Springer Geology

Series Editors

Yuri Litvin, Institute of Experimental Mineralogy, Moscow, Russia

Abigail Jiménez-Franco, Barcelona, Spain

Tatiana Chaplina, Ishlinsky Institute for Problems in Mechanics, Moscow, Russia

The book series Springer Geology comprises a broad portfolio of scientific books, aiming at researchers, students, and everyone interested in geology. The series includes peer-reviewed monographs, edited volumes, textbooks, and conference proceedings. It covers the entire research area of geology including, but not limited to, economic geology, mineral resources, historical geology, quantitative geology, structural geology, geomorphology, paleontology, and sedimentology.

Tatiana Chaplina
Editor

Processes in GeoMedia—Volume VII

 Springer

Editor
Tatiana Chaplina
Alicante, Spain

ISSN 2197-9545   ISSN 2197-9553 (electronic)
Springer Geology
ISBN 978-981-99-6574-8   ISBN 978-981-99-6575-5 (eBook)
https://doi.org/10.1007/978-981-99-6575-5

This Springer imprint is published by the registered company Springer Nature Singapore Pte Ltd.
The registered company address is: 152 Beach Road, #21-01/04 Gateway East, Singapore 189721,
Singapore

Paper in this product is recyclable.

Contents

Acoustic Wave Reflection Coefficient from Large-Scale Irregularities Weakly Non-Gaussian Sea Surface

A. S. Zapevalov⊙

Abstract Based on statistical estimates of sea surface elevations obtained from direct wave measurements, the properties of the ocean–atmosphere boundary as a surface reflecting acoustic waves are investigated. Wave measurements were carried out on a stationary oceanographic platform of the Marine Hydrophysical Institute of the RAS, located in the coastal zone of the Black Sea at a depth of 30 m. In the Kirchhoff approximation, the reflection of acoustic waves from large-scale irregularities is analyzed. In the analysis, the main attention is paid to the nonlinear effect that leads to a deviation of the distribution of surface elevations from the Gaussian distribution. It is shown that the wave nonlinearity weakly affects the average reflection coefficient at values of the Rayleigh parameter less than one. With values of the Rayleigh parameter greater than 2, the influence of wave nonlinearity grows rapidly.

Keywords Acoustic sounding · Sea surface · Nonlinear waves · Average reflection coefficient · Rayleigh parameter

1 Introduction

The interaction of acoustic waves with the sea surface is one of the key reasons of reverberation in the sea. In recent decades, when analyzing the interaction of acoustic and electromagnetic waves with the sea surface, much attention has been paid to the study of effects caused by the nonlinearity of surface waves (Callahan and Rodriguez 2004; Zapevalov 2007; Badulin et al. 2021). The nonlinearity leads to a deviation of the distributions of wave parameters (slope, elevation) from the Gaussian distribution (Phillips 1961; Longuet-Higgins 1963). Despite the fact that the fundamentals of the theory of wave scattering on the sea surface were formulated in the early 70 s (Brekhovskikh 1952; Barrick 1968; Bass et al. 1975; Valenzuela 1978), the problem has not been definitively solved. First of all, this is due to the fact

A. S. Zapevalov (✉)
Marine Hydrophysical Institute of RAS, Kapitanskaya Str., 2, 299011 Sevastopol, Russia
e-mail: sevzepter@mail.ru

© The Author(s), under exclusive license to Springer Nature Singapore Pte Ltd. 2023
T. Chaplina (ed.), *Processes in GeoMedia—Volume VII*, Springer Geology,
https://doi.org/10.1007/978-981-99-6575-5_1

1

that the currently existing models of the sea surface need to be clarified (Gao et al. 2020; Zapevalov and Garmashov 2021).

Depending on the ratio between the characteristic scales of the irregularities of the reflecting surface and the length of the probing wave, as well as depending on the angle of incidence, the scattered acoustic field is described by different methods. The main problem is that all existing methods do not allow obtaining an exact solution of the wave equation satisfying the boundary conditions on an rough surface (Bass and Fuks 1979). Approximate methods are used to calculate the wave field scattered by the sea surface. The method of small perturbations and the Kirchhoff method are most often used in problems of diffraction of acoustic and electromagnetic waves on the sea surface.

If the sounding of the sea surface is carried out at zero or close to zero incidence angle, then the average acoustic field formed as a result of reflection from large-scale irregularities can be calculated in the approximation of geometric optics (Kirchhoff method). Within this approximation, the average reflection coefficient is uniquely determined by the Rayleigh parameter and the one-dimensional probability density of wave displacements of the sea surface (Brekhovskikh and Lysanov 1978).

The purpose of this work is to analyze the variability of the average reflection coefficient due to nonlinear effects in the field of marine surface waves, as well as variations in wave heights in the coastal zone of the Black Sea.

2   Models and Methods

2.1   Average Reflection Coefficient

At the sea-air interface, an abrupt change in the density of the medium occurs. The magnitude of the density jump is so great that, when an acoustic wave falls on the sea surface from below, all its energy returns back to the water. We will consider the situation when the reflecting surface is formed by large-scale irregularities for which the condition is fulfilled.

$$k\rho_0 >> 1, \tag{1}$$

where k is the wave number of the acoustic wave, ρ_0 is a correlation radius of the surface elevations. We will assume that the probing is carried out at incidence angle close to the nadir. These assumptions allow us to describe the average acoustic field in the Kirchhoff approximation.

$$\langle \psi_1 \rangle = \vartheta \, \psi_0, \tag{2}$$

where ψ_1 is a potential velocity of the reflected wave, ψ_0 there is a potential for the velocity of a wave reflected from a flat surface of the same size as the irradiated area; ϑ is an average reflection coefficient. Here and and elsewhere the symbol $< \ldots >$ means averaging. Within the framework of these assumptions, the average reflection coefficient of acoustic waves is defined as (Brekhovskikh and Lysanov 1978).

$$
\upsilon = \int_{-\infty}^{\infty} P(\xi)\, \exp(-2\,i\,k\,\xi\, \cos\theta)\, d\xi, \tag{3}
$$

where ξ is the elevation of the sea surface; $P(\xi)$ is the probability density function of the elevation of the sea surface.

The distribution of a random variable is characterized by statistical moments.

$$
\mu_n = \langle \xi^n \rangle, \tag{4}
$$

where n is the order of the statistical moment. The first statistical moment $\mu_1 = 0$, since the elevation are counted from the level of the undisturbed surface. The second statistical moment μ_2 is the variance. Next, instead of statistical moments, we will consider cumulants. In the accepted notation the skewness A (third-order cumulant) and excess kurtosis E (fourth—order cumulant) определены как.

$$
A = \mu_3 / \mu_2^{1.5}, \tag{5}
$$

$$
E = \mu_4 / \mu_2^2 - 3. \tag{6}
$$

If a random variable is described by a Gaussian distribution.

$$
P_G(\xi) = \frac{1}{\sqrt{2\pi\,\mu_2}}\, \exp\!\left(-\frac{\xi^2}{2\mu_2}\right), \tag{7}
$$

then it follows from Eqs. (3) and (7) that with a linear description of the sea surface, the average reflection coefficient has the form.

$$
\upsilon_G = \exp\!\left(-\frac{R^2}{2}\right), \tag{8}
$$

where

$$
R = 2\,k\,\sqrt{\mu_2}\, \cos\theta \tag{9}
$$

is a Rayleigh parameter. In this case, R is a function of two parameters μ_2 and θ. If a random variable ξ is described by a non-Gaussian distribution (nonlinear wave field), then when setting the probability density function $P(\xi)$ included in Eq. (3), it is necessary to take into account the skewness and excess kurtosis of wave elevations of the sea surface.

2.2 Statistical Characteristics of the Sea Surface

For numerical modeling, the results of studies of sea waves obtained from measurements on a stationary oceanographic platform of the Marine Hydrophysical Institute RAS are used (Zapevalov and Garmashov 2021). The platform is located in the Black Sea at a depth of ~30 m. The measuring equipment, as well as the features of the wind regime in the vicinity of the platform are described in Toloknov and Korovushkin (2010); Solov'ev and Ivanov 2007). For numerical modeling, we will use the same data array that was used to analyze the variability of the skewness of sea waves in the work (Zapevalov and Garmashov 2022). The values of the statistical characteristics of the waves were calculated for measurement sessions lasting 20 min. The measurements were carried out in the period from May 2018 to January 2019.

Wave heights are usually characterized by a parameter called significant wave height H_S, which is equal to the average height of one third of the highest waves. This parameter is related to the second statistical moment by the expression $H_S = 4\sqrt{\mu_2}$.

The statistical relationship between the parameters H_S, A and E is shown in Fig. 1. Parameters H_S, A and E are weakly correlated with each other. The correlation coefficient between the parameters H_S and A, denote it as $\rho(H_S, A) = 0.44$, correlation coefficients for other parameter pairs are $\rho(H_S, E) = 0.15$ и $\rho(E, A) = 0.26$.

The consequence of the weak statistical relationship between the parameters H_S, A and E is the need to consider the variability of the average reflection coefficient in an ensemble of situations.

3 Discussion

3.1 Rayleigh Parameter Distribution

The probability density function of the Rayleigh parameter $P_R(R)$ determined for the coastal zone of the Black Sea during vertical sounding of the sea surface from below at a wavelength of 1 m is shown in Fig. 2. The calculation of the function $P_R(R)$ was carried out on the basis of histogram. The smoothed function corresponds to the histogram bin equal to 0.5, the non-smoothed function corresponds to 0.1. For our dataset, the most probable values for the Rayleigh parameter lie between 0.3 and 1.5. The Rayleigh parameter is directly proportional to the wave number of the acoustic

Fig. 1 Variability of statistical characteristics of sea surface elevations, A is skewness, E is excess kurtosis, H_S is significant wave height

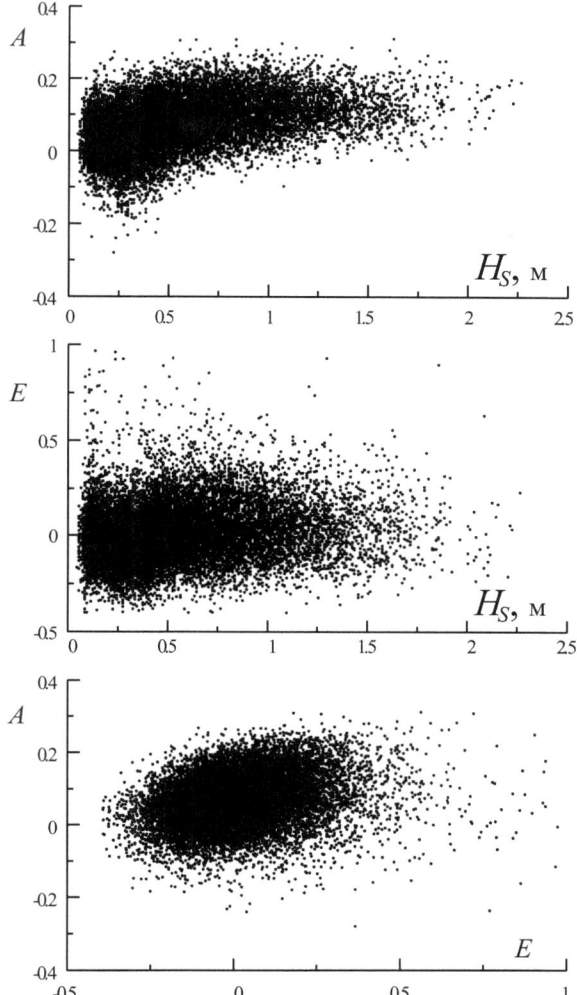

wave, so the function shown in Fig. 2 is not difficult to recalculate for sounding at other wavelengths.

3.2 Numerical Simulation of the Average Reflection Coefficient

The average reflection coefficient from a non-Gaussian surface with an asymmetric probability density function of its elevations is a complex quantity. Let's analyze the

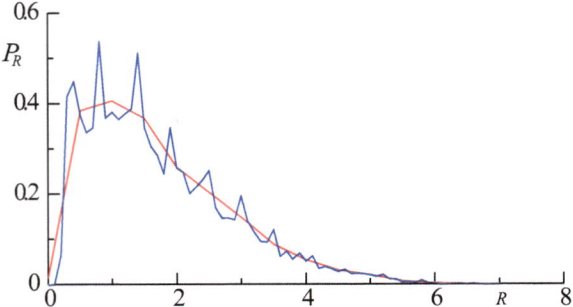

Fig. 2 Probability density function P_R of Rayleigh parameter R. The red curve is a smoothed function, the blue curve is a non-smoothed function

changes in the modulus of the average reflection coefficient in those wave situations that occur in the coastal zone of the Black Sea. As in the previous section, we will assume that the sounding is carried out at a zero incidence angle and an acoustic wave length of 1 m. The dependence of the modulus of the average reflection coefficient on the Rayleigh parameter is shown in Fig. 3.

In order to evaluate the influence of the nonlinearity of sea waves on the average reflection coefficient, consider the ratio.

$$\Psi(R) = |\upsilon_N(R)| \big/ \upsilon_L(R). \tag{10}$$

The average value of $\langle \Psi(R) \rangle$ and the standard deviation of individual estimates.

$$O(R) = \sqrt{\langle (\Psi(R) - \langle \Psi(R) \rangle)^2 \rangle}, \tag{11}$$

obtained for the intervals of the Rayleigh parameter with a width of 0.3 are shown in Fig. 4.

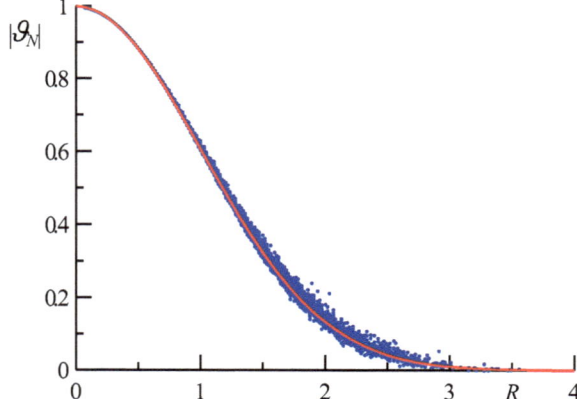

Fig. 3 Dependence of the modulus of the average reflection coefficient on the Rayleigh parameter R. The blue dots are the values $|\upsilon_N|$ calculated for the real wave field; the red curve is the theoretical dependence (8)

Fig. 4 Mean $\langle\Psi(R)\rangle$ and standard deviation $O(R)$ of the modulus of the average reflection coefficient from the linear model

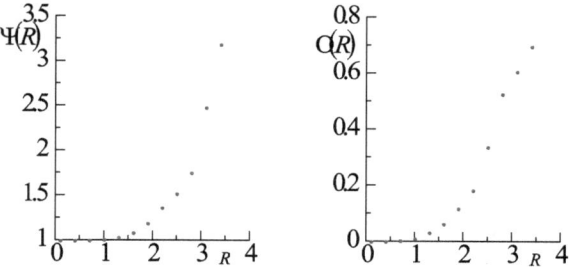

It is easy to see that with the growth of the Rayleigh parameter, the values of the modulus of the average reflection coefficient calculated for the values of skewness and excess kurtosis observed in full-scale measurements generally exceed υ_G. This excess is small at $R < 1$ and grows rapidly at $R > 2$. Changes in $O(R)$ occur similarly: at $R < 1$, the values $O(R)$ do not exceed the level of 0.01, for large values of the Rayleigh parameter $O(R = 2) = 0.15$, $O(R = 3) = 0.6$.

4 Conclusion

Based on statistical estimates of the sea surface elevation obtained from direct wave measurements, the properties of the ocean–atmosphere boundary as a surface reflecting acoustic waves are studied. The analysis is carried out for the coastal zone of the Black Sea.

It is shown that the statistical relationship between the parameters characterizing the distribution of sea surface elevations is weak. Such parameters are significant wave height H_S, skewness A and excess kurtosis E. The correlation coefficients between these parameters are equal $\rho(H_S, A) = 0.44$, $\rho(H_S, E) = 0.15$ и $\rho(E, A) = 0.26$. The consequence is the need to take into account the variability of the average reflection coefficient in a set of situations.

The probability density function of the Rayleigh parameter is constructed for sounding at a zero incidence angle and a wavelength of one meter. It is shown that the most probable values of the Rayleigh parameter lie in the range from 0.3 to 1.5. It is also shown that the deviation of the distribution of sea surface elevations from the Gaussian distribution has little effect on the average reflection coefficient at values of the Rayleigh parameter less than unity. For values of the Rayleigh parameter greater than 2, the effect of non-linearity increases rapidly.

Acknowledgements The work was carried out within the framework of the state task on the topic: FNNN-2021-0004 "Fundamental studies of oceanological processes which determine the state and evolution of the marine environment influenced by natural and anthropogenic factors, based on observation and modeling methods".

References

Badulin SI, Grigorieva VG, Shabanov PA, Sharmar VD, Karpov IO (2021) Sea state bias in altimetry measurements within the theory of similarity for wind-driven seas. Adv Space Res 68(2):978–988. https://doi.org/10.1016/j.asr.2019.11.040

Barrick D (1968) Rough surface scattering based on the specular point theory. IEEE Trans Antennas Propag 16(4):449–454. https://doi.org/10.1109/tap.1968.1139220

Bass FG, Fuks IM (1979) Wave scattering by statistically rough surface. Pergamon

Bass FG. Braude SYa, Kalmykov AI, Men' AV, Ostrovskii IY, Pustovoitenko VV, Rozenberg AD, Fuks IM (195) Radar methods for the study of ocean waves (Radiooceanography). Soviet Physics Uspekhi 18:641–642. https://doi.org/10.1070/PU1975v018n08ABEH004920

Brekhovskikh LM (1952) The diffraction of waves by a rough surface. J. Exp Theoret Phys 23:275–289

Brekhovskikh LM, Lysanov YuP (1978) Ocean acoustics. Oceanology, Ser. Ocean Phys 2:49–146

Callahan PS, Rodriguez E (2004) Retracking of Jason_1 data. Mar Geodesy 27:391–407. https://doi.org/10.1080/01490410490902098

Gao Z, Sun Z, Liang S (2020) Probability density function for wave elevation based on Gaussian mixture models. Ocean Eng 213:107815. https://doi.org/10.1016/j.oceaneng.2020.107815

Longuet-Higgins MS (1963) The effect of non-linearities on statistical distribution in the theory of sea waves. J Fluid Mech 17(3):459–480. https://doi.org/10.1017/S0022112063001452

Phillips OM (1961) On the dynamics of unsteady gravity waves of finite amplitude. Part 2. Local properties of a random wave field. J Fluid Mech 11:143–155

Solov'ev YuP, Ivanov VA (2007) Preliminary results of measurements of atmospheric turbulence over the sea. Phys Oceanogr 3:154–172

Toloknov YuN, Korovushkin AI (2010) System for Hydrometeorological Information Collection. In: MHI, 2010. monitoring systems of environment. Sevastopol: ECOSI-Gidrofizika (13):50–53. (in Russian)

Valenzuela G (1978) Theories for the interaction of electromagnetic and ocean waves. A review. Bound. Layer Meteorol 13(1–4):61–85

Zapevalov AS (2007) Evaluation of the scattering coefficient for high-frequency sound scattered from the sea surface. Acoust Phys 53(5):603–610. https://doi.org/10.1134/s1063771007050119

Zapevalov AS, Garmashov AV (2021) Skewness and kurtosis of the surface wave in the coastal zone of the Black Sea. Phys Oceanogr 28(4):414–425. https://doi.org/10.22449/1573-160X-2021-4-414-425

Zapevalov AS, Garmashov AV (2022) The appearance of negative values of the skewness of sea-surface waves. Izv Atmos Ocean Phys 58(3):263–269. https://doi.org/10.1134/s0001433822030136

Numerical Modeling of Wind Currents in the Sivash Gulf (Sea of Azov)

A. A. Polozok, V. V. Fomin, and E. V. Ivancha

Abstract In this work the current fields in the Sivash Gulf (Sea of Azov) are studied for different wind conditions. A numerical sigma-coordinate model of water circulation of the POM type with a spatial resolution of 300 m is used. Analysis of the results of numerical simulation showed that the structure of water circulation in the bay is determined by the general direction of the wind. The field of medium-depth currents consists of several eddy structures located in wide parts of the bay water area. The eddy structures are most pronounced in the southern part of the bay. With winds directed along the bay, water circulation is most intense. In these cases, a jet stream with velocities of more than 1 m/s occurs in the narrow places of the bay.

Keywords Sea of Azov · Sivash Gulf · Numerical modeling · Sigma-coordinate model · Wind currents · Vortex structures · Backflow

1 Introduction

Sivash is a shallow bay of the Sea of Azov, separated from it by a sandy spit—the Arabat Spit (Fig. 1). In the north, the bay is connected to the sea by the Tonky Strait. The water area of the bay consists of several reservoirs separated by shoals and islands. The Sivash Bay is divided by the Chongar Peninsula into two parts—western and eastern. The western part is divided into the western and middle Sivash, the eastern part—into the eastern and southern Sivash (Sovga and Shchurova 2013).

The hydrological and biogeochemical characteristics of the bay have been studied in sufficient detail. Special attention in the existing publications is paid to the study of the factors that form the water and salt regimes of the bay (Dyakov and Belogudov 2015; Lomakin 2021; Evstigneev and Eremina 2019; Eremina and Evstigneev 2020; Sovga et al. 2020). From the point of view of hydrodynamics the bay has not been studied enough. There are three works on the hydrodynamics of the Sivash Gulf. In Fomina et al. (2022), on the basis of the spectral model, the characteristics of the

A. A. Polozok (✉) · V. V. Fomin · E. V. Ivancha
Marine Hydrophysical Institute of the Russian Academy of Sciences, Moscow, Russia
e-mail: polozok.umi@gmail.com

T. Chaplina (ed.), *Processes in GeoMedia—Volume VII*, Springer Geology,
https://doi.org/10.1007/978-981-99-6575-5_2

Fig. 1 Sivash Gulf location

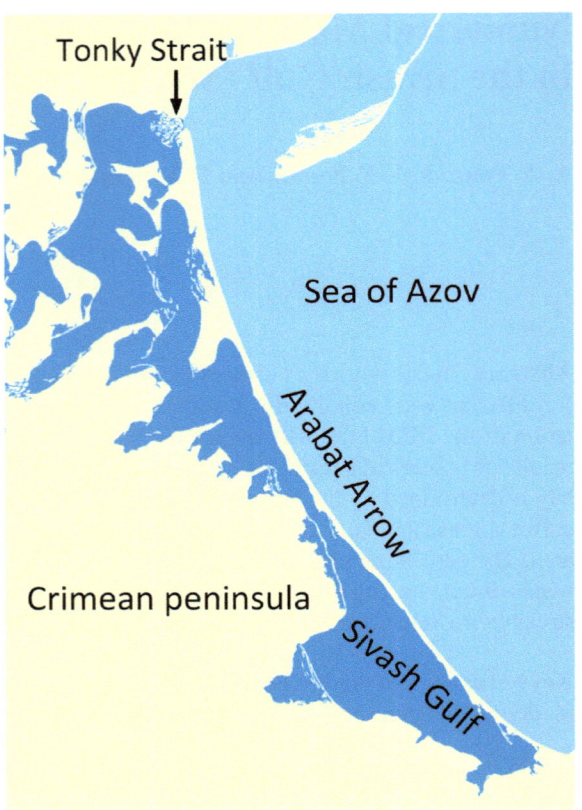

wind waves in the bay were studied. It is shown that the main factor influencing the intensity of waves is the wind speed; even during strong storms the height and period of waves in the bay do not exceed 1 m and 3 s, respectively.

In Fomin and Polozok (2022), using a numerical model, the characteristics of a freshwater plume in the area where the Salgir River flows into the Sivash Gulf are studied. The effect of various types of currents on the position, shape and size of the plume has been studied. It is shown that under the influence of river runoff a plume several hundred meters in size is formed at the exit from the river mouth. The plume is concentrated in the upper layer about 0.5 m thick.

There is a single publication (Sovga et al. 2018) devoted to modeling wind currents in the Sivash Gulf. It presents schemes of surface currents in the bay with easterly and western winds. It is shown that with easterly winds the Azov waters flow into the Sivash. The calculations were performed with a horizontal resolution of 1 km, which makes it impossible to reproduce the detailed structure of the bay water circulation.

This work aims to perform a more detailed study of the spatial structure of wind currents in the Sivash Gulf for various wind effects based on numerical simulation.

Mathematical model. The model is based on three-dimensional σ—coordinate equations of hydrodynamics in the Boussinesq approximation and hydrostatics (here summation is carried out over repeated indices α and β from 1 to 2) (Ivanov and Fomin 2008):

$$\frac{\partial}{\partial t}(Du_\alpha) + \Lambda u_\alpha + \varepsilon_{\alpha\beta} f \, Du_\beta + g D \frac{\partial \eta}{\partial x_\alpha} = \frac{\partial}{\partial x_\beta}(D\tau_{\alpha\beta}) + \frac{\partial}{\partial \sigma}\left(\frac{K_M}{D}\frac{\partial u_\alpha}{\partial \sigma}\right),$$

(1)

$$\frac{\partial \eta}{\partial t} + \frac{\partial}{\partial x_\alpha}(Du_\alpha) + \frac{\partial w_*}{\partial \sigma} = 0,$$

(2)

$$\Lambda\phi = \frac{\partial}{\partial x_\beta}(Du_\beta\phi) + \frac{\partial}{\partial \sigma}(w_*\phi), \quad \tau_{\alpha\alpha} = 2A_M\frac{\partial u_\alpha}{\partial x_\alpha},$$

$$\tau_{\alpha\beta} = \tau_{\beta\alpha} = A_M\left(\frac{\partial u_\beta}{\partial x_\alpha} + \frac{\partial u_\alpha}{\partial x_\beta}\right)$$

(3)

where $(x_1, x_2) = (x, y)$; σ—dimensionless vertical coordinate varying from -1 to 0; $(u_1, u_2) = (u, v)$—the components of the flow velocity along the axes x_1, x_2; w_*—normal to surfaces $\sigma = $ const current velocity component; $D = h + \eta$; h—the depth of the basin; η—sea level; f—the Coriolis parameter; $\tau_{\alpha\beta}$—the components of the turbulent stress tensor; A_M, K_M—the coefficients of turbulent viscosity; g—the acceleration due to gravity.

On the free surface ($\sigma = 0$) the boundary conditions have the form

$$w_* = 0, \quad \frac{K_M}{D}\frac{\partial u_\alpha}{\partial \sigma} = c_a|u|u_\alpha$$

(4)

At the bottom ($\sigma = -1$) the boundary conditions have the form

$$w_* = 0, \quad \frac{K_M}{D}\frac{\partial u_\alpha}{\partial \sigma} = c_b|u|u_\alpha$$

(5)

where c_a, c_b—friction coefficients; $|u| = \sqrt{u_1^2 + u_2^2}$.

On the solid boundaries of the computational domain the no-slip condition is set. In the Tonkii Strait the condition of free passage is used.

The bottom friction coefficient had the form: $c_a = (0.49 + 0.065W) \cdot 10^{-3}$, where $W = \sqrt{W_x^2 + W_y^2}$—the wind speed at a height of 10 m. The bottom friction coefficient was determined by the formula: $c_b = 0.16/\ln^2(\delta z/z_b)$, where δz—the vertical distance from the bottom to the point where the friction coefficient is determined; $z_b = 0{,}001$ m–the bottom surface roughness parameter.

Problem (1)–(5) was solved numerically using explicit difference schemes in horizontal coordinates and implicit difference schemes in vertical coordinates. To approximate the advection operator Λ, monotone difference schemes were used. A detailed description of the numerical algorithm is given in Ivanov and Fomin (2008).

Results of numerical simulation. Features of water circulation in the Sivash Gulfare determined by the features of atmospheric processes in the study area. According to ERA atmospheric re-analysis data for the period 1979–2020 (apps.ecmwf.int), the statistical characteristics of the surface wind speed W for the study area are as follows (Sovga et al. 2020): long-term average $\overline{W} = 7.3$ m/s; standard deviation $r = 3.6$ m/s. Figure 2 shows the wind rose for the Sivash Gulf area. As can be seen, the wind of the northeast direction has the highest frequency (23%), and the wind of the southeast direction has the smallest frequency (5%). The repeatability of wind in other directions is 10–12%. Based on the results of Amarouche and Akpınar (2021), the storm wind speed will be considered as the speed: $W \geq W_s$, where $W_s = \overline{W} + 2r = 14.5$ m/s. All calculations were carried out for this value of wind speed. The calculations used the bottom topography of the Sivash Gulf, shown in Fig. 3. A grid with a horizontal resolution $\Delta x = \Delta y = 300$ m (284 × 334 nodes) was used. 11 calculated levels were used along the vertical coordinate.

Situations were modeled when a spatially uniform wind of a given direction θ_W acts on the surface of the bay. The duration of the wind is 24 h. At the initial time $t = 0$ the wind speed $W = 0$. At t > 0 the wind speed increases linearly from zero to W_s for 3 h and then does not change. The calculations were carried out for 8 wind directions with discreteness $\Delta \theta_W = 45°$. The depth-averaged current velocity was analyzed: $U = \sqrt{\overline{u}_1^2 + \overline{u}_2^2}$, where $\overline{u}_1 = \int\limits_{-1}^{0} u_1 d\sigma$, $\overline{u}_2 = \int\limits_{-1}^{0} u_2 d\sigma$. On Figs. 4, 5, 6 and 7 for different values θ_W the speed and direction of the average depth current at $t = 24$ h were shown (large arrow—wind direction).

According to the results of calculations, the average depth velocity of the currents in the Western Sivashis predominantly directed along the wind. This is explained by the shallowness of the basin (0.3–0.4 m). The maximum values (0.85 m/s) U are

Fig. 2 Wind rose in the Sivash Gulf area

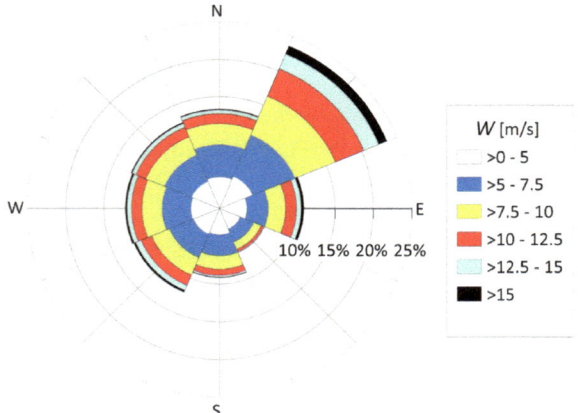

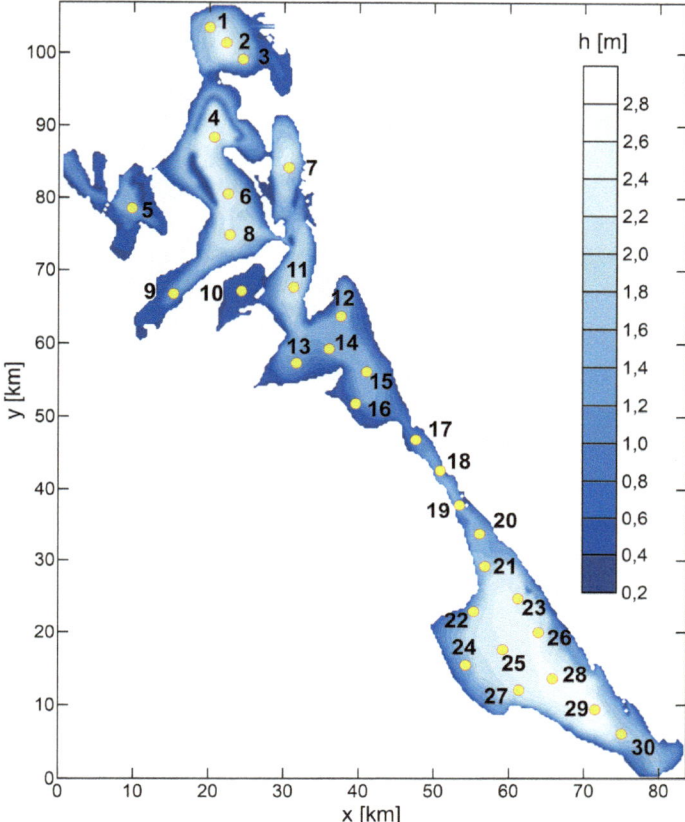

Fig. 3 Model bottom topography h (m) of the Sivash Gulf. The numbers mark the position of data output points

reached during the northern and southern winds; in the case of western and eastern winds the values of the maxima U vary from 0.6 m/s to 0.8 m/s. In the main part of the water area the values U are 0.3–0.5 m/s.

In the Middle Sivash, areas with depths of 0.5 mare most common, but in some places the depths reach 1–2 m. The directions of the velocities of currents with average depth in almost the entire water area of the Middle Sivash coincide with the direction of the wind. Only in the northern part of the Middle Sivash, with all wind directions, a vortex structure is formed, consisting of a cyclone and an anticyclone, the position of which in space changes with the wind. So, with a westerly wind, an anticyclone is located to the north, and a cyclone is located to the south. With an east wind, they change places. With southerly winds, the anticyclone is located to the west, the cyclone—to the east. With northern winds, on the contrary, there is a cyclone to the west, and an anticyclone to the east. In most of the water area of the Middle Sivash the values U are 0.3–0.4 m/s. The maxima U can reach values from

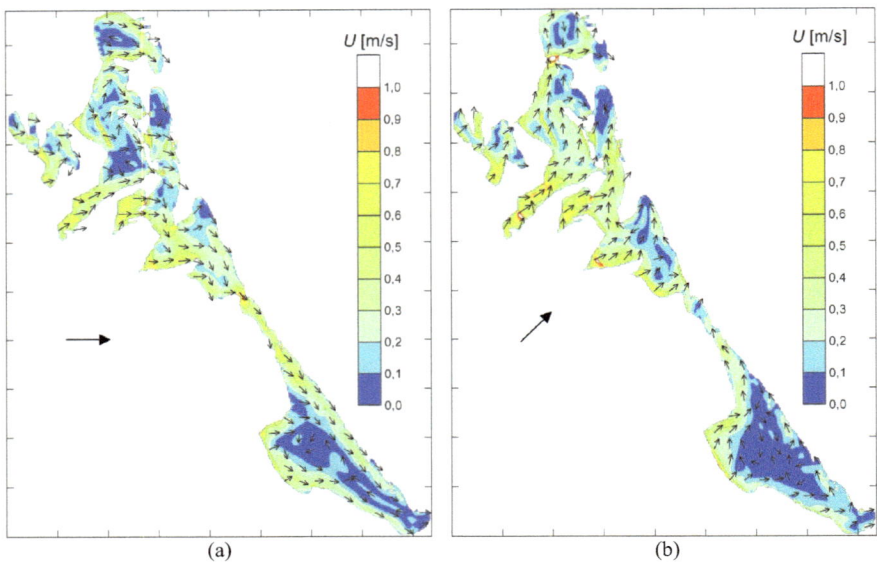

Fig. 4 Mean currents in the Sivash Gulf with west (**a**) and southwest (**b**) winds

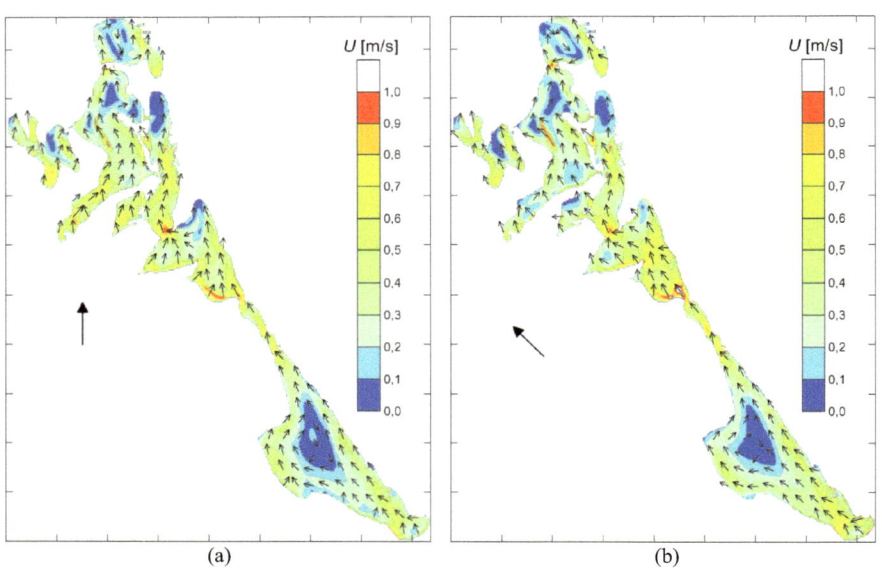

Fig. 5 Mean currents in the Sivash Gulf with south (**a**) and southeast (**b**) winds

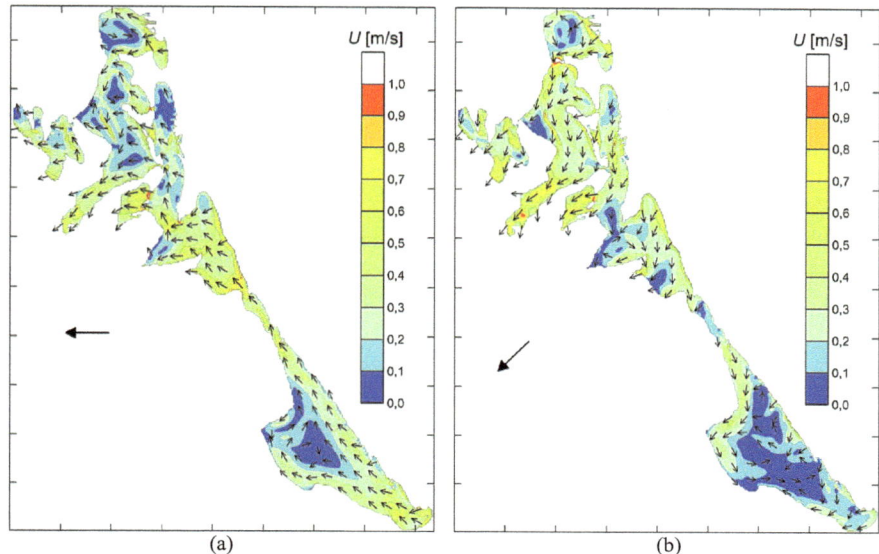

Fig. 6 Mean currents in the Sivash Gulf with east (**a**) and northeast (**b**) winds

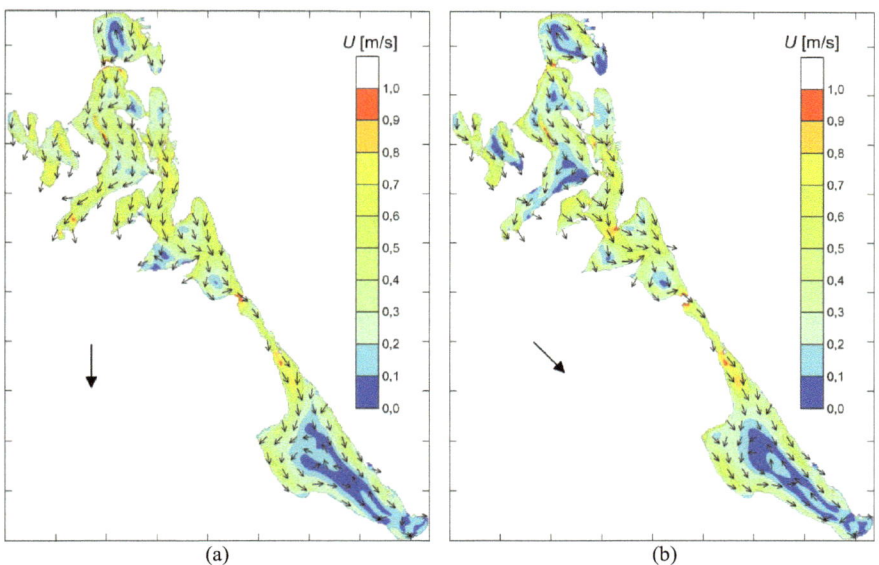

Fig. 7 Mean currents in the Sivash Gulf with north (**a**) and northwest (**b**) winds

0.6 m/s with an east wind to 1.1 m/s with a northwest wind. In straits U reaches 1.4 m/s.

In the Eastern Sivash, which directly connects with the Sea of Azov, the depths reach 2.5 m. Here, the currents, which are average in depth, are also mainly directed downwind. And only in the southern part of the Eastern Sivash, with two wind directions, the occurrence of eddy circulation is noted. With a southwest wind, a cyclone appears near the northeast coast; with a northeast wind, an anticyclone appears near the southwest coast. At the same winds, as well as at the north and south winds in the Eastern Sivash, in most of the water area, currents intensify up to 0.5–0.7 m/s, with a maximum in the straits up to 1.1 m/s. At other winds, the values U in the main part of the water area of the Eastern Sivashare 0.2–0.4 m/s.

The Southern Sivashis stretched from the northwest to the southeast. It communicates with the Eastern Sivashthrough a strait 1.8 kmwide and up to 1.5 mdeep. For the Southern Sivash, depths over 2 mare characteristic. In this part of the Sivash, the currents form two elongated eddy structures of the opposite sign. With the western wind and the winds of the northern rhumbs, an anticyclonic eddy is located near the eastern coast of the South Sivash, and a cyclonic eddy is located near the western one. With the east wind and winds of southern rhumbs in the Southern Sivash, the cyclonic eddy shifts to the eastern coast. The anticyclonic eddy moves to the western coast. At the center of the eddies, the values U do not exceed 0.1–0.2 m/s. At the periphery of eddy formations, the values U increase to 0.3–0.6 m/s. According to the wind rose (Fig. 2), the most likely pattern of currents in the Sivash Gulfwill be the pattern for the northeast wind (Fig. 6b).

Thus, the structure of medium-depth currents in the Sivash Gulf is determined mainly by the direction of the wind, which is explained by the shallowness of the basin. In the Middle and Southern Sivash, with all wind directions, there is a system of two vortex formations of the opposite sign. The most intensive circulation of water in the Sivash Gulf occurs during the northern, northwestern, southern and southeastern winds. This is due to the elongation of the Sivash Gulf from northwest to southeast.

The Table 1 for points 1–30 (Fig. 3) gives the direction of the current on the surface of the bay (θ_0) and the angle between θ_0 and the direction of the current near the bottom ($\Delta\theta$). The location of the points is shown in Fig. 3. The duration of the wind is 24 h. The direction of the wind (θ_w) and the direction of the current are measured from the x_1 axis directed to the east counterclockwise. At $\Delta\theta > 0$ the flow near the bottom is deviated to the right of the flow at the surface; at $\Delta\theta < 0$ the flow near the bottom is deviated to the left from the flow at the surface.

Table 1 Values of angles θ_0 and $\Delta\theta$ at points 1–30 of the Sivash Gulf at t = 24 h for different wind directions θ_w

№	h [m]	$\theta_w = 225°$ (NE)		$\theta_w = 315°$ (NW)		$\theta_w = 45°$ (SW)		$\theta_w = 135°$ (SE)	
		θ_0	$\Delta\theta$	θ_0	$\Delta\theta$	θ_0	$\Delta\theta$	θ_0	$\Delta\theta$
1	1.8	222	−1	290	101	67	−48	53	83
2	2.6	160	99	56	−71	343	97	316	1
3	1.8	236	−24	274	103	32	24	88	115
4	2.5	229	−10	297	96	58	−46	114	126
5	1.6	243	−36	308	154	58	−87	124	159
6	2.2	267	−32	313	2	72	−44	123	13
7	2.3	246	−96	298	60	63	−153	129	169
8	2.2	239	−19	281	108	47	−6	103	59
9	1.2	225	−1	268	51	51	−3	109	66
10	0.5	215	4	312	2	41	1	139	−7
11	2.1	231	−27	272	22	66	−21	98	22
12	1.6	249	−25	306	22	306	64	131	5
13	1.4	150	92	298	52	35	8	136	−3
14	1.6	257	−62	336	−19	62	−124	143	−5
15	1.5	249	−41	306	8	11	95	124	5
16	1.2	235	157	290	43	42	126	120	7
17	1.1	240	−51	311	2	63	−73	131	2
18	1.4	272	−64	319	−3	108	−48	141	−2
19	1.4	278	−27	307	2	101	−36	127	3
20	1.5	268	−30	296	5	90	−49	120	10
21	1.8	263	−23	297	8	87	−32	119	32
22	1.4	226	−1	289	30	43	0	80	44
23	2.3	221	131	300	14	32	138	124	20
24	1.5	275	−44	302	6	70	−28	129	4
25	2.4	207	147	286	108	52	−161	160	−113
26	2.6	205	140	309	149	8	106	127	36
27	2.4	263	122	338	141	46	−173	161	−37
28	2.6	222	176	28	148	51	−156	138	−13
29	2.8	224	178	16	−111	41	173	136	−2
30	2.0	245	−51	272	120	40	14	143	−3

Due to the complex configuration of the Sivash Gulf, there are no obvious patterns in the spatial distribution of $\Delta\theta$. However, for most points the condition is met: $|\Delta\theta| < 90°$. In general, there is a tendency to decrease $|\Delta\theta|$ with decreasing depth. For 10% of points the condition is fulfilled: $|\Delta\theta| > 90°$. This condition indicates the presence

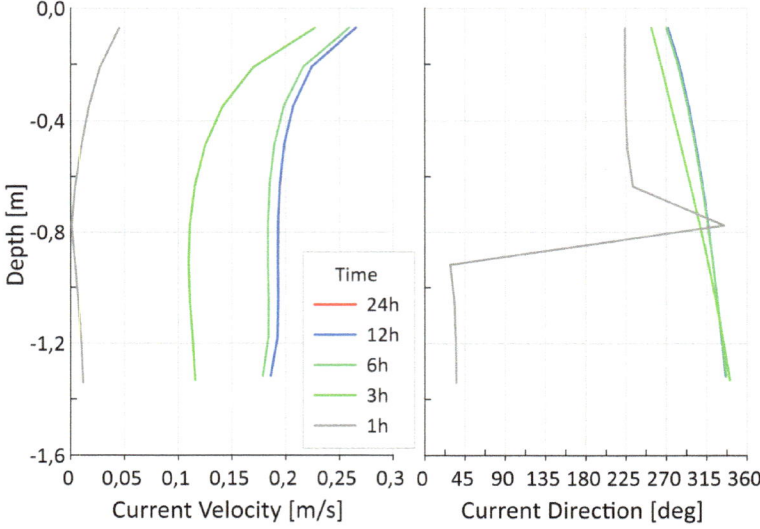

Fig. 8 Changes in depth of the speed and direction of the current at point 18 with a northeasterly wind for different values of t

of a backflow in the near-bottom layer. The points at which backflow occurs are located mainly in the South Sivash.

Below, we consider two typical examples of the evolution of vertical current profiles in the Sivash Gulf for the northeast wind direction, as the most probable one. In the first example (the strait between the Middle Sivash and the Southern Sivash, point 18, Fig. 8), under the influence of the wind, the flow speed gradually increases throughout the depth. In this case, the flow velocity profile has a monotonic character. At $t \geq 6$ h, the flow direction ceases to change with time. The angle of rotation of the current velocity vector near the bottom relative to θ_0 is less than 90°.

In the second example (the central part of the Southern Sivash, point 25, Fig. 9), the current velocity profile has a non-monotonic character. Over time, at a depth of ~0.9 m, a region of minimum current velocity is formed. The angle of rotation of the current velocity vector near the bottom relative to θ_0 reaches 147°, which indicates the presence of a backflow in the lower layer.

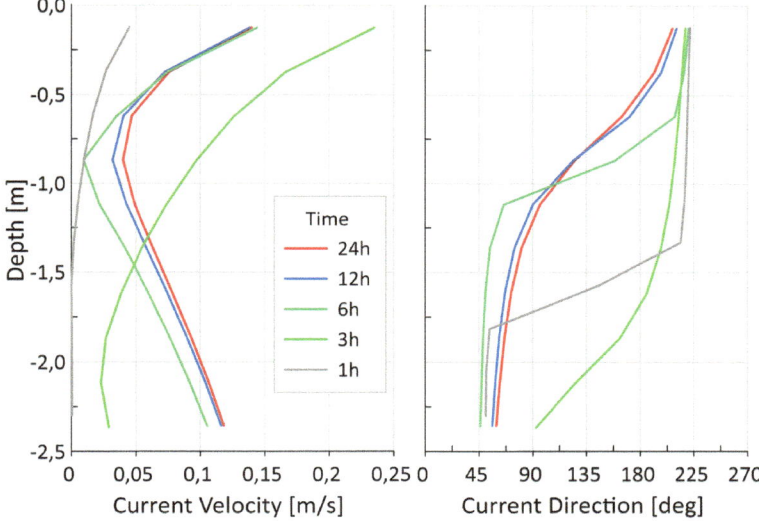

Fig. 9 Changes in depth of the speed and direction of the current at point 25 with a northeasterly wind for different values of t

2 Conclusion

On the basis of a numerical sigma-coordinate model of water circulation with a spatial resolution of 300 m, the characteristics of wind currents in the Sivash Gulf were studied for different directions of the storm wind. An analysis of the results of numerical simulation showed that the structure of water circulation in the bay is determined by the general direction of the wind. The field of medium-depth currents consists of several eddy structures located in wide parts of the bay water area. The eddy structures are most pronounced in the southern part of the bay. With winds directed along the bay, circulation becomes most intense. In these cases, in the narrowness connecting the northern and southern parts of the bay, a jet stream occurs with velocities of more than 1 m/s. The water circulation in the bay reaches a steady state after 12 h of wind action. In the deepest parts of the bay near the bottom, the formation of a backflow is possible.

Acknowledgements The study was carried out within the theme of the Marine Hydrophysical Institute no. FNNN-2021-0005 «Coastal studies». Model calculations were carried out on the MHI computing cluster (hpc-mhi.org).

References

Amarouche K, Akpınar A (2021) Increasing trend on Storm Wave intensity in the western Mediterranean. Climate 9(1):11. https://doi.org/10.3390/cli9010011

Dyakov NN, Belogudov AA (2015) Water exchange of the gulf Sivash with the Sea of Azov through the Strait Genichesk (Thin). Proceedings of the N. N. Zubov State Oceanographic Institute. Moscow: Artifex, Issue. 216, pp 240–253. (In Russian)

Eremina ES, Evstigneev VP (2020) Inter-annual variability of water exchange between the Azov Sea and the Sivash Bay through the Tonky Strait. Phys Oceanogr [e-journal] 27(5):489–500. https://doi.org/10.22449/1573-160X-2020-5-489-500

Evstigneev VP, Eremina ES (2019) Calculation of precipitation over the Sivash Bay. Ecological safety of the coastal and shelf zones of the sea, Issue. 2. pp 19–29. https://doi.org/10.22449/2413-5577-2019-2-19-29 (In Russian)

Fomin VV, Polozok AA (2022) Features of river plume formation in a Shallow Lagoon (the Case of the Sivash Bay, the Sea of Azov). Ecol Safety Coastal Shelf Zones Sea 3:28–42. https://doi.org/10.22449/2413-5577-2022-3-28-42

Fomina IN, Fomin VV, Polozok AA (2022) Wind waves in Sivash Bay according to the results of numerical modeling. Ecology. Economy. Informatics. System analysis and mathematical modeling of ecological and economic systems, Issue 7. Rostov-on-Don: SSC RAS Publishers, pp 97–102. (In Russian)

Ivanov VA, Fomin VV (2008) Mathematical modeling of dynamic processes in the sea-land zone. Sevastopol: ECOSI-Hydrophysics, 363 p. (In Russian)

Lomakin PD (2021) Features of the Oceanological values fields in the Sivash bay (The Sea of Azov). Phys Oceanogr [e-journal] 28(6):647–659. https://doi.org/10.22449/1573-160X-2021-6-647-659

Sovga EE, Shchurova ES (2013) The resource potential of Lake Sivash and the current ecological state of its water area. Ecological safety of the coastal and shelf zones and the integrated use of shelf resources. Sevastopol: ECOSI-Hydrophysics, Issue. 27. pp 276–283. (In Russian)

Sovga EE, Eryemina ES, Khmara TV (2018) Water Balance in the Sivash Bay as a result of variability of the natural-climatic and anthropogenic factors. Phys Oceanogr [e-journal] 25(1):67–76. https://doi.org/10.22449/1573-160X-2018-1-67-76

Sovga EE, Eremina ES, Latushkin AA (2020) Research expeditions performed by marine Hydrophysical Institute in the Sivash Bay Waters in Spring and Autumn, 2018. Phys Oceanogr, [e-journal] 27(2):161–170. https://doi.org/10.22449/1573-160X-2020-2-161-170

Metamorphic Conditions Carbonaceous Rocks Framing of the Larino Granite-Gneiss Dome (South Ural, Russia)

A. V. Snachev

Abstract The article briefly considers the geological structure of the Larino granite-gneiss dome. It is located within the Aramil-Sukhtelinsky zone and is the southern continuation of the Ilmenogorsk-Sysert anticlinorium. The section of metamorphic strata is a single megarhythm, in which schists and apovolcanic amphibolites from the center to the periphery of the dome are replaced by graphitic quartzites and schists of the Bulatovo Strata (S_1-D_1 bl). These rocks are intruded by granitoids of the Pervomay and Larino massifs ($\rho C_{1-2}v$) forming a two-headed structure. From granitoids to shales, a series of concentric high-gradient zones of metamorphism is observed. Mineral parageneses of amphibolite facies at a distance of several kilometers are replaced by associations of epidote–amphibolite and greenschist facies. Based on the study of amphibole-garnet paragenesis and the exothermic effect of organic carbon, it was shown that for rocks of the epidote–amphibolite facies, the metamorphism parameters were: $T = 530$–$550\,°C$ and $P = 8.0$–8.4 kbar. The manifestation of zonal metamorphism within the Larino granite-gneiss dome contributed to the migration and redeposition of gold mineralization. As a result of processing the obtained data, a very clear pattern of placement of elevated gold grades was revealed. All points with high values of gold (Malouvelskoe, Nikol'sk, Pridannikovo occurrences) are confined to the outer high-temperature subfacies of the greenschist facies. This is a very important search feature when search for gold mineralization.

Keywords Southern Urals · Bulatovo Strata · Shemetovo sequence · Carbonaceous deposits · TOC · Organic carbon · Temperature · Regional metamorphism · Gold

A. V. Snachev (✉)
Ufa Federal Research Centre of the Russian Academy of Sciences, Ufa, Russian Federation
e-mail: savant@rambler.ru

© The Author(s), under exclusive license to Springer Nature Singapore Pte Ltd. 2023
T. Chaplina (ed.), *Processes in GeoMedia—Volume VII*, Springer Geology,
https://doi.org/10.1007/978-981-99-6575-5_3

1 Introduction

Carbonaceous deposits are known to be a very favorable geochemical environment for the primary accumulation of many industrially important elements (Johnson et al. 2017; Shumilova et al. 2016; Snachev and Snachev 2018; Starostin and Yapaskurt 2007). Thus, the clarke gold content in the black shales of the world, estimated by various methods, is 0.008–0.01 g/t (Yudovich and Ketris 2015). For unaltered black shale deposits of the Bulatovo suite, the background values of gold are 0.005 g/t, which is consistent with the clark values (Snachev et al. 2015). The model of metamorphogenic-hydrothermal gold formation applicable to black shale strata suggests a complex participation in ore genesis of interrelated processes of sedimentation, tectonics, magmatism and metamorphism, with the latter playing a leading role. Thus, in the works of numerous researchers it was shown that during the processes of metasomatism and sulfidization, gold migration occurs (Korobeinikov 1985; Large et al. 2011; Nekrasov 1996; Plyusnina et al. 2004). The mechanism of its concentration is most clearly manifested when higher stages of regional, contact, and dislocation metamorphism are superimposed on carbonaceous deposits (Groves et al. 2003). In particular, on the examples of Siberian and Russian Far Eastern (Buryak 1975; Razvozzhaeva et al. 2008; Wood and Popov 2006) gold deposits and manifestations, as well as South Ural objects (Snachev et al. 2021, 2013, 2012). It is shown that gold-sulfide mineralization is associated with a high-temperature subfacies of the greenschist facies, which is considered to be a zone of gold deposition, in while higher temperature facies are zones of potential removal. The most successful example that fits into the developed model is the Larino granite-gneiss dome.

2 Research Methodology

Thermogravimetric analysis of carbonaceous rocks was carried out on a Q-1500 derivatograph (analyst T.I. Chernikova, Institute of Geology—Ufa Federal Research Center of the Russian Academy of Sciences). Heating was carried out in air from 20 to 1000 °C at a rate of 10 °C/min. For analysis, samples of the least silicified and sulfidized rocks were taken outside the zones of intrusive exocontacts and intensive tectonic processing, which made it possible to exclude their influence and reconstruct the degree of regional metamorphism.

To determine the degree of metamorphism of host rocks, garnet-amphibole parageneses were studied according to the method of Perchuk and Ryabchikov (1976). The compositions of garnets and amphiboles were analyzed by V. A. Kotlyarov (Institute of Mineralogy—Ural Branch of the Russian Academy of Sciences, Miass) on a SEMM-202 M scanning electron microscope with an LZ-5 energy-dispersive spectrometer (SiLi detector, resolution 140 eV). Accelerating voltages of 20 or 30 kV at probe currents of 4–6 nA, beam diameter of 1–2 μm (standards for garnet—garnet, for amphibole—amphibole).

3   Geological Structure of the Larino Granite-Gneiss Dome

The Larino granite-gneiss dome is located within the Aramil-Sukhteli zone and is the southern extension of the Ilmenogorsk-Sysert anticlinorium (Snachev et al. 2022) (Fig. 1). The section of metamorphic strata in the Larino dome is a single megarhythm with a thickness of more than 1500 m, in which schists and apovolcanic amphibolites from bottom to top and from the center to the periphery of the dome are replaced by garnet-micaceous (aluminous) and graphitic quartzites. These rocks are intruded by granitoids of the Pervomay and Larino massifs (C_{1-2}, according to other sources P_1), forming a two-headed structure. From granitoids to shales, a series of concentric high-gradient zones of metamorphism is observed. At a distance of a few kilometers, mineral parageneses of the amphibolite facies are replaced by associations of epidote–amphibolite and greenschist facies.

In stratigraphic terms, within the territory under consideration, according to the latest geological surveys, the following are distinguished from bottom to top: Shemetovo Sequence (O_3šm) (composed mainly of basalts), Bulatovo Strata (S_1-D_1bl) and Krasnokamensk Sequence (D_3kr). Carboniferous deposits are known and clearly predominate only in the composition of the Bulatovo Strata.

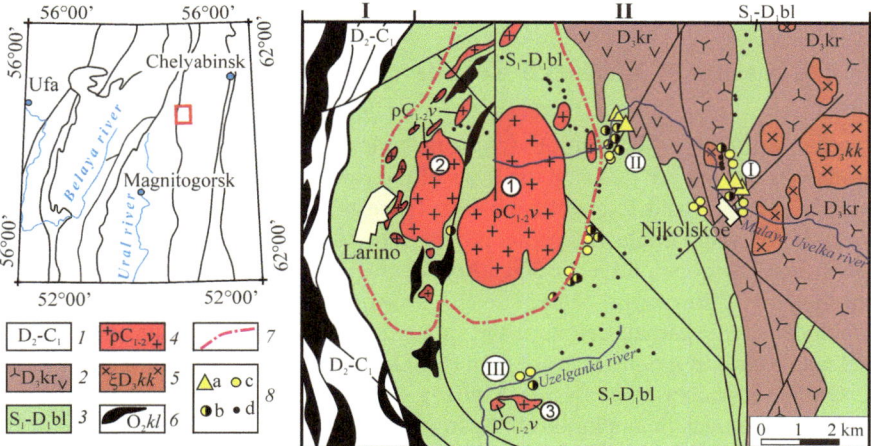

Fig. 1 Geological structure of the Larino granite-gneiss dome (according to Puzhakov et al. 2018; Snachev et al. 2022; Zhdanov et al. 2018)). Structural-formation zones: I—Uysko-Novoorenburg, II—Kochkarsko-Adamovsk. 1—volcanic-sedimentary complexes, undivided; 2—Krasnokamensk Sequence, volcanomictic sandstones and siltstones, felsic tuffs, trachybasalts and their tuffs; 3—Bulatovo Series, carbonaceous shales and siltstones; 4—Warshaw Complex, muscovite granites with garnet, granite gneisses; 5—Krasnokamensk Complex, syenites, quartz monzodiorites; 6—Kulikovo Complex, apodunite, apoharzburgite serpentinites; 7—boundary of amphibolite and epidote–amphibolite facies of metamorphism; 8—gold content (g/t): a—more than 1.0, b—0.5–1.0, c—0.1–0.5, d—Is less than 0.1. Numbers in circles: names of massifs—1—Pervomay, 2—Larino, 3—Pridannikov; occurrences of gold—I—Nikol'sk, II—Malouvelskoe, III—Pridannikovo

The Bulatovo Strata has a fairly homogeneous composition. Its characteristic sections were described by the author near the northern outskirts of the village. Nikolskoe and on the right side of the Malaya Uvelka river, where high gold grades were found. The sequence is homogeneous in composition and is composed of the so-called phtanites: carbonaceous-siliceous, carbonaceous-argillaceous-siliceous and siliceous shales. In the lower part of the section, there are sometimes interbeds of basalts and tuffaceous siltstones of the Shemetovo Sequence (O_3šm). Shale composition is dominated by quartz (90–95%), carbonaceous (graphite) matter is from 1 to 5%, sericite and biotite are present in small amounts. Shales usually contain dissemination of pyrite, pyrrhotite, and magnetite. Carbonaceous matter forms a uniform dust-like dissemination, often so dense that it makes the rock completely opaque. Its age was accepted on the basis of finds of graptolite fauna near the village of Bulatovo, which make it possible to date the sequence as Late Llandover Wenlock (Plyusnin et al. 1965). In another block of shale to the west of the village Mirny found conodonts *Ozarkodina aff. ziegleri Wall.* and others, characteristic of the Late Silurian, as well as graptolites and conodonts included in the complex of Lower Devonian remains (Puchkov and Ivanov 1989). Thus, the age of the Bulatovo Strata covers the interval from the early Silurian to the early Devonian. The total thickness of the Bulatovo Strata is more than 900 m.

4   Discussion of Research Results

Elucidation of the $P–T$ conditions for the metamorphism of the rocks of the Bulatovo Strata was carried out by us on the basis of the study of the amphibole-garnet paragenesis (Perchuk and Ryabchikov 1976; Thermo- and barometry of metamorphic rocks 1977). Amphibole and garnet monofractions were analyzed using a scanning electron microscope (Table 1). Their crystal chemical formulas were calculated by the method of Borneman-Starynkevich (1964). Analysis of Table 1 allows us to conclude that the garnets belong to the almandine type. The proportion of spessartite and andradite end members is not high (MnO = 4.03–5.46%; CaO = 1.54–4.33%).

On Fig. 2a in coordinates $X_{Mg} = Mg/(Mg + Fe + Mn)$ (values of the mole fractions of the components in amphibole and garnet), the composition points of two pairs of these minerals in plagioschists from the lower section of the Bulatovo Strata (Amf-1, Amf-2, Gr-1, Gr-2). The sample was taken in the upper reaches of the river Malaya Uvelka, on 1.0 km east of the contact of the Pervomay massif of carbonaceous shales with high gold grades (Fig. 1). It is clearly seen that the considered associations fall into the field with formation temperatures of 530–550 °C.

Calculation of the pressure from the known temperature and coefficient lnK, where $K = X_{Mg}{}^{Gr}/X_{Mg}{}^{Amf}$ (Thermo- and barometry of metamorphic rocks 1977) (Fig. 2b) showed its values of 8.0–8.4 kbar. The position of garnet-amphibole pairs of plagioschists on the petrogenetic diagram (Fig. 3) indicates an epidote–amphibolite facies of contact metamorphism.

Table 1 Composition of amphibole and garnet from rocks framing the Larino Dome (wt %)

Mineral	SiO_2	TiO_2	Al_2O_3	FeO	MnO	CaO	MgO	Na_2O	K_2O	Sum	$X_{Mg}^{Amf,Gr}$
Amf-1	44.50	0.35	12.01	18.14	0.29	11.78	9.71	1.33	0.26	98.35	0.48
Amf-2	44.08	0.30	12.42	17.78	0.29	11.27	9.84	1.54	0.28	97.81	0.49
Gr-1	37.39	–	21.12	31.76	5.42	3.21	1.94	–	–	100.83	0.08
Gr-2	36.99	–	21.18	32.76	5.46	1.59	2.23	–	–	100.20	0.09

Amf-1—$(Ca_{1.85}Na_{0.38}K_{0.05})_{2.28}(Fe_{2.22}Mg_{2.12}Mn_{0.04}Ti_{0.04}Al_{0.59})_5(Si_{6.52}Al_{1.48})_8O_{22}[O_{0.34}(OH)_{1.66}]_2$

Amf-2—$(Ca_{1.77}Na_{0.44}K_{0.05})_{2.26}(Fe_{2.18}Mg_{2.15}Mn_{0.04}Ti_{0.04}Al_{0.60})_5(Si_{6.46}Al_{1.54})_8O_{22}[O_{0.15}(OH)_{1.85}]_2$

Gr-1—$(Ca_{0.27}Fe_{2.13}Mg_{0.23}Mn_{0.37})_3Al_2Si_3O_{12}$

Gr-2—$(Ca_{0.14}Fe_{2.21}Mg_{0.27}Mn_{0.37})_3Al_2(Si_{2.99}Al_{0.01})_3O_{12}$

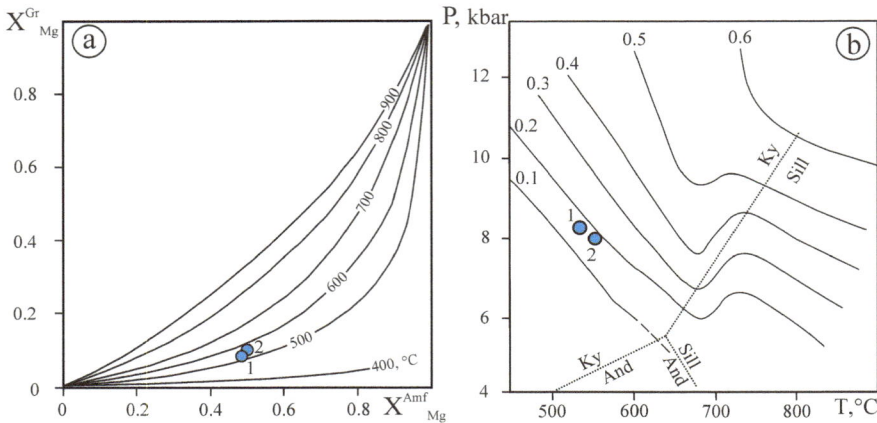

Fig. 2 Phase correspondence diagram for a amphibole-garnet paragenesis (**a**) and the position of the lines of equal values of K_{Mg}^{Gr-Amf} in the diagram T-P (Thermo- and barometry of metamorphic rocks 1977) (points 1 and 2 see to Table 2)

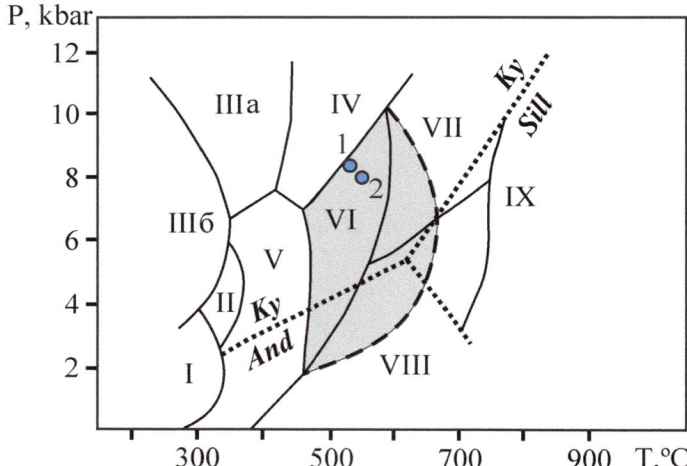

Fig. 3 Position of garnet-amphibole pairs of rocks framing the Larinsky dome on the petrogenetic diagram (Thermo- and barometry of metamorphic rocks 1977). Solid lines are facies boundaries, gray field is the stability field of staurolite, dots show the And-Sill-Ky triple diagram. Facies: I—zeolite or pumpellyite-prenitic, II—pumpelliite-actinolithic, IIIa—glaucophane-lavsonite, IIIb—glaucophane-zoisite, IV—zoisite-kyanite-quartz schist, V—greenschist, VI—epidote amphibolite, VII—almandine amphibolites, VIII—cummingtonite amphibolites, IX—granulite

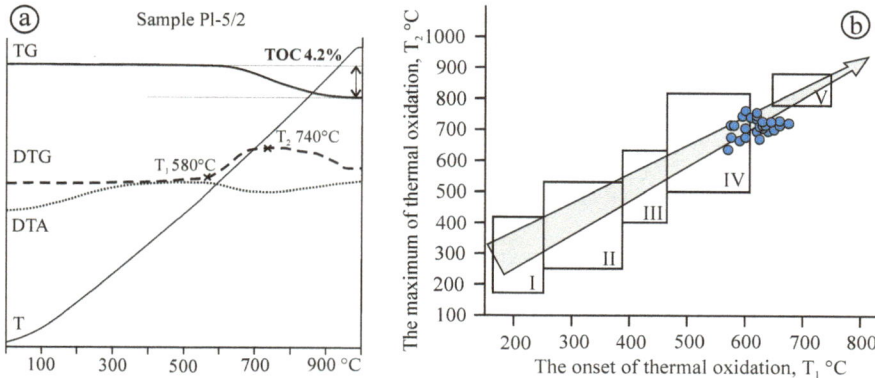

Fig. 4 Typical DTA thermograms for carbonaceous shales of the Bulatovo Strata (**a**); The position of the points of black shales of the Bulatovo Strata on the diagram of thermal stability of carbonaceous substances (**b**). *Note* burnout stages according to V.I. Silaev et al. (Silaev et al. 2012): I—modern plants, organic matter in unmetamorphosed sedimentary rocks, coprolites; II—asphalts, lower kerites; III—asphaltites, kerites; IV—higher kerites, anthraxolites, shungites; V—graphite, carbonado

To assess the degree of influence of the zonal thermogradient thermal field during the formation of the Larino gneiss-granite dome, the exothermic effect of organic carbon was studied in 105 samples of siliceous-carbonaceous shales of the Bulatovo Strata (Fig. 4, Table 2).

It is noteworthy that all sample data on the diagram of thermal stability of carbonaceous substances by Silaeva et al. (2012) (Fig. 4b) fall into field IV (higher kerites, anthraxolites, shungites) located in the central part of the trend.

Table 2 Results of thermal analysis of type carbonaceous rocks

No.	No. sample	T_1 °C	T_2 °C	TOC	No.	No. sample	T_1 °C	T_2 °C	TOC
1	Pl-5/1	600	765	5.1	14	Ms-11	630	720	3.5
2	Pl-5/3	580	720	2.1	15	Chb-3	630	730	1.3
3	Pl-7/1	645	730	0.6	16	5529/2	610	745	3.0
4	Lr-10	595	750	2.8	17	Nik-1/2	625	675	0.8
5	H-17/1	635	720	1.1	18	Nik-3/4	620	760	3.0
6	H-24/2	600	680	1.3	19	Nik-7/2	625	715	1.0
7	Uv-3/3	575	720	1.5	20	Nik-7/4	650	705	2.0
8	Uv-5/1a	575	680	3.3	21	Nik-8	630	710	1.7
9	Uv-5/1b	660	720	1.0	22	Nik-10/6	620	740	2.7
10	Ms-1/2	590	670	2.9	23	Nik-12/1	640	700	0.8

(continued)

(continued)

No.	No. sample	T_1 °C	T_2 °C	TOC	No.	No. sample	T_1 °C	T_2 °C	TOC
11	Ms-5	600	710	1.9	24	Nik-12/2	620	700	1.6
12	Ms-7	620	750	3.2	25	Nik-13	640	700	1.2
13	Ms-8/1	675	725	1.8	26	Nik-14/2	660	735	2.4

Note T_1—The onset of thermal oxidation; T_2—The maximum of thermal oxidation

It should be noted that not absolute figures are of greater interest, but the distribution of temperatures over the area. The construction of a map of burnup temperature values made it possible to identify zones with varying degrees of metamorphic transformations (Fig. 5).

According to the isotherms corresponding to the maximum carbon burnout temperatures of 700 and 650 °C, the boundaries between the amphibolite and epidote–amphibolite, as well as between the epidote–amphibolite and greenschist facies are drawn. The difference of 100 and 150 °C is formed due to the fact that the temperature of the exothermic effect (carbon burnout temperature) is determined by the peak on the thermogram (T_2), and the real temperature of metamorphism is determined by the beginning of the rise of the peak (T_1). It is believed that the graphitization process is irreversible (Bluman et al. 1974; Busek and Beissak 2014). It is noteworthy

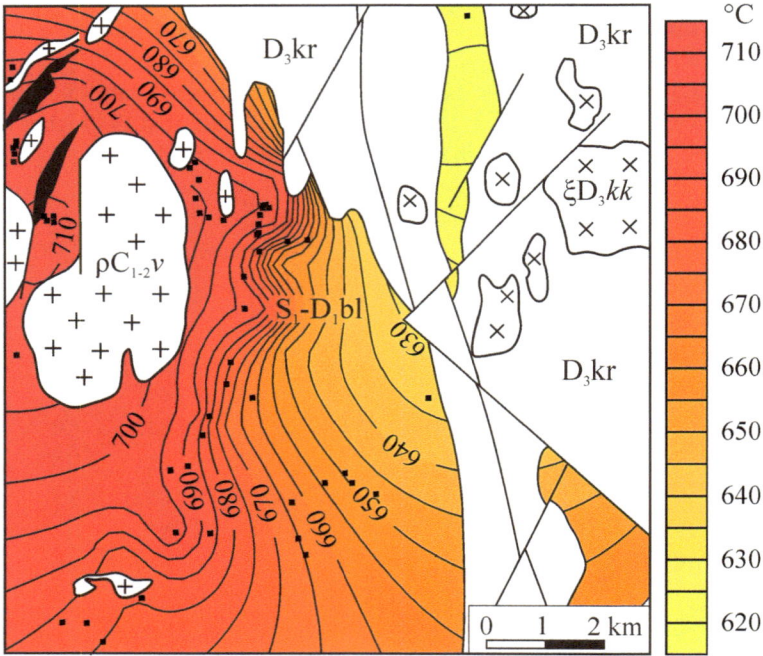

Fig. 5 Map of exothermic temperature isolines (DTA) for carbonaceous shales of the Bulatovo Strata of the eastern framing of the Larino dome (Kriking interpolation method)

that the boundaries of the amphibolite and epidote–amphibolite facies, drawn from petrographic (Fig. 1) and thermogravimetric (Fig. 5) data, satisfactorily correspond to each other.

As a result of processing the obtained data, a very clear pattern of placement of elevated gold grades was revealed. All points with industrial values of gold fall into the area of development of rocks with greenschist facies of metamorphism. Its most significant contents (Malouvelskoe, Nikol'sk, Pridannikovo occurrences) are confined to the outer high-temperature subfacies of the greenschist facies. As well as zones of intensive metasomatic processing of rocks associated with the formation of subalkaline massifs of the Krasnokamensk Somplex near the village Nikolskoe. Within the amphibolite facies, there is not a single point with a gold content higher than 0.01 g/t, all of them have values of this metal either in the region of hundredths of a gram per ton, or below the sensitivity of the method.

5 Conclusions

The geological, petrogeochemical and analytical material presented in this article made it possible to draw the following main conclusions:

1. The framing of the Larino and Pervomay massifs is composed of apovolcanic amphibolites, garnet-mica and garnet-amphibole plagioschists, graphitic quartzites, and siliceous-carbonaceous schists of the Bulatovo Strata.
2. A series of concentric high-gradient zones of metamorphism is observed from granitoids to shales. Mineral parageneses of amphibolite facies at a distance of several kilometers are replaced by associations of epidote–amphibolite and greenschist facies. For rocks of the epidote–amphibolite facies, the metamorphism parameters were: T = 530–550 °C and P = 8.0–8.4 kbar.
3. Disseminated sulfide type of mineralization, identified within the siliceous-carbonaceous deposits of the Larino dome framing, has a distinct lithological-structural control and a clear selective confinement to carbonaceous deposits. The manifestation of zonal metamorphism within the Larino dome contributed to the migration and redeposition of gold mineralization. The confinement of the latter to the outer zone of the greenschist facies of metamorphism is beyond doubt and is a very important search feature in the search for gold mineralization.

Acknowledgements The work was carried out within the framework of the State Order on the topic No. FMRS-2022-0011.

References

Bluman BA, Dyakonov YS, Krasavina TN, Pavlov MG (1974) The use of thermal and radiographic characteristics of graphite to determine the level and type of metamorphism. Notes All-Union Mineralogical Soc 103(1):95–103

Borneman-Starynkevich ID (1964) Guide to the calculation of mineral formulas. Nauka, Moscow, p 224

Buryak VA (1975) Metamorphic-hydrothermal type of gold mineralization. Geology of Ore Deposits 1:37–46

Busek PR, Beissak O (2014) From organic matter to graphite: graphitization. Elements 10(6):421–426. https://doi.org/10.2113/gselements.10.6.421

Groves DI, Goldfarb RJ, Robert F, Hart CJR (2003) Gold deposits in metamorphic belts: overview of current understanding, outstanding problems, future research, and exploration significance. Econ Geol 98:1–29. https://doi.org/10.2113/gsecongeo.98.1.1

Johnson SC, Large RR, Coveney RM, Kelley KD, Slack JF, Steadman JA, Gregory DD, Sack PJ, Meffre S (2017) Secular distribution of highly metalliferous black shales corresponds with peaks in past atmosphere oxigenation. Mineralium Deposita 52:791–798. https://doi.org/10.1007/s00126-017-0735-7

Korobeinikov AF (1985) Peculiarities of gold distribution in the rocks of black shale formations. Geokhimiya 2:1747–1757

Large RR, Bull SW, Maslennikov VV (2011) A carbonaceous sedimentary source-rock model for Carlin type and orogenic gold deposits. Econ Geol 106(3):331–358. https://doi.org/10.2113/econgeo.106.3.331

Nekrasov IYa (1996) Geochemistry, mineralogy and genesis of gold deposits. Rotterdam: Brookfield, 344 p. https://doi.org/10.1201/9780203753651

Perchuk LL, Ryabchikov ID (1976) Phase correspondence in mineral systems. Nedra, Moscow, p 287

Plyusnin KP, Plyusnina AA, Zenkov II (1965) New data on graptolitic shales of the eastern slope of the Southern Urals/Izv. Academy of Sciences of the USSR. Geolgy, No. 11. pp 121–123

Plyusnina LP, Kuz'mina TV, Avchenko OV (2004) Modeling of gold sorption on carbonaceous matter at 20–500°C and 1 Kbar. Geochem Int 42(81):755–763

Puchkov VN, Ivanov KS (1989) On the stratigraphy of black shale strata in the east of the Urals/ Yezhegodnik–1988. IGiG UFAN USSR: Sverdlovsk, pp 4–7

Puzhakov BA, Shokh VD, Schulkina NE, Shchulkin EP, Dolgova OYa, Orlov MV, Popova TA, Tarelkina EA, Ivanov AV (2018) State geological map of the Russian Federation. Scale 1:200 000 (2nd ed.) South Ural series. Sheet N-41-XIII (Plast). Explanatory note. Moscow: VSEGEI, . 205 p

Razvozzhaeva EA, Nemerov VK, Spiridonov AM, Prokopchuk SI (2008) Carbonaceous substance of the Sukhoi Log gold deposit (East Siberia). Russ Geol Geophys 49(6):371–377. https://doi.org/10.1016/j.rgg.2007.09.015

Shumilova TG, Shevchuk SS, Isayenko SI (2016) Metal concentrations and carbonaceous matter in the black shale type rocks of the Urals. Dokl Earth Sci 469(1):695–698. https://doi.org/10.1134/S1028334X16070060

Silaev VI, Smoleva IV, Antoshkina AI, Tchaikovsky II (2012) Experience of conjugate analysis of isotopic composition of carbon and nitrogen in carbonaceous substances of different origin / Problems of mineralogy, petrography and metallogeny: materials of scientific readings in memory of P.N. Chirvinsky. Perm: Publishing House of PSU. 2012. No. 15, pp 342–366

Snachev AV, Kolomoets AV, Rassomakhin MA, Snachev VI (2021) Geology and gold content of carbonaceous shale in Baikal mineralization site. Southern Ural. Eurasian Mining 1:8–13. https://doi.org/10.17580/em.2021.01.02

Snachev AV, Rykus MV, Snachev MV, Romanovskaya MA (2013) A model for the genesis of gold mineralization in carbonaceous schists of the Southern Urals. Mosc Univ Geol Bull 68(2):108–117. https://doi.org/10.3103/S0145875213020105

Snachev AV, Snachev VI (2018) Prospects of carbonaceous deposits of Novousmanovsky area on rhenium, tungsten, molybdenum (Uraltau zone). Geologicheskii Vestnik. 2:68–78. https://doi.org/10.31084/2619-0087/2018-2-5

Snachev AV, Snachev VI, Romanovskaya MA (2015) The geology, petrogeochemistry, and ore content of carbonaceous deposits from the Larinsky dome (South Urals). Mosc Univ Geol Bull 70(2):131–140. https://doi.org/10.3103/S014587521502009X

Snachev AV, Snachev VI, Rykus MV, Savelyev DE, Bazhin EA, Ardislamov FR (2012) Geology, petrogeochemistry and ore bearing of carbonaceous deposits of the Southern Urals. Ufa: Designpoligrafservice, 208 p

Snachev VI, Snachev AV, Prokofiev VYu (2022) Physicochemical conditions of the formation of the Larino granite-gneiss dome (South Ural). Georesursy = Georesources 24(1):74–83. https://doi.org/10.18599/grs.2022.1.7

Starostin VI, Yapaskurt OV (2007) Au–Cu black shale formations. Earth Sci Front 14:245–256. https://doi.org/10.1016/S1872-5791(08)60014-7

Thermo- and barometry of metamorphic rocks/ed. by V.A. Glebovitsky. Leningrad: Nauka. 1977, 207 p

Wood BL, Popov NP (2006) The giant Sukhoi Log deposit, Siberia. Russian Geol Geophys 47:315–341

Yudovich YaE, Ketris MP (2015) Geochemistry of black shales. Moscow-Berlin: Direct-Media, 272 p. https://doi.org/10.23681/428042

Zhdanov AV, Obodov VA, Makariev LB, Matyushkov AD, Molchanova EV, Stromov VA (2018) State geological map of the Russian Federation. Scale 1:200 000 (2nd ed.) South Ural series. Sheet N-40-XVIII (Uchaly). Explanatory note. Moscow: VSEGEI, 386 p

Chemical and Microbiological Features of the Coastal Waters of the Black and Azov Seas in Summer Season

N. V. Burdiyan, O. V. Soloveva, and E. A. Tikhonova

Abstract The purpose of the work was to obtain quantitative characteristics and study the features of the distribution of heterotrophic and hydrocarbon-oxidizing bacteria in surface waters and bottom sediments of coastal waters in the summer, to investigate the relationship between microbiological indicators and some characteristics of the environment (water temperature, depth, chloroform-extractable substances content, oil hydrocarbons content). The results of assessing of abundance and distribution of heterotrophic and hydrocarbon-oxidizing groups of bacteria in the surface water layer and bottom sediments of the coastal waters of the Crimea peninsula, the Caucasus and the Azov Sea in June 2020 are presented. At the same time studies of bottom sediments (habitat of bacteria) of the coastal regions of the Crimea and the Caucasus, as well as the water area in front of the Kerch Strait, were carried out. New data were obtained on the concentrations of the main organic components of origin in sea bottom sediments: chloroform-extractable substances, oil hydrocarbons. The number of heterotrophic bacteria in water ranged from 10^2 to 10^6 cells/mL, and in bottom sediments, from 10^3 to 10^6 cells/g. For hydrocarbon-oxidizing bacteria, oil and diesel fuel were used as the only source of carbon. The data obtained indicate that hydrocarbon-oxidizing microorganisms capable to transform petroleum hydrocarbons are permanent components of plankton and benthos microcenoses. A wide range of number of hydrocarbon-oxidizing bacteria was noted, due to the microzonal distribution, which depends on the hydrological conditions of the water areas and on the distribution of the substrate itself—hydrocarbons. The concentration of bacteria in bottom sediments differed from that in the aquatic environment. Hydrocarbon-oxidizing bacteria isolated from water and bottom sediments grew more abundantly on the medium with the addition of oil than on diesel fuel. According to the correlation analysis, there was no significant relationship between the water temperature index and the number of observed groups of bacteria. A positive correlation was found between the content of hydrocarbon-oxidizing bacteria in the soil and the depth of the sea at sampling stations, between the number of heterotrophic bacteria and hydrocarbon-oxidizing bacteria in the surface water layer. A slight increase in the

N. V. Burdiyan (✉) · O. V. Soloveva · E. A. Tikhonova
A.O. Kovalevsky Institute of Biology of the Southern Seas of RAS, Sevastopol, Russian Federation
e-mail: burdiyan@mail.ru

© The Author(s), under exclusive license to Springer Nature Singapore Pte Ltd. 2023
T. Chaplina (ed.), *Processes in GeoMedia—Volume VII*, Springer Geology,
https://doi.org/10.1007/978-981-99-6575-5_4

concentration of oil hydrocarbons in the bottom sediments of both the Black and Azov Seas and their share in the total amount of chloroform-extractable substances was noted. In general, the level of pollution of bottom sediments with organic substances remained unchanged compare with the data of previous years, in particular, with the data of 2016.

Keywords Coastal seawaters · Sea bottom sediments · Heterotrophic bacteria · Hydrocarbon-oxidizing bacteria · Chloroform-extractable substances · Oil hydrocarbons · Diesel fuel · The Azov Sea · The Black Sea

1 Introduction

The coastal waters of the Black Sea are experiencing a significant anthropogenic load currently. According to average international estimates, about 80 tons of mercury, 4,500 tons of lead, 2,200 tons of phenol, 200,000 tons of oil products, 140,000 tons of phosphates, 570,000 tons of nitrates, 9,800,000 tons of various organic substances enter the Black Sea (mainly with river runoff) per year (Zaitsev 2000). Oil, being the most common source of fuel in the world, is one of the most dangerous pollutants of the biosphere. Getting into the environment, it has a negative impact on all links of the biological chain. Oil pollution inhibits the development of zoo- and phytoplankton, slows down the mineralization of organic matter, increases the oxidizability of water, and changes its salt composition (Israel and Tsyban 2009). In this regard, spills of oil and oil products on the surface of the seas and oceans are a real environmental disaster. Thus, during the transportation of oil in the World Ocean, according to various sources, from 5 to 16 million tons of oil are spilled annually (Nemirovskaya 2013). Therefore, the problem of changes in marine ecosystems under the influence of anthropogenic factors is becoming more and more urgent.

   The coastal water area is the most vulnerable in the environmental aspect, since pollutants are concentrated here, falling both from the coast and from the sea. At the same time, mass contact of people with the marine environment takes place in this zone (Mironov and Mironov 2020).

   As a result of natural sorption processes occurring in water bodies, the levels of accumulation of chemicals in bottom sediments, pore water, and the near-bottom water layer are much higher than in the water column. In addition, bottom sediments are "repositories" of many pollutants. In coastal and shallow water zones, under the influence of wave and wind activity, as well as biological agitation by mollusks, bottom sediments become a source of secondary water pollution (Dauvalter 2012), Therefore they should be studied along with the water column when assessing the quality of water bodies (Gennadiev et al. 2015).

   Bacterioplankton and bacteriobenthos play the main role in the remineralization of biogenic elements and the transformation of organic matter of allochthonous and autochthonous origin. A separate group of heterotrophic bacteria that are of interest for evaluating the process of self-purification of the water area are bacteria that can use oil and oil products as the only source of carbon. Due to the high plasticity of metabolic processes, microbial populations are a powerful factor in the processes of self-purification of the marine environment from pollution by oil and oil products. The study of hydrocarbon-oxidizing microorganisms in natural ecosystems is usually associated with their pollution with oil and oil products. However, it is known that hydrocarbon-oxidizing bacteria are widely distributed in nature, their nutritional requirements are diverse, and there are no highly specialized forms among them (Shlegel 1987). This group of bacteria is a constant component of microbiocenoses, independent of the level of oil pollution. In the marine environment, oil oxidation proceeds along the following pathways: direct oxidation to CO_2, H_2O and organic substances; converting oil into surfactants such as fatty acids that emulsify the more stable components of the crude oil; oxidation to intermediate products. Products of incomplete oxidation of hydrocarbons (hydroperoxides, alcohols, ketones, alde-hydes, lipids, organic acids, amino acids, nucleotides, pigments, sugars, polysac-charides, phenol), in turn, are a substrate for the development of microorganisms (Coronelli 1996).

The purpose of the work was to obtain quantitative characteristics and study the features of the distribution of heterotrophic and hydrocarbon-oxidizing bacteria in surface waters and bottom sediments of coastal waters in the summer, to investi-gate the relationship between microbiological indicators and some characteristics of the environment (water temperature, depth, chloroform- extractable substances, oil hydrocarbons content).

2   Material and Methods

The material for this work was samples of sea water and bottom sediments taken in the coastal waters of the Caucasus and the Black Sea coast of Crimea, as well as at two stations in the Azov Sea, in June 2020 during cruise 113 of the research vessel «Professor Vodyanitsky» (Fig. 1).

Depth at the stations varied from 10 to 723 m, sea water temperature varied from 16.0 to 24.0 °C. Depth, as well as water and air temperature are presented in Table 1. A total of 17 water samples from the surface horizon and 11 samples of bottom sediments for microbiological analysis and 16 samples for chemical analysis were taken for.

Sea bottom sediments were sampled using an «Ocean-50» bottom grab-sampler. The top 5 cm layer was used for analysis. All samples were packed in special containers and labeled. Under laboratory conditions, the sediments were dried to an air-dry state, ground in a mortar, and part of the sample was sieved through sieves with a mesh size of 0.25 mm to determine the concentrations of petroleum hydrocar-bons by means of infrared spectrometry (Oradovsky 1977) on the FSM-1201 spec-trophotometer and chloroform-extractable substances by the gravimetric method. All results obtained for the concentrations of chloroform-extractable substances and petroleum hydrocarbons were recalculated per 100 g of air-dry bottom sediment

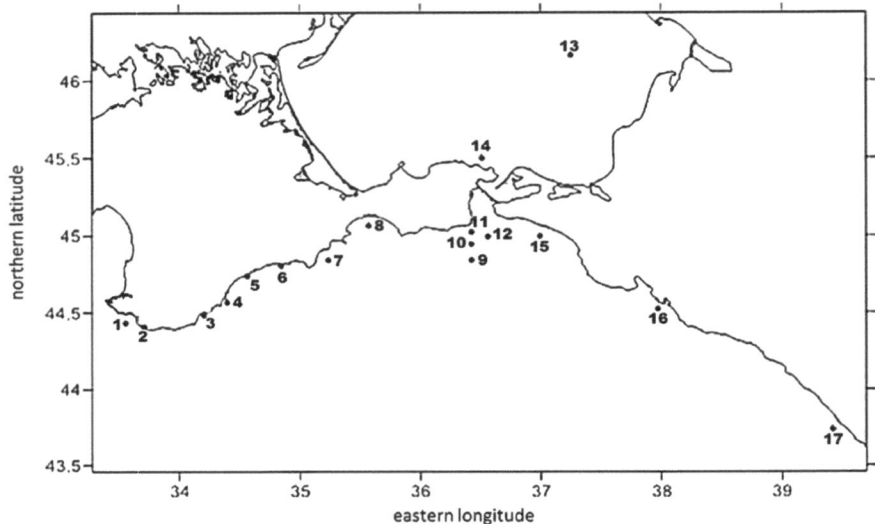

Fig. 1 Scheme of microbiological sampling stations during cruise 113 of the research vessel «Professor Vodyanitsky»

Table 1 Water temperature, air temperature, depth at the sampling stations

Station №	Depth (m)	Temperature	
		Water (°C)	Air (°C)
1	87	16,0	17,9
2	51	16,0	17,9
3	39	18,7	19,1
4	82	20,6	21,3
5	59	20,6	21,3
6	47	22,1	22,2
7	50	21,8	21,9
8	23	20,6	19,8
9	50	21,3	23,8
10	37	21,4	24,8
11	15	21,4	24,8
12	26	21,4	21,9
13	10	23,1	21,2
14	12	23,7	23,6
15	25	23,4	22,1
16	723	24,0	23,1
17	115	24,6	24,3

(air-dry. b.s.). The correlation coefficient was calculated at $P = 0.05$ in Microsoft Excel.

Microbiological work was carried out in the ship's laboratory, immediately after taking samples of water and bottom sediments. Samples of sea water from the surface horizon were taken from the bathometer into sterile glass test tubes with a volume of $21 cm^3$. Sampling of bottom sediments was carried out from the Ocean-50 grab into sterile jars. In the studied samples, the number of heterotrophic and hydrocarbon-oxidizing bacteria was determined.

The determination of the number of bacteria was carried out by the method of limiting tenfold dilutions using elective media. For heterotrophic bacteria, a medium with peptone (g) was used: peptone, 10.0; yeast extract -1.0; sea water -1 L (Mironov et al. 1988). For hydrocarbon-oxidizing bacteria, the Voroshilova-Dianova medium (Voroshilova and Dianova 1952) was used with the following composition (g): NaCl, 18.0; $MgSO_4 \times 7H_2O$ -0.2; KCl -0.7; $KH_2PO_4 - 1.0$; $K_2HPO_4 - 1.0$; $CaCl_2 - 0.02$; $FeCl_3 - 2$ drops; distilled water -1.0 L, pH 7.2–7.4.

As the only source of carbon and energy for hydrocarbon-oxidizing bacteria, depending on the tasks set, sterile oil or diesel fuel was added to each test tube (1% of the volume of the medium). When preparing the media, the salinity of sea water was taken into account. The most probable number of microorganisms per unit volume was calculated using the McCready table (in triplicate), based on the method of variation statistics (Netrusov 2005). The cultures were incubated at room temperature.

3 Results and Discussion

Surface waters. The results of microbiological studies showed that the number of heterotrophic bacteria in the surface water layer varied from 10^2 to 10^6 cells/mL (Fig. 2a).

Starting from the water area of Cape Khersones to the Feodosia Bay (inclusive), the number of heterotrophic bacteria ranged from 10^2 to 10^6 cells/mL. The maximum number was observed once—in the water area of the Karadag nature reserve (station 7). In the Feodosiya Bay, the number of heterotrophic bacteria was 10^4 cells/mL; at other stations on the Crimean coast, the quantitative indicators of heterotrophic bacteria varied within 10^2–10^3 cells/mL. In the zone of active navigation, which is the region near the Kerch strait, the number of heterotrophic bacteria varied within 10^3–10^4 cells/mL. In the region of the Azov fore-strait and the Taganrog Bay (Azov Sea), the number of heterotrophic bacteria did not exceed a thousand bacterial cells per mL of water. In the coastal area of the Caucasus coast, the largest number of heterotrophic bacteria (10^5) was observed at st. 15, at other stations the heterotrophic bacteria indicators were within the third and fourth order.

Hydrocarbon-oxidizing bacteria, using crude oil as the sole source of carbon and energy, were isolated from all seawater samples. The number of hydrocarbon-oxidizing bacteria varied in a fairly wide range: from 1 to 4.5×10^3 cells/mL

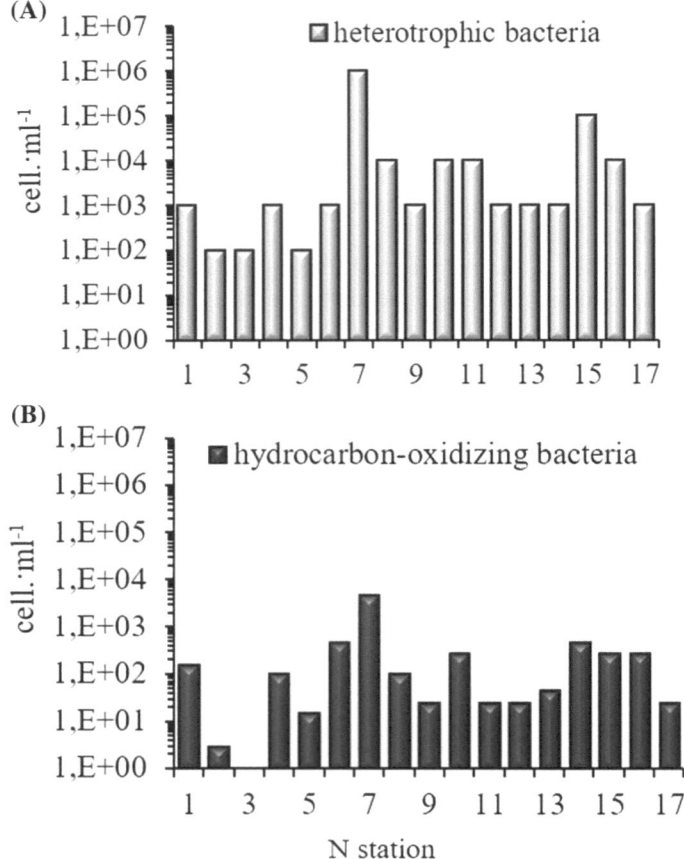

Fig. 2 Abundance (cells/mL) of heterotrophic (**a**) and hydrocarbon-oxidizing bacteria (**b**) in the surface water layer of the coastal waters of the Black and Azov Seas

(Fig. 2b). In 53% of the samples, the indicators of the number of hydrocarbon-oxidizing bacteria varied from 95 to 4.5×10^3 cells/ml. The maximum was recorded in the Karadag area, high values of the hydrocarbon-oxidizing bacteria abundance were observed both at the stations closest to Karadag and in the Feodosiya Bay. Near the Kerch strait, the abundance of hydrocarbon-oxidizing bacteria ranged from 25 to 250 cells/mL. In turn, in the Azov Sea near the Kerch strait, the content of hydrocarbon-oxidizing bacteria was 450 cells/mL, and in the Taganrog Bay, it was 45 cells/mL. In the water area of the Caucasus, the largest number of hydrocarbon-oxidizing bacteria −250 cells/mL was noted at stations 15 and 16. Single indicators (1–3 cells/mL) were found only in the water area of Laspi Bay and travers of Yalta. It should be noted (Burdiyan and Doroshenko 2021) that in April 2019, the number of hydrocarbon-oxidizing bacteria in the surface water layer of Laspi Bay was the smallest in the coastal waters of the Crimean Black Sea coast, and did not exceed one bacterial cell per ml of water. The results of our studies in 2016 showed slightly elevated concentrations of petroleum hydrocarbons (about 0.5 MPC) in the area of the Laspi bay and on the traverse of the Yalta city. At other stations, including those in the Caucasus coast and the Azov Sea of in 2016, the concentrations of petroleum hydrocarbons in the water indicated the absence of oil pollution.

In 53% of the total number of water samples obtained, the number of bacteria using diesel fuel as the only source of carbon and energy was determined. The number of hydrocarbon-oxidizing bacteria in 55% of the samples ranged from 1 to 450 cells/ mL, in the rest of the samples zero values were recorded. The maximum indicator was singled out once at the entrance to the Balaklava Bay, where small-sized motor vessels often ply. In the rest of the samples, the number of hydrocarbon-oxidizing bacteria ranged from 1 to 25 cells/mL.

A close relationship was found between the abundance of heterotrohic and hydrocarbon-oxidizing bacteria in the surface water layer (R = 0.98; $\alpha < 0.05$). A similar relationship between the numbers of these groups of bacteria in water was noted in the Kola Bay (Litvinova et al. 2012).

There is no reliable relationship between the water temperature in the range from 16.0 to 24.6 (°C) and the number of observed groups of bacteria. It is known that water temperature, as an essential physical factor of the environment, affects the quantitative indicators of marine bacteria only within those temperature limits that favor the growth of microorganisms in general. Only low temperatures make the number of bacterial populations in the sea linearly dependent (Nemirovskaya 2004).

The absence of a relationship between the number of heterotrophic bacteria and temperature is apparently explained by the state of the phytoplankton community. As is known, outbreaks in the development of bacterioplankton follow, with some delay, outbreaks in the development of phytoplankton. As a rule, the seasonal succession of the Black Sea bacterioplankton has two peaks: early spring and autumn, caused by the development of phytoplankton. The death of phytoplankton leads to the enrichment of water with organic substances and, indirectly, to an increase in the number of heterotrophic microorganisms (Muravyeva and Gaponyuk 2004).

Bottom sediments. The number of heterotrophic bacteria in bottom sediments varied from 10^3 to 10^6 cells/g (Fig. 3a). It should be noted that the sea bottom sediments samples were silty with an admixture of sand and shell rock; in some samples, a slight smell of hydrogen sulfide was felt. The described type of bottom sediments is the most common in the study region and is capable of accumulating not only natural organic matter, but also compounds of petrogenic origin.

In the bay of Laspi and on the traverse of Yalta, the number of heterotrophic bacteria was 10^4 cells/g, at the same time, the chloroform-extractable substance's values in the sea bottom sediments of the Crimean coast ranged from 28 to 225 mg/100 g air-dry. b.s. (Fig. 4a). The highest indicators of chloroform-extractable substances for the Crimean water area were noted in the area of Opuk Cape, the lowest in the bottom sediments of the st. 1 (near Sevastopol city). In the water area of the Karadag nature reserve, the concentration of chloroform-extractable substances was

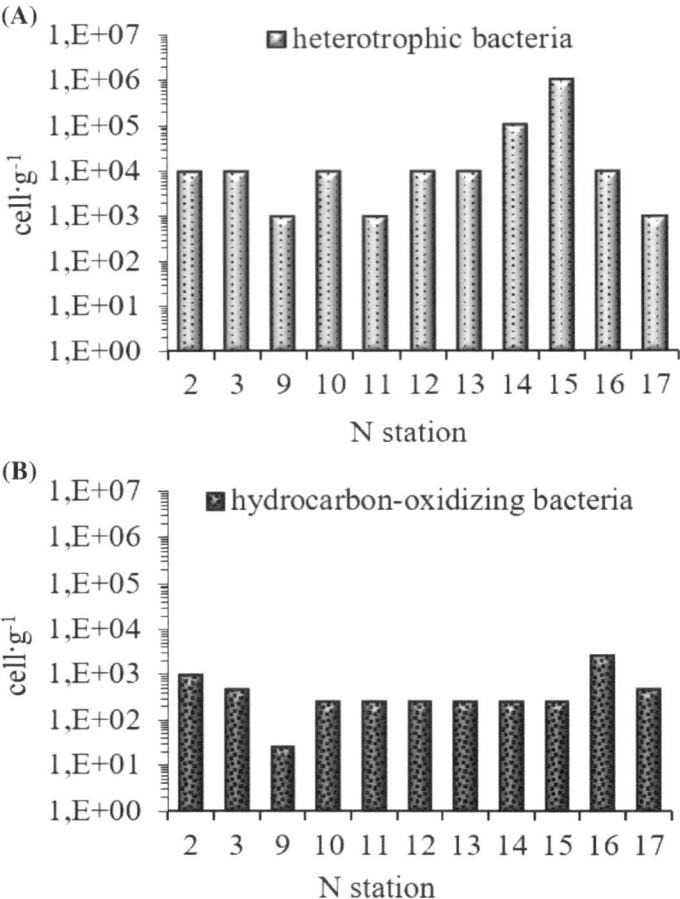

Fig. 3 Abundance (cells/g) of heterotrophic (**a**) and hydrocarbon-oxidizing (**b**) bacteria in bottom sediments of the coastal waters of the Black and Azov Seas

120 mg/100 g air-dry. b.s. Quantitative indicators of heterotrophic bacteria in the near the Kerch strait varied within the third-fourth order of magnitude, in the Azov water area near the strait—an order of magnitude higher. At the time of the study, the concentration of chloroform-extractable substances in the region of the Kerch strait was 70 mg/100 g before. At the time of the study, the concentration of CEI in the region of the Kerch strait was 70 mg/100 g air-dry. b.s. At st. 13 (Azov Sea), the number of heterotrophic bacteria did not exceed 10^4 cells/g. In the bottom sediments of the Azov Sea, the chloroform-extractable substances values varied in the range from 85 to 125 mg/100 g air-dry. b.s. The maximum number of heterotrophic bacteria was allocated once per st. 15. At two stations in the water area of the Caucasian coast, the number of heterotrophic bacteria was equivalent and amounted to 10^3 cells/g. The

chloroform-extractable substances values in the bottom sediments of the Caucasus coast ranged from 20 to 110 mg/100 g air-dry. b.s.

According to the regional scale for the assessment of organic pollution and its impact on macrozoobenthos (Tang et al. 2012), five levels of pollution of bottom sediments were determined according to the content of mg/100 g air-dry. b.s. in them: the first—less than 50 mg/100 g air-dry. b.s.; the second −50–100 mg/100 g air-dry. b.s.; third −100–500 mg/100 g air-dry. b.s.; fourth −500–1000 mg/100 g air-dry. b.s.; and the most dangerous fifth—above 1000 mg/100 g air-dry. b.s. Most samples of sea bottom sediments of the Black Sea coast of Crimea belong to the II-nd level of pollution with an average content of chloroform-extractable substances of 72 mg/100 g air-dry. b.s. Conditionally clean (I-th level), according to the content of

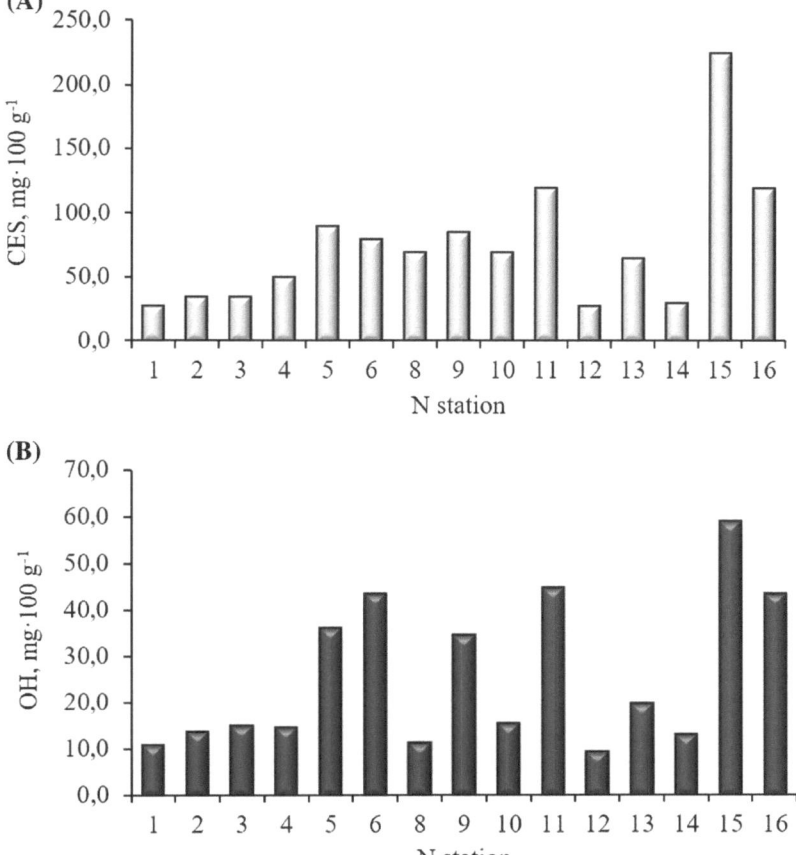

Fig. 4 Concentrations of chloroform-extractable substances (CES) (**a**) and total petroleum hydrocarbons (TPH) (**b**) in sea bottom sediments sampled during the 113th cruise of the R/V «Professor Vodyanitsky»

chloroform-extractable substances, is Laspi Bay, the coastal water area of Opuk Cape and the Parthenid village. Sea bottom sediments classified as pollution level III were determined at station 1 and in the Karadag area. The concentration of chloroform-extractable substances in the sea bottom sediments of the Caucasian coast were somewhat lower than near the Crimean coast. The sea bottom sediments of the noted areas of the Crimean coast were previously classified as conditionally clean. However, for example, in the surface and near-bottom layers of water in the coastal waters of the Laspi Bay reserve, elevated concentrations of hydrocarbons were previously noted (Soloveva et al. 2019, 2020), while the level of pollution of sea bottom sediments with organic substances remained quite low. In previous years of research (2016–2017), the concentration of chloroform-extractable substances was 42 mg/100 g, at present it is 40 mg/100 g air-dry. b.s. But, in general, based on a comparison of this indicator over a long period of research, the content of chloroform-extractable substances in the sea bottom sediments of the Black Sea coast of Crimea can be considered characteristic of the region (Tikhonova et al. 2020; Mironov et al. 1992). It should be noted that the obtained data on the concentration of chloroform-extractable substances in the bottom sediments of the Caucasian coast are somewhat lower than those near the Crimean coast. This is also consistent with the data on the content of hydrocarbons in water, where this indicator in the sea water of the Caucasian coast was lower than that of the Crimean coast (Mironov and Mironov 2020). Also, lower concentrations could be associated with different composition of bottom sediments (st. 15 shell rock with sand admixture) and sampling depth (st. 16) were deep-water with depths of 723 m. In the bottom sediments of the Azov Sea, the chloroform-extractable substances values varied from 85 to 125 mg/100 g air-dry. b.s., which is somewhat lower than previously noted. In 2016, the maximum recorded values (187 mg/100 g air-dry. b.s.) corresponded to those obtained by us in 2010 (186 mg/100 g air-dry. b.s.) (Tikhonova et al. 2021). Nevertheless, the concentrations of chloroform-extractable substances did not exceed the values typical for the study area (20 mg/100 g air-dry. b.s. for shell specimens, up to 230 mg/100 g air-dry. b.s. for pelitic sediments (Tikhonova et al. 2016)). The results obtained correspond to those previously described, and sea bottom sediments could be classified as naturally clean according to this indicator (Tikhonova et al. 2021; Soloveva et al. 2018). Previously (Tikhonova et al. 2016; Mironov 1996), trace amounts of oil hydrocarbons (up to 5 mg/100 g air-dry. b.s.) were noted in the sea bottom sediments in both the Black and Azov Seas at the greatest number of stations. Whereas in the present study, the indicators were somewhat higher: in the Black Sea area of the Crimean coast—from 9.3 to 59 mg/ 100 g air-dry. b.s., in the Caucasus—from 5.3 to 27 mg/100 g air-dry. b.s. In the Azov water area in 2010, at 65% of stations, the oil hydrocarbons concentrations in bottom sediments were less than 5 mg/100 g air-dry. b.s. (Tikhonova et al. 2021). In 2016 (Tikhonova et al. 2016) and in 2020, at all stations, the amounts were higher than trace amounts (with an average value of 13.5 mg/100 g air-dry. b.s.). The obtained indicators may indicate the inflow of fresh oil products into the studied water areas. First of all, this was due to shipping activity and coastal runoff, including the increasing recreational load on the Crimean coast. Thus, for the Black Sea, the share of their inflow with runoff, including domestic, industrial, storm and river, is more than 95%

(Lebedev 2008). The most unfavorable conditions were noted in the runoff area of the city of Sochi. According to modern satellite observations, the formation of an oil film on the sea surface is not always caused by human activity; it can also be a natural factor—the functioning of oil and gas seeps, mud volcanoes (Lavrova and Mityagina 2013). The total volume of oil products entering the Black Sea is estimated at 270 thousand tons per year (Zagranichny 2014). At the same time, the level of oil pollution of both the Crimean and Caucasian coasts did not exceed the values typical for clean and slightly polluted water areas of the Black Sea. In the Azov Sea, an undulating increase and decrease in the concentrations of oil products in bottom sediments was also noted earlier (Pavlenko et al. 2018). On the one hand, the frequency of occurrence of their content of more than 1 g/kg, at which sublethal effects are possible, had decreased to 0–6% since 2002 (Pavlenko et al. 2018). On the other hand, in the places of sampling of sea bottom sediments (for example, the Taganrog Bay), in most cases, the normative concentrations of oil hydrocarbons in the water were exceeded (Stepanyan 2019). And one of the sources of pollutants entering the bottom sediments is their entry from the water column of the sea.

Thus, the maximum oil hydrocarbons values, as well as chloroform-extractable substances concentrations, were recorded in the coastal water area of Sevastopol (59 mg/100 g air-dry. b.s.) and Karadag (45 mg/100 g air-dry. b.s.). If compare with the standards of the so-called "Dutch Lists", then the obtained values did not exceed the normative ones, with the exception of the coastal areas of the city of Sevastopol. At the entrance to the Kerch Strait at all stations, the mg/100 g air-dry. b.s. concentration did not exceed 18 mg/100 g air-dry. b.s., which is a characteristic indicator for this navigable water area (Mironov et al. 1992; Tikhonova et al. 2015). The maximum load falls on the water area of the Kerch Strait itself and the port. Here, along with active shipping, dumping and hydraulic construction, the currently prohibited reloading of fuel from small oil tankers to large-capacity tankers is practiced (Nemirovskaya 2013). But, in these areas, sampling of bottom sediments was not carried out in this work. The minimum amounts, close to trace ones, were noted in the coastal waters of the Caucasian coast.

The percentage of oil hydrocarbons from chloroform-extractable substances, which indicates the degree of hydrocarbon content (Mironov et al. 1992), fluctuated within the following limits: in the Black Sea region of Crimea, from 21% (pre-strait zone) to 55% (Malorechensk, the water area near the Mayak temple), up to 50% in the water area of the village Partenid. In other water areas the percentage was below 44%; the coastal water area of the Caucasus—from 16 to 35%, the Azov water area—from 11 to 15%. Compared to the data of previous years, the percentage of oil hydrocarbons from chloroform-extractable substances slightly increased: in 2016, in the Azov Sea bottom sediments it did not exceed 6%, in the Black Sea − 31%. Previously, indicators of more than 50% were not recorded in any sample, at the present—in two samples out of 19. These indicators indicate a constant supply of allochthonous hydrocarbons to the environment, the source of which may be coastal runoff. Since samples of bottom sediments were taken mainly in the coastal water area, and the second most intense source of this class of substances (after oil spills) is their entry from the shore (Soloveva et al. 2017).

A close correlation was also noted between the content of chloroform-extractable substances and oil hydrocarbons: in the bottom sediments of the Crimean coast, the correlation coefficient was 0.89, in the Caucasian coast, 0.88, and in the Azov water area, 0.8. Previously, such a close dependence was not noted (r = 0.5) (Tikhonova et al. 2016).

Hydrocarbon-oxidizing bacteria were found in 100% of the samples, their number varied in a wide range: from 25 to 2.5×10^3 cells/g (Fig. 3b). The maximum values of hydrocarbon-oxidizing bacteria -2.5×10^3 cells/g were found at the maximum depth (723 m), near the Caucasian coast. In the bottom sediments near the Kerch strait, the hydrocarbon-oxidizing bacteria number varied from 25 to 250 cells/g, oil hydrocarbons concentration was 18 mg/100 g air-dry. b.s. At other stations, the abundance of hydrocarbon-oxidizing bacteria varied from 250 to 950 cells/g. The maximum values of oil hydrocarbons, as well as chloroform-extractable substances, were recorded at st. 1 (59 mg/100 g air-dry. b.s.) and in the sea bottom sediments of the Karadag Nature Reserve (45 mg/100 g air-dry. b.s.). The minimum concentration of hydrocarbons, close to trace levels, was noted in the bottom sediments of the Caucasian coast (Fig. 4b).

In 53% of the total number of bottom sediment samples taken by us, bacteria were identified that use diesel fuel as the only source of carbon and energy. The number of hydrocarbon-oxidizing bacteria varied from 1 to 2.5×10^3 cells/g. The content of hydrocarbon-oxidizing bacteria in the seddiments of the Caucasian coast was expressed as zero and single values. The number of hydrocarbon-oxidizing bacteria in the area of parking of large-tonnage vessels did not exceed 25 cells/g. For comparison, at two stations in the Azov Sea, including the Azov fore-strait, the number of hydrocarbon-oxidizing bacteria was 250 cells/g. The highest indicator of the abundance of hydrocarbon-oxidizing bacteria -2.5×10^3 cells/g was observed in the bottom sediments of Laspi Bay. On the traverse of Yalta, the content of this group of bacteria was two orders of magnitude lower and did not exceed 100 cells/g.

A positive correlation was found between the number of hydrocarbon-oxidizing bacteria and depth at sampling stations (R = 0.95; $\alpha < 0.05$), while the correlation analysis showed a weak relationship between the initial parameter and the number of heterotrophic bacteria (R = $-0, 13$; $\alpha < 0.05$).

In the majority of samples, heterotrophic bacteria grew more abundantly on the medium with the addition of oil than on diesel fuel (Fig. 5a, b).

A wide range of hydrocarbon-oxidizing bacteria abundance in water and bottom sediments was noted. It is known that the mesoscale distribution of hydrocarbon-oxidizing bacteria in the Black Sea is determined by water mixing due to circulation, surge phenomena, as well as periodic current fluctuations and turbulence (Rozman and Tarkhova 1984). The mosaic nature of the hydrocarbon-oxidizing bacteria abundance distribution is also explained by the distribution of the pollutant itself, its spatial heterogeneity, temporal variability, and pronounced seasonal variations in concentrations (Nemirovskaya 2013).

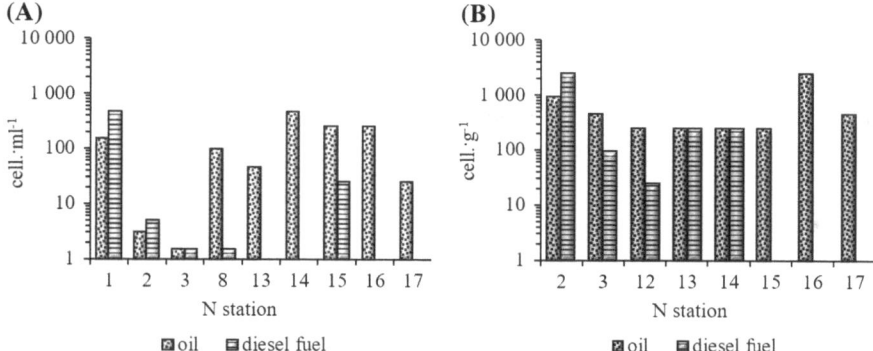

Fig. 5 Number of hydrocarbon-oxidizing bacteria isolated from water samples (**a**) and bottom sediments (**b**), obtained by cultivation on a medium with the addition of oil and diesel fuel

4 Conclusion

The results of microbiological studies carried out during the summer period during cruise 113 of the research vessel «Professor Vodyanitsky» showed that the number of heterotrophic bacteria in water varied from 10^2 to 10^6 bacterial cells per ml of water, in bottom sediments—from 10^3 to 10^6 bacterial cells per gram of the bottom sediments. The highest number of heterotrophic bacteria in the surface water layer was found in the water area of the Karadag nature reserve, and in bottom sediments in the area of the Taman Peninsula.

The data obtained indicate that hydrocarbon-oxidizing microorganisms capable of transforming petroleum hydrocarbons were constant components of plankton and benthos microcenoses. Their share in the total number of heterotrophic bacteria varies from 2.5 to 45%. The number of hydrocarbon-oxidizing bacteria using oil as a source of nutrition in the surface layer of water varied from 1 to 4.5×10^3 cells/ml, in bottom sediments from 25 to 2.5×10^3 cells/g. In the surface layer of water, the highest indicators of the abundance of hydrocarbon-oxidizing bacteria were recorded in the water area of Karadag, in bottom sediments—in the region of the Caucasian coast.

Quantitative indicators of hydrocarbon-oxidizing bacteria capable of using diesel fuel as a food source varied from 1 to 4.5×10^2 cells/mL in the surface water layer, and from 1 to 4.5×10^3 cells/mL in bottom sediments. An increase in hydrocarbon-oxidizing bacteria was not detected in all samples of water and bottom sediments. High number of hydrocarbon-oxidizing bacteria in the surface water layer was observed in the area with active navigation of small boats (the entrance to the Balaklava Bay). In the bottom sediments, the highest abundance of hydrocarbon-oxidizing bacteria was found in Laspi Bay. Quantitative indicators of bacteria that oxidize oil in most were cases higher than the number of bacteria grown on a medium with the addition of diesel fuel.

For the group of hydrocarbon-oxidizing bacteria, a wide range of numbers was noted, due to the microzonal distribution, which, as a rule, depends on the hydrological conditions of the water areas and on the distribution of the substrate itself—hydrocarbons. The concentration of bacteria in bottom sediments differs from that in the aquatic environment.

There was no reliable relationship between the water temperature in the range from 16.0 to 24.6 ($^\circ$C) and the number of bacteria groups observed. A positive correlation was found between the amount of hydrocarbon-oxidizing bacteria in the sea bottom sediments and the depth of the sea at sampling stations (R $= 0.95$; $\alpha < 0.05$), between the number of heterotrophic bacteria and hydrocarbon-oxidizing bacteria in the surface water layer (R $= 0, 98$; $\alpha < 0.05$).

The analysis of sea bottom sediments for the content of chloroform-extractable substances and oil hydrocarbons in them allowed us to conclude that the sea bottom sediments of the Crimean and Caucasian coasts of the Black Sea, as well as the Azov coast, taken during the summer expedition, had properties typical of marine sediments of the studied region. This indicated the favorable state of the studied water areas as a whole. In accordance with the regional classification of pollution of bottom sediments according to the chloroform-extractable substances concentrations, the maximum indicators, both for the Black Sea and for the Azov coast, belong to the III level of pollution (23% of the studied samples). At the same time, the concentrations of chloroform-extractable substances in the bottom sediments of the Caucasian and Azov coasts were somewhat lower than those near the Crimean coast. The maximum recorded concentrations of chloroform-extractable substances related to the III level of pollution were noted in the bottom sediments of the water area of Sevastopol (225 mg/100 g air-dry. b.s.), the Karadag region (120 mg/100 g air-dry. b.s.), the southern part of the Azov Sea (125 mg/100 g air-dry. b.s.) and the city of Tuapse (110 mg/100 g air-dry. b.s.).

In 46% of all analyzed samples, sea bottom sediments (the waters of the Feodosiya Bay, Malorechensk, Yalta, the pre-strait part of the Kerch Strait, the Taganrog Bay (Azov Sea), the coastal part of the Caucasian coast belonged to the II-nd level of pollution with an average chloroform-extractable substances content of 72 mg/100 g air-dry. b.s.

The rest of the water areas (31%) could be attributed to conditionally pure level I (less than 50 mg/100 g air-dry. b.s.). These were the water areas of Opuk Cape, Laspi Bay, Partenida vilage and the coastal zone of the Caucasian coast.

A slight increase in the concentration of hydrocarbons in the sea bottom sediments of both the Black and Azov Seas, as well as their share growth in the total amount of chloroform-extractable substances was noted. In general, the level of pollution of bottom sediments with organic substances remained unchanged compared with the data of previous years, in particular, with the data of 2016.

Acknowledgements Present work has carried out within framework of the State assignment of the IBSS on the theme "Molismological and biogeochemical foundations of the marine ecosystems homeostasis" (№ 121031500515-8). The studies were carried out during cruise 113th of the research vessel «Professor Vodyanitsky» (center of collective usage R/V «Professor Vodyanitsky»).

References

Burdiyan NV, Doroshenko YuV (2021) The number and distribution of heterotrophic and hydrocarbon-oxidizing bacteria in the coastal waters of the Crimea and the Caucasus in the spring period (based on the materials of the 106th cruise of the R/V "Professor Vodyanitsky"). Study of aquatic and terrestrial ecosystems: history and modernity: tez. dokl. International Scientific Conference, dedicated. The 150th anniversary of the Sevastopol Biological Station—Institute of Biology of the South Seas named after A. O. Kovalevsky and the 45th anniversary of the NIS "Professor Vodyanitsky", September 13–18, 2021, Sevastopol, Russian Federation, 500–501. Sevastopol: IBSS. (in Russ.)

Coronelli TV (1996) Principles and methods of intensification of biological destruction of hydrocarbons in the environment. Appl Biochem Microbiol 32(6):579–585 (in Russ.)

Dauvalter VA (2012) Geoecology of Lake Sediments, p. 242. Murmansk: izd-vo MGTU. (in Russ.)

Gennadiev AN, Pikovskii YI, Tsibart AS, Smirnova MA (2015) Hydrocarbons in soils: orgin, composition, and behavior (Review). Eurasian Soil Sci 48(10):1076–1089

Israel YuA, Tsyban AV (2009) Anthropogenic ecology of the ocean. Flinta, Moscow, p 529 (in Russ.)

Lavrova OYu, Mityagina MI (2013) Satellite monitoring of oil slicks on the black sea surface. Izvestiya Atmosph Oceanic Phys 49(9):897–912

Lebedev SA (2008) Model calculations of background values of anthropogenic pollution with petroleum products and assimilation capacity of the black Sea (using Remote Sensing Data). Engineering Ecology 5:41–51 (in Russ.)

Litvinova MYu, Ilyinsky VV, Semenenko MN, Peretrukhina IV (2012) Distribution and potential activity of hydrocarbon-oxidizing bacteria in the water of the middle and northern tribes of the Kola Bay. Proceedings of the Murmansk State Technical University "Bulletin of MSTU" 15(3), 533–540. Murmansk, MSTU. (in Russ.)

Mironov OG (1996) Sanitary-biological characteristic of the Azov Sea. Gidrobiologicheskiy Zhurnal [hydrobiological Journal] 32(1):61–67 (in Russ.)

Mironov OA, Mironov OG (2020) Current level of oil hydrocarbons in Russian coastal waters of the black Sea and Azov Sea. South Russia Ecol Dev 15(3):77–85 (in Russ.)

Mironov OG, Milovidova NYu, Shchekaturina TL (1988) Biological aspects of oil pollution of the marine environment, p 248. Kiev: Naukova dumka. (in Russ.)

Mironov OG, Kiryukhina LN, Divavin IA (1992) Sanitary and biological studies in the black sea, p115. Saint-Petersburg: Girdometeoizdat. (in Russ.)

Muravyeva IP, Gaponyuk TO (2004) Some factors affecting the self-purification of seawater. Ecologiya Morya 66:79–81 (in Russ.)

Nemirovskaya IA (2004) Hydrocarbons in the ocean (snow-ice-water-suspension-bottom sediments). Scientific World, Moscow, p 328 (in Russ.)

Nemirovskaya IA (2013) Oil in the Ocean (Pollution and Natural Flow), p 432. M.: Scientific World. (in Russ.)

Netrusov AI (2005) Practical guidelines for microbiology. Akademiya, Moscow (in Russ.)

Oradovsky SG (1977) Guidelines for methods of chemical analysis of sea waters, 118–131. L.: Gidrometeoizdat. (in Russ.)

Pavlenko LF, Skrypnik GV, Klimenko TL, Anokhina NS, Ekylic VS, Sevostyanova MV, Barabashin TO (2018) Long-term dynamics of oil pollution of hydrobionts environment in the Azov Sea. Prob Fisheries 19(4):534–544 (in Russ.)

Rozman LD, Tarkhova EP (1984) The effect of water mixing on the distribution of oil-oxidizing microorganisms in the surface layer of the Black Sea. Ecologiya Morya 16:60–64 (in Russ.)

Shlegel G (1987) General microbiology. Mir, Moscow, p 567 (in Russ.)

Soloveva OV, Tikhonova EA, Mironov OA (2017) The concentrations of oil hydrocarbons in coastal waters of crimea. Scientific Notes of V.I. Vernadsky Crimean Federal University. Biol Chem 3(3):147–155 (2017). (in Russ.)

Soloveva OV, Tikhonova EA, Mironov OA, Zakharchenko DA (2018) Monitoring of oil hydro-carbons concentrations in the coastal waters of the crimea. Water Chem Ecol (4–6):19–24. (in Russ.)

Soloveva O, Tikhonova E, Burdiyan N (2019) Catastrophe of a tanker and its traces in the ecosystem of the strait (on Example of the Accident in the Kerch Strait). In: BAS, 2019. 19th international multidisciplinary scientific Geoconference SGEM 2019: conference proceedings. 30 June–6 July 2019, Albena, Bulgaria. Sofia. 19(5.2), 203–208

Soloveva O, Tikhonova E, Mironov O (2020) Total petroleum hydrocarbons in the coastal waters of Crimean Peninsula. In: BAS, 2020. 20th international multidisciplinary scientific geoconference SGEM 2020: conference proceedings. 18–24 August 2019, Albena, Bulgaria. Sofia 20 (5.1), 857–862

Stepanyan OV (2019) Impact of oil Spils on coastal waters and aquatic plants. Environ Protect Oil Gas Complex 2:12–17 (in Russ.)

Tang J, Lu X, Sun Q, Zhu W (2012) Aging effect of petroleum hydrocarbons in soil under different attenuation conditions. Agr Ecosyst Environ 149:109–117

Tikhonova EA, Burdiyan NV, Soloveva OV, Doroshenko JV (2015) Chemical and microbiological parameters of the Kerch strait sea bottom sediments after the accident of "Volgoneft-139" Ship. Environ Protect Oil Gas Complex 4:12–16 (in Russ.)

Tikhonova EA, Kotelyanets EA, Soloveva OV (2016) Evaluation of the contamination level of sea bottom sediments on the Crimean coast of the black and Azov Seas. Principles Ecol 5:56–70 (in Russ.)

Tikhonova EA, Soloveva OV, Mironov OA, Burdiyan NV (2020) Sanitary and biological charac-teristics of the Laspi Reserve Coastal Waters (the Black Sea). Ecological Safety of Coastal and Shelf Zones of Sea 3, 95–106. (in Russ.)

Tikhonova EA, Kotelyanets EA, Soloveva OV (2021) Sea bottom sediments pollution of the Crimean coast (the Black and Azov Seas). In: Chaplina T (ed) 2021. Progress in GeoMedia 2, 199–211. Cham, Switzerland: Springer Nature Switzerland AG

Voroshilova AA, Dianova EV (1952) Oil–oxidizing bacteria—indicators of the intensity of biological oxidation of oil in natural conditions. Microbiology 21(4):408–415 (in Russ.)

Zagranichny KA (2014) To the question of sources and volumes of oil components, input into black sea. Eng J Don 1:80–92 (in Russ.)

Zaitsev YuP (2000) The Black Sea: the state of the ecosystem and ways to improve it, p 45. V. I. Vernadsky Youth Ecological Center. (in Russ.)

On the Issue of "True Polar Wander" Phenomenon and Its Alternative Physical Interpretation Based on Galactic Model

A. A. Barenbaum

Abstract The phenomenon of "true polar wander (TPW)" is discussed, which consists in a change in the orientation of Earth's rotation axis, which is judged by a change in the position of Earth's magnetic poles, determined by paleomagnetic methods. The TPW phenomenon is used to study the movements of lithospheric plates in Earth's geological past, as well as to explain their periodic association into supercontinents in different hemispheres of Earth. It is believed that TPW phenomenon is caused by a change in the direction of Earth main inertia axis due to endogenous mantle processes. However, this explanation is not sufficiently substantiated physically. Based on the physical mechanism of the cyclic formation of supercontinents by galactic comets previously proposed by the author, a hypothesis is proposed that considers lithosphere with lithospheric plates as a convective shell of our planet, autonomously rotating in asthenosphere relative to mantle. A model has been constructed that explains the TPW phenomenon by a change not in Earth's axis orientation, but in the rotation axis of its lithospheric shell. This model takes into account for the first time the Coriolis force influence on continental and oceanic plate's movement. Based on this model, a conclusion was made about a higher rate of the shell rotation compared to rotation of mantle and Earth as a whole. Using Pangaea supercontinent breakup over the past 250 Ma as an example, it is shown that Atlantic Ocean formation and appearance of transform faults on Atlantic mid-ocean ridge can be explained by Coriolis force action. The lithospheric shell, in which the "old"—continental and "young"—oceanic plates are formed and moved, is considered as a consequence of the convection of cooling surface layer of rocks, which were completely melted of 4.6 billion years ago during the Moon formation.

Keywords True polar wander · Supercontinental cyclicity · Galactic model · Galactic comets · Coriolis force

A. A. Barenbaum (✉)
Oil and Gas Research Institute RAS, Moscow, Russia
e-mail: azary@mail.ru

© The Author(s), under exclusive license to Springer Nature Singapore Pte Ltd. 2023   49
T. Chaplina (ed.), *Processes in GeoMedia—Volume VII*, Springer Geology,
https://doi.org/10.1007/978-981-99-6575-5_5

1 Introduction

"True polar wander (TPW)" is a loop-like change in the direction of Earth's axis relative to mantle (Sager and Koppers 2000) or lithosphere (Mitchell et al. 2021) as a result of processes occurring inside our planet (True polar wander 2023).

It is known from mechanics that the rotation of rigid bodies can be characterized by three axes of inertia moments passing through the body center of mass, and by the angular momentum vector of the body. In a free state, all rotating bodies tend to take a position in which the axis of their greatest (principal) inertia moment is their rotation axis and coincides in direction with the angular momentum vector. If the latter condition is not met, then during the transition to a stable state, the body rotation axis can perform a complex reciprocating motion "polhode" (Greenwood 1988).

Although our planet is not a "solid" body, such a theoretical model is used to establish the absolute movement of lithospheric plates (Steinberger and Torsvik 2008; Piper 2018; Gordon 2018; Goldrreich and Toomre 1969; Mitchel et al. 2021; Dubrovin et al. 2012; Kravchinsky et al. 2014; Grand 1988). The conditions for applying the model to our planet are: (1) the constancy of Earth's mechanical rotation moment and (2) a change in the Earth's main inertia moment axis as a result redistribution of density inhomogeneities in mantle (Goldrreich and Toomre 1969). It is believed that geodynamic processes in mantle can lead to a mismatch between the direction of main inertia axis and Earth angular momentum vector. At that the transition of our planet to a new stable rotation state can be accompanied by a complex movement of its rotation pole—the TPW effect.

The traditional approach to estimating TPW effect (Mitchel et al. 2021; Dubrovin et al. 2012) is to compare the average position of Earth's magnetic pole, which is determined by magnetic methods and attributed to any one plate, with the movement of this plate in the frame of reference corresponding to Earth's mantle. The average magnetic pole position is identified with Earth's rotation axis. Next, for this plate, the trajectory of Earth's rotation axis motion is found, which is called "apparent polar wander path (APWP)". The APWP motion is considered to be a superposition of two motions: (1) the plate absolute motion relative to mantle, and (2) the rotation (of mantle and lithosphere) around Earth's axis (TPW). By subtracting the absolute motion of the plate (1) from the motion APWP, the change in time of the position of the Earth's axis (2) in the mantle reference frame is determined, which gives an estimate of speed and direction TPW motion.

Determining the coordinates of Earth's rotation pole, especially its longitude, and establishing the trajectories of the lithospheric plates movement is a very not a simple theoretical task (Kravchinsky et al. 2014). Calculations show (Steinberger and Torsvik 2008; Dubrovin et al. 2012; Kravchinsky et al. 2014) that many lithospheric plates repeatedly changed speeds and directions of movement and experienced turns both in one direction and in the other. However, these conclusions can be considered sufficiently reliable only for the last 300 Ma, since the errors and inaccuracies in the calculations increase rapidly with rocks age.

In this regard, we note that such calculations do not take into account the possibility of changing the angular momentum and Earth's rotation speed, and also neglect other factors that can significantly affect the movement of lithospheric plates. In addition, the change in direction of Earth's main inertia axis over a time of $\sim 10^7$ years (Goldrreich and Toomre 1969), assumed in TPW, is very doubtful from a physical point of view. Therefore, explaining the movement of continents by the mechanism of plate tectonics (Piper 2018; Grand 1988), it is proposed to transfer the matter convection from the mantle to the asthenosphere (Hofmeister 2020).

2 A New Approach to the TPW Phenomenon Interpretation

The author develops a new approach to interpretation of TPW phenomenon, eliminating difficulties of its physical explanation. The approach take into account the important astrophysical phenomenon of the jet outflow of gas and dust from nuclear disk of our Galaxy (Barenbaum 2002), which was not previously taken into account in astronomy and geology (Barenbaum 2010). According the phenomenon, during orbital motion in Galaxy, the Sun cyclically crosses jet streams and Galaxy spiral arms, and at such moments Earth and other planets are subjected to powerful bombardments by galactic comets. A galactic model (Barenbaum 2010) has been developed and tested, linking main events of Earth's geological history with processes in Galaxy and Solar System.

An optimized version of galactic model (Barenbaum and Titorenko 2020) is used in this article to substantiate an alternative interpretation of TPW phenomenon proposed by the author. According to the representations being developed, the TPW phenomenon is due to a change in the inclination not of Earth's rotation axis, but of rotation axis of our planet lithospheric shell. It is assumed that Earth's lithospheric shell, in which the plates (continental and oceanic), interacting with each other, move in asthenosphere can rotate with an angular velocity different from Earth's rotation rate as a whole. In this case, the axes of Earth rotation and its lithospheric shell do not necessarily coincide.

A simple phenomenological model has been constructed, which, from the standpoint of such representations, makes it possible to give TPW phenomenon a more adequate, in our opinion, physical interpretation. The model uses the results of the author's previous works, the conclusions from which are summarized in this article. These questions include:

- Information about the used version of galactic model.
- General information about galactic comets.
- Formation of supercontinents by galactic comets.
- Precession of Earth's rotation axis
- Action on lithospheric plates of Coriolis force.
- Origin of continental and oceanic plates and lithospheric shell.

The constructed phenomenological model is tested on two examples using empirical data. These examples, in the author's opinion, make it possible to explain: (1) succession of Pangaea supercontinent breakup in the last 250 Ma and (2) specifics of Atlantic Mid-Ocean Ridge structure and its transform faults that arose during Pangaea breakup.

3 Galactic Model

The model implements the representations (Barenbaum 2010), according to which, after formation in Galaxy, Sun revolves around galactic center in an elliptical orbit, from time to time crossing jet streams and Galaxy spiral arms. At these moments, Solar System is subjected to intense bombardments by galactic comets, which are formed at place intersections of jet streams with galactic arms. Such cometary bombardments are the main factor that forms the surface topography on all planets (Barenbaum 2010, 2015, 2016; Barenbaum and Titorenko 2020; Barenbaum and Shpekin 2016, 2018, 2019).

In particular, on Earth, galactic comets split and move lithospheric plates (Barenbaum et al. 2002, 2003, 2004; Barenbaum and Yasamanov 2004), and also initiate cycles of global biotic and climatic processes, the climaxes of which are recorded by geologists in the geochronological scale by its boundaries of different ranks. The rank of these boundaries, according to the galactic model, is determined by the power of cometary bombardments (Barenbaum 2010; Barenbaum and Titorenko 2020).

A general idea of the used galactic model (Barenbaum and Titorenko 2020) is given in Fig. 1. In Fig. 1a in the projection onto Galaxy plane shows the current Sun position relative to 4 arms and 2 jet streams, and in Fig. 1b change in this position on Phanerozoic scale in the last 600 Ma.

On Fig. 1b, it can be seen that powerful cometary bombardments define boundaries of eras, less powerful ones determine the boundaries of systems, and even weaker ones determine the boundaries of epochs of Phanerozoic scale. The boundaries of epochs (not shown) correspond to Sun's hits into jet streams, the boundaries of systems correspond to Sun's hits simultaneously to jet streams and galactic arms, while the same hits, but at a distance of corotation radius $R*$ from Galaxy center, are the boundaries of eras and eons. The boundaries of eras and eons fall on Sun's orbit parts coinciding with corotation radius $R*$, where formation of stars and comets in galactic arms is sharply activated (Barenbaum 2002). In the last 600 Ma, the most powerful cometary bombardment took place in Crux-Scutum (IV) arm at V/Pz boundary, which is also Phanerozoic boundary corresponding to highest rank of geochronological scale.

We note the high calculating accuracy of Phanerozoic boundaries, which indicates the adequacy of our galactic model. These qualities of model are achieved by optimizing the Galaxy design parameters (Barenbaum and Titorenko 2020) and taking

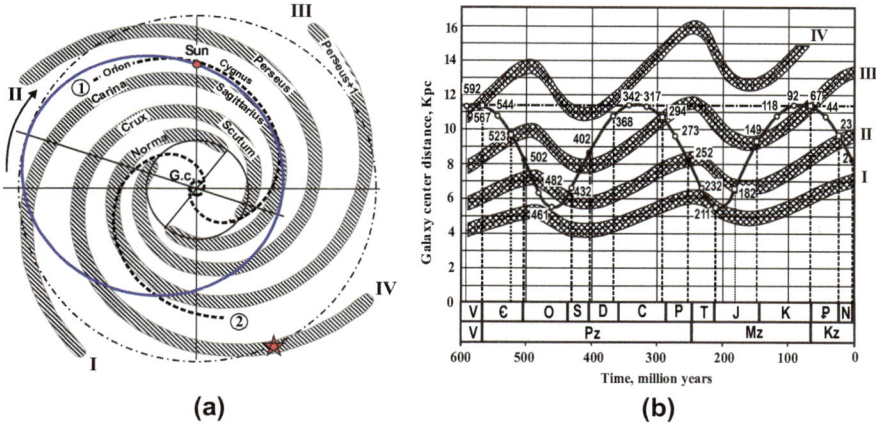

Fig. 1 **a** Sun's position in orbit (ellipse) in relation to 4 arms (Roman numerals) and 2 jet streams (Arabic numerals) of Galaxy. The small circle is nuclear disk, the middle circle is Galaxy isothermal core radius, the large circle is corotation radius R*. The arrow indicates Sun's movement on orbit, the direction rotation of the line of apsides of its orbit (straight line), as well as arms and nuclear disk of Galaxy. The asterisk is the place where Sun was formed in arm IV; **b** Change in Sun's position when moving along an orbit (sinusoid) relative to Galaxy center, as well as its arms and jet streams. Numbers are the time (in millions of years) of Sun entry into jet streams (circles) and simultaneous entry into Galaxy arms (squares). The black icons are the moments when jet streams cross Galaxy plane. The dashed-dotted line is corotation radius R*. Below is Phanerozoic scale indicating the boundaries of systems and epochs

into account the participation of Sun motion in parametric resonance (Molchanov 1966) with rotation of Galaxy arms and its nuclear disk. So, for one revolution of the line of apsides of solar orbit (with period $T_\alpha = 2.0$ billion years), Sun makes 8 revolutions in orbit and 9 revolutions around galactic center, and Galaxy and its nuclear disk make 10 and 80 revolutions. In this case, Sun makes 80 small-amplitude oscillations athwart Galaxy plane.

The galactic model also made it possible to establish that since Solar System formation, Galaxy spiral structure and Sun orbit in it have not changed much. Therefore, the sequence of cometary bombardments of different intensity in Phanerozoic (Fig. 1b) is also characteristic of Precambrian. Moreover, in Precambrian, the most powerful bombardments of Earth by comets occurred when Sun was in galactic arms at a distance R* from Galaxy center (Table 1).

Table 1 shown that the most powerful cometary bombardments occurred alternately in each of Galaxy arms and were repeated with a period of $T_B = 500$ Myr. Moreover, the power of bombardments in different Galaxy arms differed. In Carina-Sagittarius (I) and Norma-Perseus + 1(III) arms, bombardments were weaker than in Perseus (II) arm, and even more so in Crux-Scutum (IV) arm, in which the both main stages of Solar System planets formation took place (Barenbaum 2002).

Table 1 Relationship between Precambrian mega-events and galactic arms (Barenbaum 2022a)

Galactic arm	Event time, billion years	Precambrian boundaries of eons and eras
I. Carina-Sagittarius	1.067	Neoproterozoic
	3.067	–
II. Perseus	1.567	Mesoproterozoic
	3.567	Paleoarchean
III. Norma-Perseus + 1	2.067	–
	4.067	Archean (Eoarchean) (conditional border)
IV. Crux-Scutum	0.567	Phanerozoic (Paleozoic)
	2.567	Proterozoic (Paleoproterozoic)
	4.567	2nd cycle of planet formation
	6.567	Solar System formation

4 Galactic Comets

Galactic comets represent a special class of large cosmic bodies, which, as already noted, falling like an avalanche to Earth, play an extremely important role in all geological processes taking place on our planet. However, these comets are not currently detectable by astronomy means. Therefore, everything that is known about them today is obtained on the basis of studying the geological consequences of their falls to Earth and other Solar System planets, as well as the results of their collisions with the bodies of asteroid belt. Let us briefly give some information about galactic comets (Barenbaum 2010).

The Galaxy has comets 1) jet streams and 2) galactic arms. The former move relative to Sun at speeds of ~400–500 km/s, while the latter move at speed an order of magnitude lower Those and others consist of condensed gas-and-dust matter. However, comets of galactic arms contain higher abundances of chemical elements with average atomic weights than comets of jet streams. Depending on the astronomical situation (Fig. 1b), during one bombardment, $\sim 10^4 \div 10^7$ galactic comets can fall on Earth.

The most studied are the Orion-Cygnus jet stream comets (Fig. 1a), which bombarded Solar System at the turn of Neogene and Quaternary in period from ≈ 5.0 to 0.7 Ma. These comets had a matter density of ~1.0 g/cm^3 and consisted mainly of water ice. Diameter of comet nuclei was 0.1–3.5 km, nuclei mass was $\sim 10^{12}–10^{17}$ g, and their kinetic energy was $\sim 10^{20}–10^{25}$ J. The speed of their collision with Earth was ≈ 450 km/s. These comets mainly bombarded the southern hemisphere on all planets of the terrestrial group.

The work (Barenbaum and Shpekin 2021) determined the flux density of galactic comets of Orion-Cygnus, which was 5×10^{-10} year^{-1} km^{-2}. This value was used to estimate the number, energy and frequency of galactic comet falls on terrestrial planets and the Moon in period from 5 to 0.7 Ma. In particular, 2.6×10^5 comets of all sizes fell to Earth at that time with a total energy of 1.3×10^{27} J. The estimates showed that falls of Orion-Cygnus jet stream comets can explain the phenomena of "recent uplifts of crust" and "young volcanism" on Earth, emergence of shield volcanoes on Venus, formation of continental and marine hemisphere on Mars, as well as the origin of large craters, seas and mascons on the Moon, Mars and Mercury (Barenbaum 2010, 2015).

In connection with these conclusions, we will make two important remarks concerning the properties of galactic comets. The first remark is related to the 62° inclination of the ecliptic plane, in which the planets revolve around Sun, to Galaxy plane, in which Sun and galactic comets move. Since Earth's rotation axis little deviates from perpendicular to the ecliptic, then when Sun moves along galactic orbit, comets alternately bombard the southern and northern hemispheres of our planet.

The second remark refers to the physical mechanism of high-speed galactic comets interaction with planets (Barenbaum 2010, 2015, 2016; Barenbaum and Shpekin 2016, 2018, 2019). This mechanism on planets without an atmosphere and with an atmosphere turns out to be different. If on the Moon, Mercury and Mars, these comets form craters with a diameter of 10 to 180 km, and when crater funnels are superimposed, they also form basalt seas with a diameter of hundreds of kilometers. Whereas in Earth and Venus atmospheres, the comet nuclei of evaporate, turning into a gas jet, which, when falling to the surface, forms a narrowly directed hypersonic shock wave that deep penetrates into bowels of planets. Calculations show that small galactic comets (~100 m) heat lithosphere under the continents to depths of ~10–20 km and melt rocks at depths of ~1–3 km. In large comets (~3 km) the heating zone of rocks for large comets reaches depths of ~ 250–300 km, corresponding to the asthenosphere. In this case, up to a depth of ~6 km, evaporation of rocks occurs, and below, in the depth interval of ~6–40 km, complete melting of rocks occurs, as a result of which a channel can form, through which magma from the asthenosphere can pour out to the surface for a long time.

It is important to note that with this mechanism, most of galactic comets kinetic energy is converted into heat, which is involved in the melting and heating of rocks in deep layers of lithosphere. At that, depending on the lithosphere structure, both crustal and mantle rocks undergo melting, which affects the composition and further behavior of the resulting magmas. This causes powerful convective processes in the asthenosphere, which initiate various tectonomagmatic and geodynamic processes on Earth's surface.

When Sun moves in Galaxy, the zone of maximum heating of rocks by comets migrates between the south and north Earth's poles, activating tectonic–magmatic and geodynamic processes at corresponding latitudes. By heating of asthenosphere rocks, cometary bombardments cause a local "swelling" of Earth's surface, which is later replaced by its subsidence and alignment of the emerging mountain structures. These processes in the tectonosphere occur against the background of lateral movement of continental masses, the formation of a new oceanic crust, and extensive magmatic outpourings.

Thus, in contrast to the well-known hypotheses that explain the asthenosphere origin as a result of heating this layer of rocks with heat coming "from below" from supposed deep endogenous sources, the considered mechanism explains the heating of the asthenosphere rocks "from above" by shock waves that are created by galactic comets disintegrating in Earth's atmosphere (Barenbaum 2010, 2015). Note that the asthenosphere existence is not a exclusively terrestrial phenomenon, but it also inherent in Mars and the Moon (Barenbaum 2016, 2012).

5 Supercontinental Cyclicity

The TPW method has established that continental plates move and from time to time, colliding with each other, form supercontinents. This process, as well as the formation of lithospheric plates themselves, has been going on since the Archean (Palin et al. 2020).

Although the concept of "supercontinent" is interpreted by geologists quite freely, the existence of 10 supercontinents that formed with a period of $T_S = 400$ Ma has been revealed to date with varying degrees of reliability (Bozhko 2009). In the life of each supercontinent, one can distinguish an assembly stage lasting ~150 Ma, when it is formed from relatively small plates, and a decay stage ~250 Ma, when it breaks up into parts. The first stage proceeds under conditions of high plume activity, and at the second stage, the disintegration of supercontinents occurs under the influence of progressive rifting. The collapse can be complete or partial. In the latter case, the heavily destroyed part is alternately located either in the northern or in the southern hemisphere of Earth (Bozhko 2011).

To date, hypotheses have been proposed (Palin et al. 2020; Daiziel 1997; Doucet, 2019, Doucet et al. 2019; Dien et al. 2019) that explain the supercontinental cyclicity by the mechanism of lithospheric plate tectonics under one or the other model of mantle convection. However, these hypotheses do not answer 2 key questions: (1) what is energy source for moving plates and (2) what is the physical mechanism that causes the formation of supercontinents in different Earth's hemispheres.

The author showed (Barenbaum 2022a, b) that cometary bombardments of Earth in Galaxy arms are such a mechanism. The essence of the mechanism is that supercontinents are formed in the subpolar zones of Earth as a result of a very high density of comet falls, and in the middle latitudes they cease to exist under conditions of a much lower frequency of their falls. In this mechanism, it does not matter on how many parts the supercontinent break up under the action of cometary bombardments, and also in what directions and at what speeds their breakaway parts—terranes—move. All these processes are random.

The only non-random process is the assembly and breakup of supercontinents, which repeats itself with period $T_S = 400$ Myr. This period clearly differs from T_B period of powerful cometary bombardments (Table 1) and, therefore, other factors also influence its value. The most important of them is the inclination angle of Earth's axis to Galaxy plane, in where galactic comets move (Barenbaum 2010). In galactic model, this angle is determined by precession of Solar System ecliptic plane with a period $T_E = 2000$ Myr (Barenbaum 2022b), which is exactly equal to rotation period T_α of the line of apsides of solar orbit in Galaxy.

Therefore, the supercontinental cyclicity period can be calculated as $T_S = (T_B^{-1} - T_\alpha^{-1})^{-1} = 400$ Myr. This means that supercontinental cyclicity is a phenomenon caused not by endogenous processes in Earth's body, but by resonant processes in Galaxy and in Solar System. The main role in this phenomenon is played by the falls of galactic comets, which alternately bombard either the southern or the northern hemisphere of Earth during periods of time when Earth's rotation axis coincides with direction of galactic comets falls.

The times of supercontinents existence in a "merged" state according to Bozhko (2009, 2011) in Fig. 2 (Barenbaum 2022b) are compared with the periods of cometary bombardments in galactic arms according to (Table 1). We see that supercontinents were formed under conditions when Sun was in Galaxy arms at a distance $R*$ from its center. So, supercontinents Vaalbara and Rodinia originated in Carina-Sagittarius (I) arm, Sebakwia and Gothia in Perseus (II) arm, and Columbia in Norma-Perseus + 1 (III) arm. All other supercontinents originated in powerful Crux-Scutum (IV) arm. At the same time, Kenoria and Pannotia were formed when Sun was at a distance $R*$ from Galaxy center, and Yatulia and Pangaea—at a distance $R < R*$.

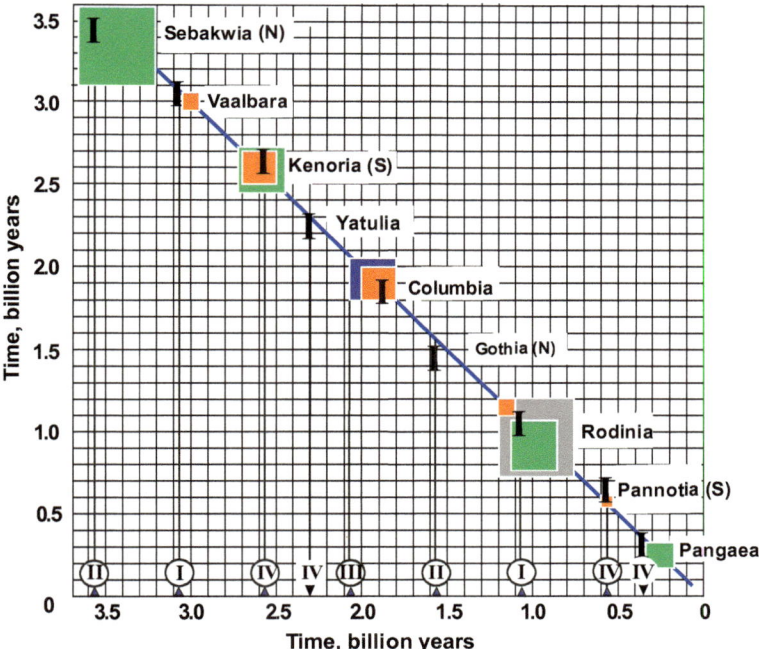

Fig. 2 Periods of supercontinents existence (vertical axis) in comparison with times of cometary bombardments in Galaxy arms (I–IV) at $R = R*$ (circled) and $R < R*$ (without circles). Black icons are data (Bozhko 2011). Multi-colored squares are the data of other authors. Icons (S) and (N) indicate supercontinents formation in the southern and the northern Earth's hemispheres

Figure 2 also shows the polarity of supercontinents location on the globe, determined by us taking into account the orientation of Earth's rotation axis with respect to direction of galactic comets incidence. As a result, it was found that supercontinents Pannotia and Kenoria, which arose in arm IV, were located in the southern hemisphere. While supercontinents Sebakwia and Gothia, which arose in the II arm of 1 billion years earlier and later than Kenoria, were located in the northern hemisphere, and supercontinents Vaalbara, Columbia and Rodinia should have been located in the central latitudes.

The paleoreconstructions of Pannotia, Kenoria, Sebakwia, and Gothia supercontinents indicate that during their formation Earth's rotation axis of was almost strictly oriented in the direction of comets motion. At the same time, supercontinents Pannotia and Kenoria were in the Southern hemisphere, and Sebakwia and Gothia were in the Northern hemisphere, which is noted in Fig. 2 icons (S) and (N). As for supercontinents Vaalbara, Columbia, and Rodinia, they were located in central latitudes and, apparently, did not form a single supercontinent.

We emphasize that the paleoreconstruction of supercontinents and the determination of trajectories of movement their parts before assembly and after the supercontinents destruction by the TPW method is a very not an easy task. The complexity of this task is due to the fact that its correct solution requires taking into account at least two other factors the influence of which is neglected. These include the precession of Earth's rotation axis and the Coriolis force effect on the movement of lithospheric plates.

6 Precession of Earth's Rotation Axis

Earlier, we noted that the applicability of TPW phenomenon for studying the motion of lithospheric plates is based on constancy of the magnitude and direction of the angular momentum of Earth's rotation over time. However, as a result of powerful bombardments of Solar System by galactic comets, this requirement for the planets is not fulfilled. Falls of high-energy comets can change the direction and magnitude of the angular momentum of planetary rotation. If these changes are small, then the planet's rotation axis can start precessing relative to the original direction with a certain opening angle, depending on direction of comets' arrival.

The proof of Earth's axis precession, caused by precession of Solar System ecliptic plane as a whole with a period of $T_E = 2000$ Myr, can be the polarity of the supercontinental cyclicity considered above.

Our other example is the precession of Earth's rotation axis discovered by Hipparchus. The period of this precession is 25.765 thousand years, and the angle between Earth's axis and the perpendicular to the ecliptic plane is 23.44°. Mars (25.2°), Saturn (26.73°) and Neptune (28.32°) have close inclination angles of rotation axis to ecliptic plane. If we also take into account that the axial rotation periods of Earth (24 h) and Mars (24 h 39 min) practically coincide, then the similarity of the tilt angles of the axes of the listed planets and Mars and Earth periods rotation can be associated with the last bombardment of Solar System by galactic comets.

In the galactic model (see Fig. 1), these were the comets of the Orion Cygnus jet stream, which in the period from 5.0 to 0.7 Ma bombarded all planets at an angle of 62° to the ecliptic plane. So for 0.7 Ma after the bombardment end, the rotation parameters of indicated planets could hardly change significantly.

Of course, in Earth's geological history, there were many similar and many times more powerful comet bombardments. Therefore, the question of influence on Earth's axis precession on the movement of lithospheric plates remains unexplored and open.

7 The Coriolis Force

The Coriolis force is the force of inertia acting on objects moving in a frame of reference rotating about an inertial frame of reference. Under the action of this force in a rotating reference frame, objects receive acceleration.

$$\alpha = 2[V \times \omega] = 2V\omega\sin(\varphi), \tag{1}$$

where: V is velocity vector of object in the rotating reference frame, ω is vector of angular rotation velocity of the rotating frame relative to the inertial reference frame, φ is angle between V and ω vectors. The Coriolis force is directed perpendicular to the vectors V and ω plane.

The Coriolis force action caused by Earth's rotation is a firmly established fact. In Earth's northern hemisphere, Coriolis force deflects objects to the right of their movement direction, and in southern hemisphere—to the left.

The question of the importance of taking into account Coriolis force influence on motion of lithospheric plates was raised by us in Barenbaum (2018) when analyzing the results of Khutorskoy and Teveleva (2018). The authors (Khutorskoy and Teveleva 2018), when studying heat fluxes along nine geotraverses crossing Mid-Ocean Ridges (MOR) in the Atlantic, Indian, and Pacific oceans at different latitudes, found that average heat fluxes on western and eastern flanks of geotraverses in northern and southern of Earth's hemispheres are differ. In southern hemisphere, the western flank of geotraverses has a higher average heat flux, and in the northern hemisphere, the eastern flank. The authors explained this fact by action of Coriolis force, which deflects ascending magma flow in the MOR zones in the corresponding direction.

However, this explanation is wrong. Since Coriolis force should deflect the ascending magma flow in northern and southern hemispheres to the west, and not in opposite directions, as was obtained by the authors (Khutorskoy and Teveleva 2018). We have shown (Barenbaum 2018) that Coriolis force does not act on magma flow, but on oceanic plates, which deviate to the right in northern hemisphere and to the left in southern hemisphere relative to their movement direction. Despite strong variations in heat fluxes and geomorphologic differences in the ocean floor in their measurements areas (Khutorskoy and Teveleva 2018), the ratio of average values of

Fig. 3 The ratio of heat fluxes between the western and eastern wings of geotraverses (Barenbaum 2018), constructed according to data of Khutorskoy and Teveleva (2018). Designations: Atlantic Ocean (triangles), Pacific Ocean (circles), Indian Ocean (squares). Numbers are numbers of geotraverses (Khutorskoy and Teveleva 2018). The solid line is the 1–sinφ dependence. The dotted line is the right envelope of measurement results

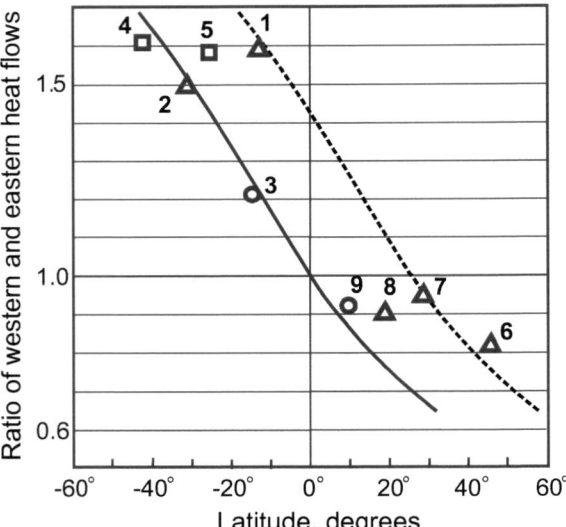

heat fluxes in the three oceans on both sides of MOR increases in the same way with increasing latitude of geotraverses (Fig. 3).

Figure 3 shows that the ratio of heat fluxes lie in latitude interval of 25°, and the curve enveloping the experimental values field on the left corresponds to Coriolis force formula (1). When explaining (Fig. 3), we assumed (Barenbaum 2018) that under the MOR there is an asthenospheric (or mantle) protrusion that determines the average heat flux at the ocean floor. The movement of oceanic plates above such protrusions under the action of the Coriolis force leads to a change in heat fluxes measured on the wings of geotraverses.

In the MOR wing, the heat flux decreases in the direction of oceanic plate motion, while in the opposite wing it increases. In addition, when oceanic plates move from north to south in southern hemisphere, the average heat flux in the eastern geotraverses flank becomes less than in the western flank, while in northern hemisphere, on the contrary, it becomes larger. If the plates begin to move from south to north, the picture will be reversed.

However, the question of lithospheric plate's motion on asthenosphere relative to mantle requires more detailed consideration. This consideration led the author to the following conclusions:

1. The Coriolis force acts not only on "young" oceanic plates involved in bottom spreading and moving on asthenosphere relative to mantle at high velocities, but also on "old" continental plates, containing ancient cratons, which move relative to mantle at a much lower velocity.
2. All lithospheric plates, both continental and oceanic, form a common lithospheric shell on our planet, which was in a completely melted state 4.6 billion years ago (Barenbaum 2010), and today remains in a state of partial convective mixing.

3. This lithospheric shell, consisting of plates of different ages interacting with each other and moving at different velocities, apparently, starting from Archean, rotates as a whole in the asthenosphere with its own angular velocity, which is different from rotation velocity of Earth's solid body.
4. The angular rotation velocity of lithospheric shell, in particular, in Phanerozoic was higher than mantle and the shell rotation axis did not necessarily coincide with mantle rotation axis.
5. Below we construct the simplest phenomenological model reflecting these conclusions, which will be used further to substantiate the possibility of an alternative explanation of the TPW phenomenon.

8 Theoretical Model (Building and Testing)

Let us represent the plate velocity relative to mantle as a vector sum of velocities $\tilde{V} = V' + V''$, where V' is the plate velocity vector relative to lithospheric shell, and V'' is the shell velocity vector relative to mantle. Further, we will consider the lithospheric shell with a coordinate grid applied to it as a rotating reference frame, and we will identify the mantle, where hot spots are located, with Earth as a whole. Let us fix Earth's axis in the mantle and take mantle as an inertial frame of reference. Let us denote the angular velocity of shell rotation as ω_0, and angular velocity of mantle rotation as ω_m.

The first question that needs to be answered is: what rotates faster—the mantle or the lithospheric shell? In the first case, the plates lag behind the rotation of the mantle, moving east relative to its hot spots, and in the second case, they are ahead of it, moving west. The facts show that, at least in Phanerozoic, lithospheric shell rotated faster than mantle, i.e. $\omega_0 > \omega_m$.

For simplicity, we will assume that lithospheric shell and mantle rotate around a common axis with angular velocities $\omega_0 > \omega_m$. Then, in the coordinate system associated with lithospheric shell, plates will move relative to mantle in a westerly direction with a speed of:

$$V = R(\omega_0 - \omega_m) \cos(\varphi), \tag{2}$$

and in longitude under Coriolis force action they will acquire acceleration:

$$\alpha = 2R\omega_0(\omega_0 - \omega_m) \cos(\varphi)sin(\varphi), \tag{3}$$

where: R is Earth's radius, φ is the place latitude.

From formulas (2) and (3) it can be seen that velocity of plates relative to mantle V'' is maximum at the equator ($\varphi = 0°$) and decreases to 0 at the north and south Earth's poles ($\varphi = \pm 90°$). At the same time, Coriolis force influence on the plates movement is maximum at latitudes $\varphi = \pm 45°$, and at the equator and at Earth's poles, the force $F_k = 0$. In this case, under influence of Coriolis force (F_k), in the northern hemisphere, plates will move along the meridian to North and in southern hemisphere to South. These conclusions are illustrated by the diagram in Fig. 4a.

If we also assume that plates, together with lithospheric shell, move at a constant speed $V' \neq 0$ along the meridian to North (Fig. 4b), then in northern hemisphere, under the force $F_k(\varphi)$ action, their velocity along the latitude will slow down, and in southern hemisphere, on the contrary, accelerate. When the plates cross the equator, the force $F_k(\varphi)$ changes sign.

This simple model is obviously applicable to movement of the plates in lithospheric shell at different speeds and in arbitrary directions.

So, in summary, we note that Coriolis force action on lithospheric plates leads to three important consequences that can be subjected to empirical verification:

- Earth's equator is a special zone in which the plates experience tensile stresses (Fig. 4a) and shear (Fig. 4b);
- Movement of plates in latitude when crossing the equator can change to opposite;
- The influence of Coriolis force on the plates reaches its maximum at middle latitudes, while this force does not act at the equator and at Earth's poles.

We also note that if rotation axes of lithospheric shell and mantle do not coincide, the models (Fig. 4a, b) can be combined into one inclined model for individual plates.

Let us now discuss the effect of Coriolis force on motion of oceanic and continental lithospheric plates using two specific examples. The first example is the structure features of Atlantic Mid-Ocean Ridge and its transform faults (Fig. 5), and the second is the plate's movement reconstructed by the TPW method after the Pangaea supercontinent breakup that caused Atlantic MOR formation (Fig. 6).

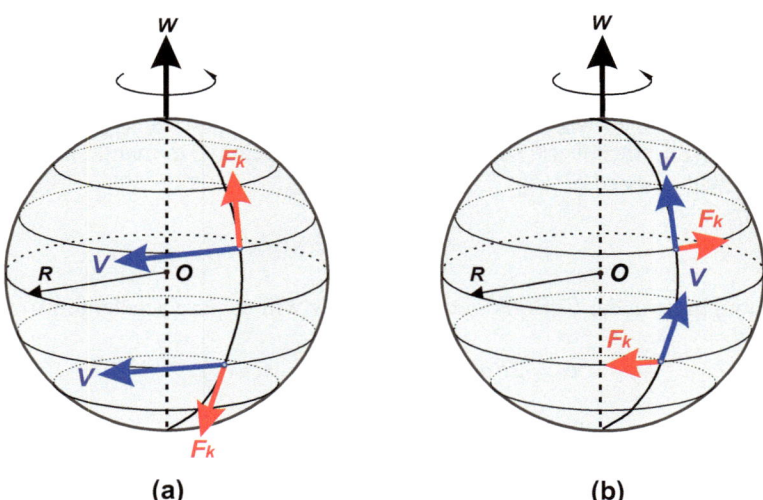

(a)   **(b)**

Fig. 4 The Coriolis force action on lithospheric plates moving at a speed V in latitude (**a**) and longitude (**b**) in northern and southern hemispheres

Fig. 5 Map of the Atlantic Ocean bottom

Both figures show important details indicating that Coriolis force may have played a key role in both cases. Such details in Fig. 5 are:

- Bending of central line on Atlantic mid-ocean ridge in both hemispheres to the west in the middle latitudes. We associate this bending with a higher speed of these MOR sections movement under Coriolis force action (see Fig. 4a). We also believe that the force gradient $F_k(\varphi)$ creates shear stresses that deform and tear the center line of Atlantic mid-ocean ridge, thereby causing the transform faults formation on it.

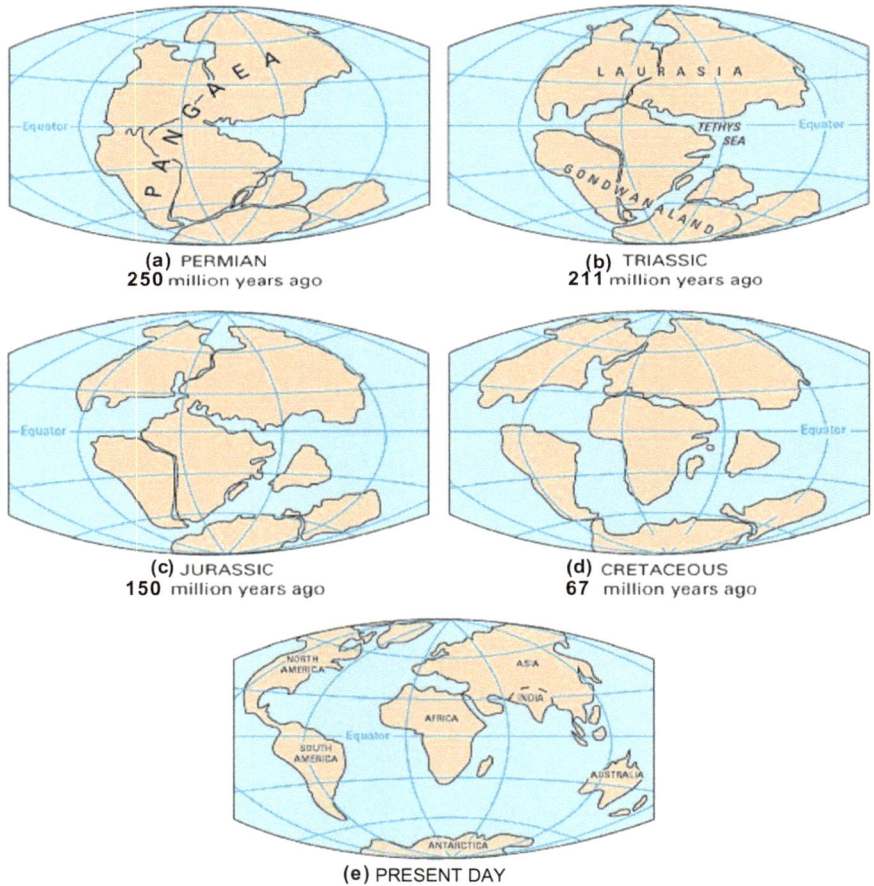

Fig. 6 The sequence of supercontinent Pangaea break up in the last 250 Ma

– Underdevelopment of transform faults of Atlantic mid-ocean ridge in high
  northern latitudes. We explain this fact by a decrease in Coriolis force in Earth's
  northern polar zone at $\varphi \sim 90°$.
– Change in the angle of inclination of the transform faults of Atlantic mid-ocean
  ridge to the equator line, as well as to the MOR axis on opposite sides of the
  equator. If at the equator the transform faults go along the line $\varphi = 0°$, then with
  an increase in latitude, the inclination angle of faults to the equator in northern
  hemisphere increases in the positive direction, and in southern hemisphere—in the
  negative ones. We explain this feature by the multidirectional action of Coriolis
  force on the plates in southern and northern hemispheres (see Fig. 4b).
– The nature of change in inclination angles of transform faults on Atlantic MOR
  (Fig. 5) leads us to the conclusion that due to general movement of lithospheric

shell and its continents to North (see Fig. 4b), at beginning formation of Atlantic ocean, the equator was located to South than at present.

This conclusion is confirmed by the breakup of the Pangaea supercontinent (Fig. 6). In the Permian (Fig. 6a), this supercontinent occupied the region of latitudes from the South Pole to the North, and Earth's equator then passed through the junction of four large plates: African, Eurasian, North American and South American. We associate the breakup of Pangaea and its subsequent fragmentation with the intense bombardments of Earth by comets of the galactic arms (see Fig. 1) and with the influence of the Coriolis force on the movement of plates (see Fig. 4).

We assume that as a result of a powerful cometary bombardment in galactic arm of Perseus (II), at the border of Permian and Triassic 250 Ma, a single Pangaea (Fig. 6a) began to be divided into two parts: northern—Laurasia and southern—Gondwana. The split occurs just along the equatorial line, where the greatest discontinuous and shear stresses (Fig. 4b) take place. Laurasia and Gondwana under the influence of the Coriolis force (Fig. 4a) move away from the equator to the north and south, respectively.

The next powerful cometary bombardments in Triassic (Barenbaum 2010) and at the Triassic–Jurassic boundary (211 Ma) led to the birth of the Tethys Sea (Fig. 6b). During this period, Pangaea is actively destroyed, continuing to move north. Laurasia breaks up into the North American and Eurasian continents and Gondwana into the South American and African continents. The North American and South American continents are moving west. At the site of their split, a plate spreading center arises and the Atlantic mid-ocean ridge is formed.

In Jurassic and Cretaceous (Fig. 6c, d), the Pangaea supercontinent practically disintegrated. The Antarctic continent separates from Gondwana, which remains in the south polar zone, while the Australian and Indian continents move north. In the Cretaceous period, the maximum impact of galactic comets falls on the equatorial latitudes of Earth, which causes the Alpine tectonic cycle (Barenbaum 2010).

We emphasize that all cometary bombardments and especially very powerful ones at the boundaries of periods of Phanerozoic geochronological scale (see Fig. 1b), crushed lithospheric plates, changing the directions and speeds of their movement in the lithospheric shell relative to mantle.

9   Formation of the Convective Shell and Lithospheric Plates

The presence on Earth of an autonomous convective lithospheric shell, in which continental and oceanic plates arise, move and interact, we closely associate with the origin of the Moon. The problem of Moon is solved by the author from the standpoint of the "Open Solar System (COSS)" cosmogony concept (Khain 2001), which takes into account the powerful bombardments of Solar System by galactic comets.

According to COSS data, the planetary system of our Sun was formed in two main stages (Barenbaum 2002). At the first stage, ~6.6 billion years ago, the Sun and planets arose; in the second, ~4.6 billion years ago, caused by destruction of the planet Phaeton, the giant planets acquired gas shells, satellites of the planets, including the Moon, arose, and a belt of asteroid bodies was formed. By this time, the Earth already had a solid–liquid Fe–Ni core, a thick anorthosite crust, and a system of differentiated mantle silicate shells, similar to the one currently existing (Barenbaum 2010).

Calculations show (Barenbaum 2010) that due to the fall of large fragments of Phaethon to Earth, our planet lost 245 ± 10 km of the surface layer of rocks, consisting on ~1/3 of the primary anorthosite crust and on ~2/3 of mantle rock. Part of the matter torn off Earth was lost forever, ~18% was ejected into near-Earth orbit and became part of Moon, and about the same amount fell back to Earth, melting it to a depth of ~100 km. After cooling, these molten rocks differentiated into crustal and mantle parts. But due to the loss of ~2/3 of the primary Earth's crust, the remaining refractory substance was enough to cover with continents only ~1/3 of the globe surface.

This deficiency of the primary crustal matter that arose after the formation of the Moon is now manifested in the presence of oceanic and continental hemispheres on Earth, the nature of evolution of Earth's tectonosphere as a whole (Khain 2001), and in existence of continental (old) and oceanic (young) plates, moving along the plastic layer rocks of asthenosphere. Under the lithospheric shell of our planet, we define the upper layer of rocks of such a "secondary" Earth's crust. We believe that this rocks layer has not yet completely cooled down and continues to be in a state of partial melting and convective mixing.

An very important role in maintaining the lithosphere in a convective state is played by cyclic bombardments of Earth by galactic comets, which cause split and movement of cooling lithospheric plates, and also lead to strong uneven heating of the underlying layer of asthenosphere rocks (Barenbaum 2010, 2015, 2016) driving oceanic and continental plates, as well as the lithospheric shell as a whole.

10 Discussion and Conclusions

The simplest phenomenological model, on basis of which the TPW scheme of the Pangaea supercontinent breakup (Fig. 6) is interpreted, allows us to propose a more adequate physical explanation of TPW phenomenon, alternative to the generally accepted one.

We focus on the main provisions of our explanation.

- The so-called "True polar wander" refers not to rotation axis and magnetic poles of Earth, but to rotation axis of its lithospheric shell (or a significant part of it). Lithospheric shell, in which continental and oceanic lithospheric plates move and interact, apparently, it can autonomously rotate in asthenosphere relative to mantle, changing the direction and speed of its movement under the influence of cometary bombardments.

- The spatial orientation of rotation axis, as well as the axis of Earth's dipole magnetic field, when averaged over a time interval of $\sim 10^6$–10^7 years, very weakly depends on both dislocation of lithospheric plates on surface and movement of mantle substance. The direction of Earth's rotation axis is determined by vector of Earth's rotational moment, which can change the spatial orientation under powerful cosmic impacts. The reaction to such influences most often becomes precession of Earth's rotation axis.
- Bombardments by galactic comets are the main reason for maintaining the Earth's asthenosphere rocks in a convective state, as well as for the splitting and movement of lithospheric plates. Alternately bombarding the southern and northern hemispheres of the Earth, comets cause the cyclic formation of supercontinents in these hemispheres.
- When paleoreconstructions of plate movements, the accuracy of solving this problem by the TPW method depends on the reliability of the choice of reference system, for which the trajectory of "apparent polar wander (APW)" of Earth's rotation pole is theoretically calculated, using paleomagnetic data. Such calculation does not take into account a number of important factors, in particular, the Coriolis force influence on movement of plates, which introduces large errors in the final result.

The approach proposed by the author to the interpretation of the TPW phenomenon, which assumes that Earth has an autonomously rotating lithospheric shell, makes it possible to eliminate the difficulties of a physical explanation of the TPW phenomenon, as well as to take into account a number of factors that affect the movement of lithospheric plates, in particular, the Coriolis force.

The author believes that taking into account the Coriolis force will open up the possibility to significantly improve the accuracy of paleoreconstructions of lithospheric plate's movement in geological history of Earth.

However, this possibility undoubtedly needs more detailed study and theoretical substantiation.

References

Barenbaum AA (2002) Galaxy, Solar system, Earth: Subordinate processes and evolution. PH: GEOS, Moscow, 393 p

Barenbaum AA, Khain VE, Yasamanov NA (2004) Large-scale tectonic cycles: analysis from the galactic concept positions. Vestnik MGU. Geol 3:3–16

Barenbaum AA (2010) Galactocentric paradigm in geology and astronomy. PH: LIBROKOM. Moscow, 544 p

Barenbaum AA (2012) On the asymmetry of the surface terrestrial planets, caused by the falls of galactic comets. Vestnik Otdeleniya Nauk o Zemle RAN, 4, NZ9001, https://doi.org/10.2205/2012NZ_ASEMPG

Barenbaum AA (2015) Geological structures created by falls of galactic comets. J Phys Conf Ser 653:012073. https://doi.org/10.1088/1742-6596/653/1/012073

Barenbaum AA (2016) Formation of the asthenosphere by galactic comets as a new direction in tectonophysics. Tectonophysics and topical issues of Earth sciences. Mater conference Moscow: Schmidt institute physics of Earth RAS. V.2. Sec. 5:430–438. http://www.ifz.ru/lab_204/konfer encii/chetvertaja-tk-2016/materialy-konf

Barenbaum AA (2018) Displacement of heat fluxes in mid-ocean ridges under the action of the Coriolis force. Proceedings of XIX international conference "Physical-chemical and petrophysical research in Earth sciences" (September 24–28, 2018) Moscow: IGEM RAS. 28–31

Barenbaum AA (2022a) Supercontinental cyclicity as a result of bombardment of Earth by galactic comets in Galaxy spiral arms. Proceedings of All-Russian Annual Seminar on Experimental Mineralogy, Petrology and Geochemistry (RASEMPG-2022) Moscow: GEOKHI RAN, pp 258–264

Barenbaum AA (2022b) Measuring the precession period of Solar System ecliptic plane using Galactic model. The 13 Moscow Solar System Symposium (October 10–14, 2022) Book of abstracts. Moscow: Space Research Institute RAS. 13MS3-GP-PS-04, 211–214

Barenbaum AA, Shpekin MI (2016) To the development of the mechanism of interaction of galactic comets with the terrestrial planets. J Phys Conf Ser 774:012096. https://doi.org/10.1088/1742-6596/774/1/012096

Barenbaum AA, Shpekin MI (2018) Problem of lunar mascons: an alternative approach. J Phys Conf Se. 946:012079. https://doi.org/10.1088/1742-6596/946/1/012079

Barenbaum AA, Shpekin MI (2019) Origin and formation mechanism of craters, seas and mascons on the Moon. J Phys Conf Ser 1147:012057. https://doi.org/10.1088/1742-6596/1147/1/012057

Barenbaum AA, Shpekin MI (2021) Estimation of the flux density of galactic comets in the Orion-Cygnus branch based on number of shield volcanoes on Venus, craters on Mars and marine basins on the Moon. J Phys Conf Ser 1787:012021. https://doi.org/10.1088/1742-6596/1787/1/012021

Barenbaum AA, Titorenko AS (2020) Galactic model of geological cyclicity: parameter optimization and testing based on geology and astronomy data. Proc. Rus. An. Seminar on Experimental Mineralogy, Petrology and Geochemistry (RASEMPG-2020), Vernadsky Inst. Geochemistry and Analytical Chemistry RAS, Moscow, 210–215

Barenbaum AA, Yasamanov NA (2004) Tectonic cycles of Wilson, Bertrand and Stille as result of Earth's bombardments by galactic comets. Evolution of tectonic processes in Earth's history (Proc. of XXXVII Tectonic Conf., Novosibirsk, February 10–13, 2004) PH: Siberian Branch RAS, Novosibirsk, V.1, 38–41

Barenbaum AA, Gladenkov YuB, Yasamanov NA (2002) Geochronological scales and the astronomic time (State of the Art). Stratigraphy and Geological Correlation. V.10, No. 2, 3–14

Barenbaum AA, Yasamanov NA (2003) Galactic comets as one of leading factors in terrestrial planets tectonic evolution // Tectonics and Geodynamics of Continental Lithosphere. Proceedings of XXXVI tectonic conference (Moscow, February 04–07, 2003) V.1, PH: GEOS, Moscow, 24–26

Bozhko NA (2009) Supercontinental cyclicity in Earth evolution. Vestnik MGU. Geol. No 2, 13–28

Bozhko NA (2011) On two types of supercontinental cyclicity. Vestnik MGU. Geol. No 5, 36–61

Daiziel IWD (1997) OVERVIEW: neoproterozoic-Paleozoic geography and tectonics: review, hypothesis, environmental speculation. GSA Bull 109(1):16–42

Dien HGE, Doucet LS, Li Z-X (2019) Global geochemical fingerprinting of plume intensity suggests coupling with the supercontinent cycle. Nat Commun

Doucet LS, Li Z-X, Ernst RE et al (2019) Coupled supercontinent–mantle plume events evidenced by oceanic plume record. Geology 48(2). https://doi.org/10.1130/G46754.1

Dubrovin PV, Steinberger B, Torsvik TKh (2012) Absolute plate motions in a reference frame defined by moving hot spots in the Pacific, Atlantic, and Indian oceans. J Geophys Res Atmosph 117(B9):9101. https://doi.org/10.1029/2011JB009072

Goldrreich P, Toomre A (1969) Some more remarks on polar wandering. J Geophys Res 74(10):2555–2567

Greenwood DT (1988) Principles of dynamics. Prentice-Hall, 552 p

Hofmeister AM (2020) Heat Transport and Energetics of the Earth and Rocky Planets. Elsevier.https://doi.org/10.1016/C2018-0-04206-1

Khain VE (2001) Tectonics of continents and oceans (2000). PH: Nauchnyy mir, Moscow, 606 p

Khutorskoy MD, Teveleva EA (2018) Heat flow asymmetry at mid-ocean ridges in the Northern and Southern hemispheres of the Earth. Georesourcy 20(2):122–132

Le Grand HE (1988) Drifting Continents and Shifting Theories. Cambridge University, 313 p

Leu W, Kravchinsky VA (2014) Derivation of paleolongitude from the geometric parameterization of apparent polar wander path: Implication for absolute plate motion reconstruction. Geophys Res Lett, 4503–4511. https://doi.org/10.1002/2014GL060080

Mitchell RN et al (2021) A late cretaceous true polar wander oscillation. Natu Commun 12(1). https://doi.org/10.1038/s41467-021-23803-8

Molchanov AM (1966) Resonances in multifrequency oscillations. DAN 168(2):284–287

Palin RM, Santosh M, Cao Wentao et al (2020) Secular change and the onset of plate tectonics on Earth. Earth Sci Rev 207:103–117

Piper JDA (2018) Dominant Lid Tectonics behavior of continental lithosphere in Precambrian times: Palaeomagnetism confirms prolonged quasi-integrity and absence of supercontinent cycles. Geosci Front 9(1):61–89

Sager WW, Koppers AAP (2000) Late cretaceous polar wander of the pacific plate: evidence of a rapid true polar wander event. Science 287(5452):455–459.https://doi.org/10.1126/science.287.5452.455

Steinberger B, Torsvik TH (2008) Absolute plate motions and true polar wander in the absence of hotspot tracks. Nature 452(7187):620–623

True polar wander. Available at: https://en.wikipedia.org/wiki/True_polar_wander. Accessed 1 Mar 2023

Woodworth D, Gordon RG (2018) Paleolatitude of the Hawaiian hot spot since 48 Ma: evidence for a mid-Cenozoic true polar Stillstand followed by late Cenozoic true polar wander coincident with northern hemisphere glaciation. Geophys Res Lett 45(21):11632–11640. https://doi.org/10.1029/2018GL080787

The Black Sea Deep-Water Circulation Under Climatic and Anomalous Atmospheric Forcing

N. V. Markova and S. G. Demyshev

Abstract The purpose of this work is to study the behavior of the Black Sea deep-water currents below the permanent pycnocline under regular (climatic) and anomalous atmospheric forcing. The annual variability of Black Sea current fields at horizons deeper than 300 m and the response of deep-water currents to the atmospheric anomalous quasitropical cyclone in September 2005 in the southwestern Black Sea are analyzed. Thus, the variability of deep-water currents is assessed both when the climatic fluxes of heat, moisture and wind at the sea surface change quite smoothly throughout the year and when the circulation is forced by an extreme atmospheric cyclone during its 5-day passage over the sea surface and one month after the cyclone leaves the sea basin. It is shown that the velocities of Black Sea currents during anomalous cyclone forcing can increase several times compared to their typical values, and the relaxation of the current field after the quasitropical cyclone crosses the sea edge can take up to 4–5 weeks in the deep sea layer.

Keywords Black Sea · Deep-water circulation · Numerical experiment · Simulation · Climatic currents · Undercurrents · Anomalous cyclone · Relaxation

1 Introduction

Knowledge of the structure and variability of ocean currents is important for studying the processes of mixing of sea waters, the transfer of impurities of various nature, the distribution of nutrients, etc. Two layers are distinguished in the Black Sea current fields with a significant difference in their dynamics. This is due to the presence of a permanent (main) pycnocline, the core of which is located at depths of 50–100 m, and the lower boundary is about 300 m (Ivanov and Belokopytov 2013). Having a blocking effect on the processes of vertical mixing and propagation of the atmosphere fluxes downward, the pycnocline contributes to a significant weakening

N. V. Markova (✉) · S. G. Demyshev
Marine Hydrophysical Institute of RAS, Sevastopol, Russia
e-mail: n.v.markova@mail.ru

© The Author(s), under exclusive license to Springer Nature Singapore Pte Ltd. 2023   71
T. Chaplina (ed.), *Processes in GeoMedia—Volume VII*, Springer Geology,
https://doi.org/10.1007/978-981-99-6575-5_6

of the currents with depth and the formation of specific features of the current fields in the deep-water (subpycnocline) layer.

To date, the dynamics of the upper layer of the Black Sea, primarily due to the availability of observations, seems to be quite clear. It is characterized by the dominance of the Rim Current, which propagates in a cyclonic direction along the continental slope, periodically forming eastern and western sub-basin cyclonic gyres, as well as a number of quasistationary anticyclonic eddies between the Rim Current and the coast (Ivanov and Belokopytov 2013). The actual configuration of the Rim Current jet is sensitive to atmospheric fluxes forcing the circulation; average velocities in its core reach 30–50 cm/s.

At the same time, the Black Sea deep-water currents below the main pycnocline, at horizons of more than 300 m, have been studied very poorly due to the lack of the required amount, unevenness and heterogeneity of field observations. The existence and interannual variability of mesoscale eddies and currents in the deep part of the Black Sea were discovered only in the early 1990s, thanks to data obtained on international R/V cruises. Then, with the launch of the first autonomous profiling floats, an assumption was made for the first time about the seasonal variability of deep-water currents (Korotaev et al. 2006). The velocities obtained from the data of the floats turned out to be higher than those previously known. Thus, the background values were indeed only up to 2 cm/s, but in deep-water currents and eddies they could reach 5–8 cm/s and even 10–12 cm/s. The basin-scale Countercurrent under the Rim Current (Neumann 1943; Bulgakov and Bulgakov 1995) was not found. Later, these conclusions were confirmed on the basis of a more complete array of ARGO float data (Markova and Bagaev 2016). In this paper, we have tried to improve our understanding of the structure and variability of the Black Sea deep-water currents using a modern numerical model with different atmospheric forcing.

2 Numerical Simulation

Numerical modeling should be considered as an effective method for reconstructing the Black Sea circulation, taking into account the need to compensate for the lack of deep-water observation data. Modern eddy-resolving models of the Black Sea dynamics are verified mainly based on results of their reconstruction of the best-studied upper sea layer circulation. Simulation results are also validated using available observational data. Over that, the use of observational data assimilation in the course of numerical calculations makes it possible to obtain a solution that most realistically describes the structure of hydrophysical fields modeled. In this study, we analyze the results of simulations based on a nonlinear z-coordinate model of the Black Sea dynamics (Demyshev and Korotaev 1992) developed at the Marine Hydrophysical Institute of the Russian Academy of Sciences.

To calculate the climatic circulation of the Black Sea and analyze the annual variation of hydrophysical fields, the method proposed in Korotaev et al. (2000) is used. At the boundaries of the basin, regular fluxes of heat, moisture, wind, river inflow, and water exchange through the straits are set, calculated from the data of long-term observations for each day of the climatic year. At the nodes of the model grid (5 km horizontally and 38 levels vertically), climatic data on temperature and salinity are also assimilated (Belokopytov 2004), because in general, there are significantly more observations of temperature and salinity than observations of currents. The bottom relief is set on the basis of EMODNet bathymetry (www.emodnet-bathymetr y.eu). As a result of model numerical integration, hydrophysical fields consistent with climatic stratification are reconstructed, including the fields of currents (horizontal velocity components) necessary for the study. The calculated data are recorded for each day of the climatic year. Note that the climatic values are rather estimates, and in real synoptic situations, the current velocities may be higher.

3   Climatic Deep-Water Currents

Analysis of the simulation results showed the presence of all the main features of the thermodynamics of the upper layer of the sea in the model fields, as well as the features of deep-water circulation known to date. It was found that at depths of more than 300 m, the general cyclonic direction of circulation is preserved, but there is no single gyre circulation. In the summer season, the western branch of the Rim Current can be detected to the horizons of 400–500 m. In winter, the current velocities increase at all horizons, which is obviously associated with an increase in the wind forcing by the northeastern winds prevailing over the region. As a result, the lower boundary of the Rim Current in the western part of the sea during this period deepens to depths of 700–800 m.

The calculated velocities of deep-water currents are, on average, an order of magnitude lower than at the sea surface. The highest velocities up to 10–15 cm/s are determined in mesoscale, predominantly cyclonic eddies characteristic of the inner abyssal region of the Black Sea. The most intense eddies at depths of more than 300 m are up to 200–250 km in size. In the eastern part of the sea along the continental slope, smaller anticyclones up to 50–100 km in diameter often form. In the centers of deep-water cyclonic eddies, positive temperature anomalies are formed, and in the centers of anticyclones, negative ones. This effect is a consequence of the features of the thermal structure of the Black Sea waters—an increase in temperature from the core of the Cold Intermediate Layer (Ivanov and Belokopytov 2013) to the bottom—and is a distinctive feature of the Black Sea deep-water eddies from the ocean ones.

The presence of narrow non-stationary deep-water currents of the anticyclonic direction (undercurrents) was revealed. They opposite to the currents in the upper layer and most often occurring in the northeastern part of the sea (Fig. 1). The undercurrents spread along the continental slope at horizons from 250–300 m to 1700 m, their lifetime can reach several weeks. The formation of undercurrents near the North Caucasian coast occurs on the background of the anticyclonic vorticity in the velocity fields along the continental slope and a number of mesoscale eddy structures in the upper 200–300 m layer, which form in this area or move here from

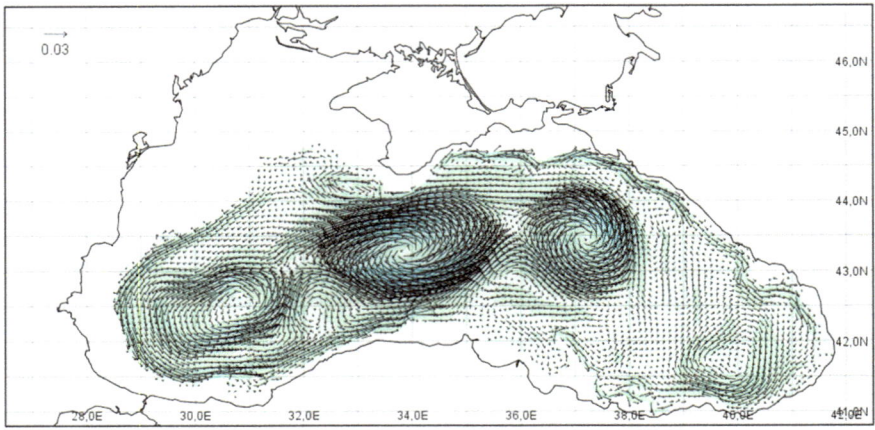

Fig. 1 The field of currents at a depth of 500 m, simulated for April 11 of the climatic year

the southeastern part of the sea. The presence of undercurrents can explain the vertical velocity shear, which was sometimes detected in field data and which contributed to the now rejected hypothesis of the Countercurrent below the Rim Current (Neumann 1943; Bulgakov and Bulgakov 1995).

4 Transformation of Currents Caused by an Anomalous Atmospheric Cyclone

In this section, in addition to the above analysis of the typical structure of deep-water currents, an assessment is made of their response to extreme atmospheric forcing. A prominent example of extreme atmospheric impact is the passage of a quasitropical cyclone, anomalous for the Black Sea region, characterized by wind velocity up to 33–35 m/s and causing a strong storm and destruction in the coastal zone of the southwestern part of the Black Sea on September 25–29, 2005. The data of this cyclone simulation in the atmosphere (Efimov et al. 2004) are used as the boundary conditions for modeling the Black Sea hydrodynamics during the cyclone passing above the sea and one month more to follow the relaxation of the current fields. Unlike the calculation of climatic fields, this simulation is of prognostic character, i.e., no assimilation of seawater temperature and salinity data is applied.

The simulation results allowed us to show the following. Such an intense atmospheric cyclone caused a response cyclonic gyre in the southwestern part of the Black Sea, which extended to a great depth and captured the water column under the atmospheric anomaly. On the sea surface, the jet current was accelerated up to 150–160 cm/s and the sea level extremely rose up to 30–35 cm along the southwestern coast. A powerful upwelling was developed in the centre of the cyclonic gyre generated, which resulted in water with a temperature of + 8.5...10 °C (against the climatic values of about +20 °C) rose to the surface. The characteristics of the simulated cold anomaly are consistent with the SST data from the NOAA-17 satellite, which observed the Black Sea surface during the period under consideration (Fig. 2, http://dvs.net.ru/mp/data/200509bs_sst_ru.shtml).

The horizontal velocity values in the anomaly cyclone region increased rapidly and even in the deep layer they rose from 1–3 cm/s to 15–20 cm/s (Fig. 3). It is also calculated that under the influence of an atmospheric anomaly, the vertical velocities of seawater can increase up to 2 orders of magnitude. The dissipation of hydrophysical fields transformed under the influence of the atmospheric anomaly occurred differently in the upper (to depths of 200–300 m) and lower layers. Thus, after the atmospheric cyclone left the sea, the relaxation of the current field in the upper layer took about 2 weeks, while in the deep layer the structure of the currents recovered much longer and was completed only by the end of October.

The subsequent atmospheric synoptic processes (of a regular nature) had the most impact in the upper layer of the sea, and the velocity field changes occurred rapidly here. At the same time, the velocities of deep-water currents weakened much slowly, and the deep-water velocity field anomalies were detected even a month after the quasitropical cyclone left the basin.

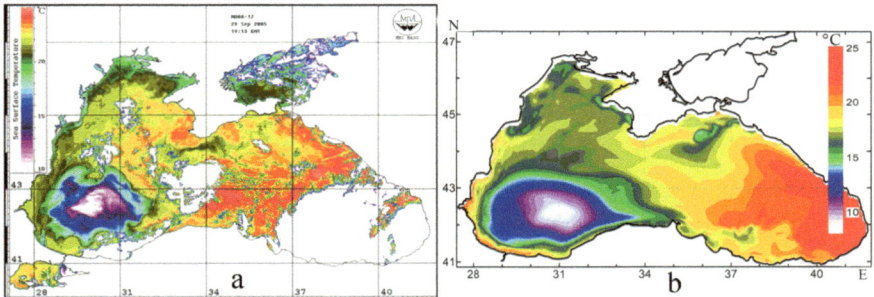

Fig. 2 Sea surface temperature on September 29, 2005: **a** NOAA-17 SST data, **b** simulation result

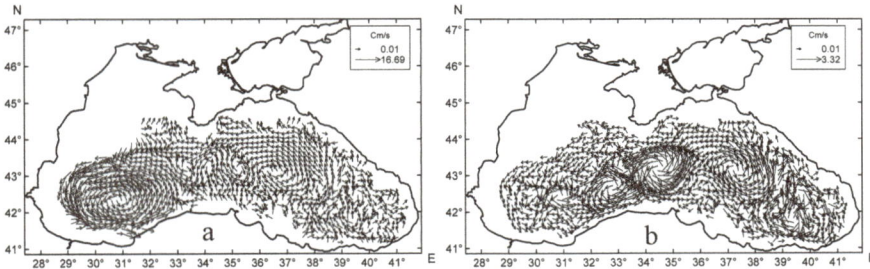

Fig. 3 Daily-mean currents simulated for September 28 at a depth of 1000 m: **a** in 2005, **b** climatic

5 Conclusions

Analysis of the features of the Black Sea currents simulated under regular (climatic), quite smooth atmospheric conditions, as well as the anomalous cyclone forcing has shown that the deep-water currents are also essential affected by the atmospheric processes. Despite the vertical exchange blockade by the permanent pycnocline, velocities of deep-water currents can reach 10–15 and even 20 cm/s with an increase in wind forcing. Under conditions of anomalous cyclonic forcing on the sea surface, the structure of currents has changed significantly both in the upper and lower layers, and the vertical and horizontal exchange has greatly increased. The subsequent relaxation of the velocity field in the deep layer was longer than in the upper layer in direct contact with the atmosphere. Winter intensification of the Black Sea deep-water currents is also observed in the climatic fields with an increase in cyclonic wind over the sea. Given the small number of in-situ deep-water velocity observations, further studies are likely to be conducted using numerical analysis of hydrodynamics for real synoptic situations at different time periods and comparison of the simulation results with relevant available observational data in order to refine the characteristics of deep-water currents in the Black Sea and clarify the mechanism of generation of deep-water circulation mesoscale features.

Acknowledgements The study was supported by the Russian Science Foundation Grant № 22-17-00150. The hydrophysical fields of the Black Sea for the autumn of 2005 were calculated according to the State task theme No FNNN-2021-0003.

References

Belokopytov VN (2004) Thermohaline and hydrologic-acoustical structure of the Black Sea waters. PhD Thesis, MHI NASU. Sevastopol: ECOSY-Gidrofizika. p 160 (in Russian)

Bulgakov NP, Bulgakov SN (1995) Manifestation of the Black Sea Countercurrent in the seawater density and hydrostatic pressure. Morskoy gidrofizicheskiy zhurnal (in Russian) 4:63–76

Demyshev SG, Korotaev GK (1992) Numerical energy-balanced model of the baroclinic currents in ocean with uneven bottom on a C-grid. In: Numerical models and results of calibration calculations of currents in the Atlantic Ocean: Atmosphere—Ocean—Space. The program "Sections". Institute of Numerical Mathematics RAS, Moscow, pp 163–231 (in Russian)

Efimov VV, Shokurov MV, Yarovaya DA (2007) Numerical simulation of a quasi-tropical cyclone over the Black Sea. Izv Atmos Ocean Phys 43:723–743

Ivanov VA, Belokopytov VN (2013) Oceanography of the Black Sea. Sevastopol: ECOSY-Gidrofizika 210

Korotaev GK, Demyshev SG, Knysh VV (2000) Three-dimensional climate of the Black Sea. In: Black Sea ecosystem processes and forecasting. Operational database management system. NReport of the workshop and project evaluation meeting (Istanbul, May, 15–16, 2000). IMS METU, Erdemli, pp 1–20

Korotaev G, Oguz T, Riser S (2006) Intermediate and deep currents of the Black Sea obtained from autonomous profiling floats. Deep-Sea Res II Top Stud Oceanogr 53(17–19):1901–1910. https://doi.org/10.1016/j.dsr2.2006.04.017

Markova NV, Bagaev AV (2016) The Black Sea deep current velocities estimated from the data of ARGO profiling floats. Phys Oceanogr 3:23–35. https://doi.org/10.22449/1573-160X-2016-3-23-35

Neumann G (1943) Uber den Aufbau und die Frage der Tiefenzirkulation des Schwarzen Meeres. Ann Hydrogr Und Marit Meteorol 71(1):1–20

Determination of the Initial Cs140 Concentration Field in the Black Sea After the Chernobyl Accident

V. S. Kochergin◉ and S. V. Kochergin◉

Abstract The work is devoted to the identification of the initial Cs140 concentration field in the Black Sea after the Chernobyl accident from contact measurement data. The assimilation procedure is based on an estimation method based on the method of adjoint equations. As input parameters for the transport-diffusion model, the values of current velocities obtained using a MGI circulation model of the Black Sea are used. The horizontal step is 1.6 km, and the wind stress is set according to SKIRON data. The method of adjoint equations is used to find fallout zones. The contact measurement data used were obtained on the 33rd flight of the NIS "Akademik Vernadsky" in June-July 1986. The results are compared with data from the National Centers for Environmental Prediction (NCEP) on the wind situation and precipitation intensity and information from the IRSN (Institute for Nuclear and Radiation Safety of France) on the dynamics of radioactive contamination from the Chernobyl nuclear power plant.

Keywords Cs140 concentration · Numerical modeling · Identification of pollution sources · Adjoint problem · Radioactive contamination · Impurity transfer

1 Introduction

The problem of identifying initial isotope concentration data from information distributed over space and time is an important task. The main fallout of radioactive contamination from the atmosphere after the Chernobyl accident come from 1 to 5 May 1986. Measurements of the isotopes Cs140 and Cs137 were made starting from June 15, 1986. In the first period, measurements were get only on the surface of the sea. The concentration of Cs140, due to its half-life (about two years), is an important characteristic of new radiation incoming in the sea.

V. S. Kochergin (✉) · S. V. Kochergin
Marine Hydrophysical Institute RAS, Kapitanskaya Str. 2, 299011 Sevastopol, Russia
e-mail: vskocher@gmail.com

© The Author(s), under exclusive license to Springer Nature Singapore Pte Ltd. 2023   79
T. Chaplina (ed.), *Processes in GeoMedia—Volume VII*, Springer Geology,
https://doi.org/10.1007/978-981-99-6575-5_7

The initial isotope concentration fields Cs140 are determined based on the method of adjoints equations (Marchuk 1982; Shutyaev 2019). It give possibilities to determine and find the spatial structure of the influence functions of the initial isotope concentration field on the values in a given region (Kochergin 2011, Kochergin and Kochergin 2017; Demyshev et al. 2018). In the work, the measurement data are distributed over time and space, the MGI model (Demyshev 2012, Demyshev and Dymova 2013) with real wind stress is used as a dynamic model. 27 horizons by the vertical coordinate are set in the model, the time step is 1.5 min. The dynamics of isotope Cs140 in Black Sea is modeled with the help of the transfer-diffusion equation of a passive impurity. At this time interval of model integration, the use of such a model is correct. In (Demyshev et al. 2023), similar calculations were performed to estimate the initial Cs137 fields after the Chernobyl accident. In this paper the same task solved for Cs140 isotope.

2 Isotope Transfer-Diffusion Model

Let the model of the transfer-diffusion has the form:

$$\frac{\partial C}{\partial t} + \frac{\partial(uC)}{\partial x} + \frac{\partial(vC)}{\partial y} + \frac{\partial(wC)}{\partial z} = A_H \nabla^2 C + \frac{\partial}{\partial z} A_V \frac{\partial C}{\partial z}, \tag{1}$$

where C is the impurity concentration; A_H is the coefficient of horizontal turbulent diffusion; A_V is the coefficient of vertical turbulent diffusion.

On the surface ($z = 0$), flow is absence:

$$A_V \frac{\partial C}{\partial z} = 0. \tag{2}$$

At the solid boundaries, the bottom is also set to zero flow. At the initial moment of time $C(x, y, z, 0) = C_0(x, y, z)$.

The horizontal coefficient of turbulent diffusion $A_H = 10^7$ cm^2/s, and the vertical coefficient of turbulent diffusion is set in the range of 2.5–0.03 cm^2/s (Demyshev et al. 2001).

3 Adjoint Problem

Equation (1) with boundary conditions (2) and initial data we will put it in the corresponding the following adjoint (Marchuk 1982) problem:

$$-\frac{\partial C^*}{\partial t} - \frac{\partial(uC^*)}{\partial x} - \frac{\partial(vC^*)}{\partial y} - \frac{\partial(wC^*)}{\partial z} = A_H \nabla^2 C^* + \frac{\partial}{\partial z} A_V \frac{\partial C^*}{\partial z}, \tag{3}$$

$$z = 0 : A_V \frac{\partial C^*}{\partial z} - wC^* = 0$$

$$z = H : A_V \frac{\partial C^*}{\partial z} = 0 \qquad (4)$$

$$\Gamma : A_H \frac{\partial C^*}{\partial n} = 0$$

$$C^*_{t_L} = h$$

In (4) h has not yet been determined, and Γ is the boundary of the Black Sea. From (1), (3), (4) taking into account the continuity equation and boundary conditions, we can obtain:

$$\int_M h C_T dM = \int_M C_0 dM C_0^* \qquad (5)$$

Let be $\Omega \in M$ some domain. Define h as

$$h = \begin{cases} \frac{1}{m(\Omega)} - in\ \Omega \\ 0 - outside\ \Omega \end{cases} \qquad (6)$$

where m is the measure of the domain Ω. Then, on the left in expression (5) we have the average concentration $\overline{C}_T$ in Ω at a finite time:

$$\overline{C}_T = \int_\Omega C_0 \cdot C_0^* d\Omega. \qquad (7)$$

Thus, with the help of formula (7), it is possible to find the concentration from the initial data using the solution of the adjoint problem.

4 Modular Dynamic, Meteorological and Radiation Situation in the Black Sea

The current velocities for the period from 01.05.1986 to 04.07.1986 were obtained by the MGI model (Demyshev 2012, Demyshev and Dymova 2013) with a space step of 1.6 km in latitude and longitude, taking into account the real wind stress for 1986 (according to ERA-5 data with a spatial resolution of 0.25°) (Kallos et al. 1997). The reanalysis information (Climate Data Store, Korotaev et al. 2016), interpolated onto the model grid, were used as the initial data. During the period under review in 1986, the circulation in the sea was quite typical for this season and described in detail in Demyshev and Dymova (2013).

Analysis of National Centers for Environmental Prediction (NCEP) data on the wind situation after the Chernobyl accident shows that, since May 2, air masses have southeasterly general direction of moving. In the period from 01.05.1986 to 05.05.1986, according to NCEP data on precipitation intensity (https://psl.noaa.gov/data/gridded/data.ncep.html), there were rains in the Black Sea area with values from 10 to 22 kg/m2. Radioactive isotopes with precipitation began to inflow the surface layer of the Black Sea starting from May 2. According to the Institute of Nuclear and Radiation Safety of France (https://www.irsn.fr/EN/global_partner/Radiation_protection) the radiation trail from the Chernobyl got the northeastern part of the sea on May 2. These data relate to the concentration of Cs137, but they can also be used to determination the zones of deposition of other isotopes after the Chernobyl accident. The maksimum of concentrations are observed on May 3–4 in the north-western, central and north-eastern parts of the sea. After, the radioactive cloud leave the Black Sea in the Anatolian coast and the Balkans direction. During this several days in the conditions of meaningful precipitation in these domains of the Black Sea, initial fields of radioactive isotopes were formed. The aim is to construct an initial concentration field Cs140 based on the method of adjoint equations.

5 Results of Numerical Experiments

The first data measurements of Cs137 and Cs140 concentrations in the Black Sea after the Chernobyl accident were made from June 15, 1986 to July 4, and the next series of measurements from November 26 to November 30 (Bayankina et al. 2021). In this paper, the measurements from the first raid are used. No measurements were carried out until June 15, and the background values of isotopes in the Black Sea were small. For each of the measurement points, the adjoint problems (3)–(4) were solved with the velocity fields obtained by the MGI model as input information. As a result, according to the solutions of the problem (3)–(4), it is possible to determine in which areas radioactive fallout had a greater impact on the Cs140 concentration at a given point. An domain of increased isotope concentration was found as a result. It is shown in Fig. 1. The concentration values in the areas of influence were given by constant value in the corresponding area. From Fig. 1, it can be seen that the maksimum means of concentration with values of about 800 Bq/m3 is mostly located off the coast of the Caucasus, the Kerch Peninsula, west of the Crimean Peninsula, and Taman Peninsula. These zones of big isotope concentration correlate well with zones of intense precipitation and domains of radioactive cloud passage. Taking into account the half-life, the concentration of the isotope before the accident will be considered small in relation to the radioactive fallout. Therefore, the initial concentration value before the accident was considered to be zero. The integration of the associated tasks was performed for each measurement point for the corresponding period of time. TVD (Total Variation Diminishing) approximations for advective

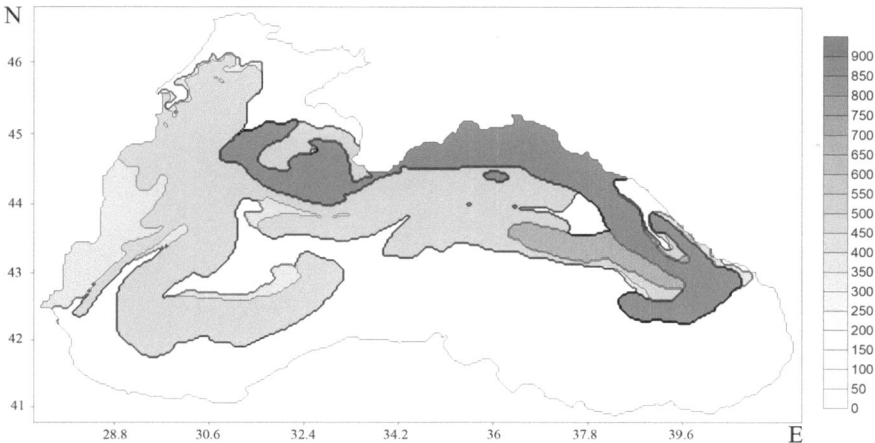

Fig. 1 The initial concentration field of Cs140 on May 4, 1986 (Bk/m^3)

terms of the equations were used to solve the main problem and related problems. In the difference approximation of these terms, Superbee (SB) discretization was used. This approach makes it possible to integrate the model with initial fields having a strongly gradient structure. This is especially important when solving conjugate problems when specifying perturbations in the initial data (4), (6). Preliminary test calculations have shown that the SB scheme turns out to be more preferable, since it gives a monotonic solution and preserves the initial extremes well. The SB scheme gives a more "compact" shape of the spot, which more correctly reflects the physics of the simulated process.

After calculating the transfer-diffusion model with a given initial field in the upper layer, we have a Cs140 concentration field for July 4, shown in Fig. 2. The model isotope concentration values have increased values near the coast of the Kerch and Taman peninsulas. The highest concentrations are located near the southern coast of Crimea and southwest of it, where the Main Black Sea Current passes.

Due to application of the adjoint equation method, the initial field Cs 140 concentration was obtained, which consistent with the measurement data and model. At the same time, the transfer model and its corresponding adjoint problem used as a space–time interpolant of concentration fields in the model integration region for the entire time interval.

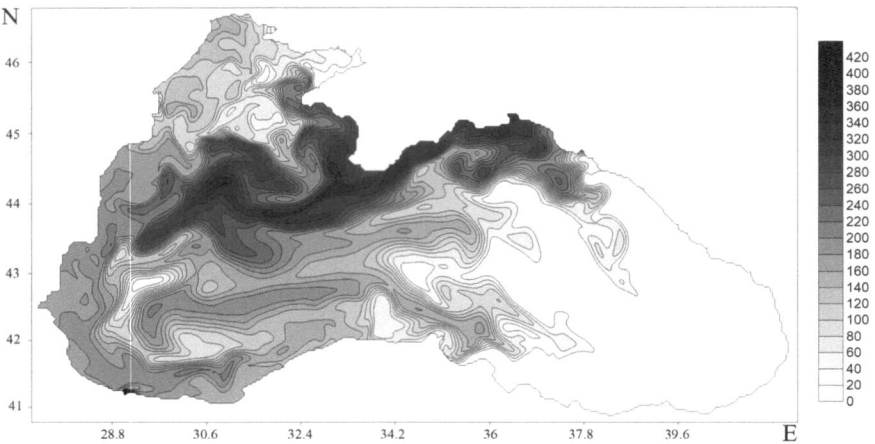

Fig. 2 Cs140 concentration field on July 4, 1986 (Bq/m^3)

6 Conclusions

Numerical experiments show that, the initial concentration field of the isotope Cs140 was found with the help of used method. This field was obtained as a result of assimilation of measurement data that were made in the Black Sea after the Chernobyl accident. The model estimates of concentration are consistent with the data of measurements distributed over space and time. Value of the functional characterizing the deviations of model estimates from measurements has significantly decreased. The found initial field is in good agreement with Black Sea dynamic and IRSN and NCEP data. Du to integration of adjoint equations, the task of determining the initial concentration field is solved.

Acknowledgements The work was carried out within the framework of the state task on the topic FNNN -2021-0005 "Complex interdisciplinary studies of oceanological processes that determine the functioning and evolution of ecosystems of the coastal zones of the Black and Azov Seas" (code "Coastal research").

References

Bayankina TM, Godin EA, Zhuk EV, Ingerov AV, Isaeva EA, Vetsalo MP (2021) Information resources of marine hydrophysical institute, RAS: current state and development prospects. In: Processes in GeoMedia–Volume 2. Hardcover, pp 187–197. ISBN 978–3–030–53520–9, Series ISSN 2197–9545. https://doi.org/10.1007/978-3-030-53521-6_22
Climate Data Store https://cds.climate.copernicus.eu/cdsapp#!/dataset/reanalysis-era5-single-lev els?tab=overview

Demyshev SG, Dymova OA (2013) Numerical analysis of the mesoscale features of circulation in the Black Sea coastal zone. Izv Atmos Ocean Phys 49:603–610. https://doi.org/10.1134/S00 01433813060030

Demyshev SG, Zapevalov AS, Kubryakov AI, Chudinovskikh TV (2001) Evolution of the Cs^{137} concentration field in the Black Sea after the passage of the Chernobyl cloud. Meteorol Hydrol 10:49–61

Demyshev SG (2012) A numerical model of online forecasting Black Sea currents. Izv Atmos Ocean Phys 48(1):120–132. https://doi.org/10.1134/S0001433812010021

Demyshev SG, Kochergin VS, Kochergin SV (2018) Using the variational approach and adjoint equations method under the identification of the input parameter of the passive admixture transport model. In Physical and mathematical modeling of earth and environment processes–3rd Inter-national Scientific School for Young Scientists, Ishlinskii Institute for Problems in Mechanics of Russian Academy of Science. Springer International Publishing AG, part of Springer Nature, pp 51–61. https://doi.org/10.1007/978-3-319-77788-7

Demyshev SG, Dymova OA, Kochergin VS, Kochergin SV (2023) Initialization of the Cs^{137} field concentration in the black sea after the chernobyl accident based on solving of adjoint problems. In: Chaplina T (ed) Processes in geomedia—volume VI. Springer geology. Springer, Cham. https://doi.org/10.1007/978-3-031-16575-7_45

Kallos G et al (1997) The regional weather forecasting system SKIRON: An overview. In: Proceedings of the symposium on regional weather prediction on parallel computer environments. Athens, Greece, pp 109–122

Kochergin VS (2011) Determination of the field of passive impurity concentration based on initial data based on the solution of conjugate problems. Environ Saf Coast Shelf Zones Integr Use Shelf Resourc 2(25):270–376

Kochergin VS, Kochergin SV (2017) Identifying the parameters of the instantaneous point pollution source in the Azov Sea based on the adjoint equation method. Phys Oceanogr (1):62–67. https://doi.org/10.22449/1573-160X-2017-1-62-67

Korotaev GK, Ratner YB, Ivanchik MV, Kholod AL, Ivanchik AM (2016) Operational system for diagnosis and forecast of hydrophysical characteristics of the Black Sea. Izv Atmos Ocean Phys 52(5):542–549

Marchuk GI (1982) Mathematical modeling in the problem of the environment. Nauka, Moscow

Shutyaev VP (2019) Methods of assimilation of observational data in problems of atmospheric and ocean physics. Izv RAS Phys Atmos Ocean 1(55):17–34

Analytical Solution of the Evolutionary Model of Wind Currents

V. S. Kochergin⊙, S. V. Kochergin⊙, and S. N. Sklyar⊙

Abstract The article presents analytical solutions for the stationary and evolutionary case for the problem of wind circulation. Analytical solutions were obtained with a fairly general wind forcing. The presence of such exact solutions allows us to analyze and select and use various numerical schemes and methods. The schemes found in this way can later be used for the numerical solution of such problems in complete models of marine hydrodynamics. In such schemes and approaches, it allows you to obtain more accurate solutions using less computing resources.

Keywords Dimensionless problem · Wind currents · Test problem · Analytical solution · Flow function · Evolutionary problem

1 Introduction

Dynamic numerical models occupy a special place in solving various oceanological problems. They make it possible to obtain and use flow fields in solving practical problems, including environmental ones. The quality of such dynamic fields depends on various factors—this is both the information entered and the difference schemes and algorithms used when integrating the model. It is better to choose such discretizations not by comparing different solutions to each other, but by comparing them with an exact analytical solution. Dynamic ocean models are complex (Marchuk and Sarkisyan 1988), so exact solutions exist only for the simplest statements, for example, for the Stommel model (Stommel 1948, 1965a, b). In (Eremeev et al. 2002; Kochergin and Dunets 2001), a similar problem is solved on the basis of the inversion method of the dynamic operator. In (Kochergin et al. 2020, 2021), a three-dimensional model was considered, which made it possible to analyze various approaches to calculating

V. S. Kochergin (✉) · S. V. Kochergin
Marine Hydrophysical Institute RAS, Kapitanskaya Str. 2, 299011 Sevastopol, Russia
e-mail: vskocher@gmail.com

S. N. Sklyar
American University of Central Asia, Abdumomunova Str.205, 720040 Bishkek, Kyrgyzstan

the vertical component of velocity. This paper presents an analytical solution of the stationary and evolutionary problem obtained by the authors in Kochergin et al. (2022); Kochergin et al. 2023) for variable wind forcing.

2 The Stationary Problem in Dimensionless Form

For a rectangular parallelepiped reservoir, its surface in the x0y plane is:

$$\Omega_0 = [0, r] \times [0, q].$$

Its constant depth $H > 0$. The axes of coordinate system are follows: Ox–to the east, Oy–to the north, Oz–vertically down. In the domain:

$$\Omega = \{(x, y, z) | (x, y) \in \Omega_0, 0 \le z \le H\}.$$

Consider the rsystem of stationary equations of motion in dimensionless form:

$$\begin{cases} -lv = -\frac{\partial P^s}{\partial x} + \frac{\partial}{\partial z}\left(k\frac{\partial u}{\partial z}\right) \\ lu = -\frac{\partial P^s}{\partial y} + \frac{\partial}{\partial z}\left(k\frac{\partial u}{\partial z}\right), t > 0, (x, y, z) \in \Omega^0 \\ \frac{\partial U}{\partial x} + \frac{\partial V}{\partial y} + \frac{\partial W}{\partial z} = 0 \end{cases} \tag{1}$$

with boundary conditions:

$$\begin{aligned} \{z = 0, (x, y) \in \Omega_{0,}^0\} &: k\frac{\partial u}{\partial z} = -\tau_x, k\frac{\partial V}{\partial z} = -\tau_y, w = 0; \\ \{z = H, (x, y) \in \Omega_{0,}^0\} &: k\frac{\partial u}{\partial z} = -\tau_x^b, k\frac{\partial V}{\partial z} = -\tau_y^b, w = 0; \\ \{0 \ge z \ge H, (x, y) \in \partial\Omega_0\} &: U \cdot n_x + V \cdot n_y = 0, \end{aligned} \tag{2}$$

in (2) the full streams are defined as follows:

$$U(x, y) = \int\limits_0^H u(x, y, z)dz, V(x, y) = \int\limits_0^H v(x, y, z)dz, \tag{3}$$

and in (1) the following parametrization of bottom friction is used:

$$\tau_x^b = \mu \le U, \tau_y^b = \mu \cdot V, \mu \equiv const > 0. \tag{4}$$

Due to the Stommel model, let:

$$l = l_0 + \beta \cdot y, k = const. \tag{5}$$

3 Analytical Solution for the Stationary Model

We exclude pressure gradients from the first two equations by cross-differentiation:

$$\begin{cases} \mu U - lv = -H\frac{\partial P^s}{\partial x} + \tau_x, \\ lU + \mu V = -H\frac{\partial P^s}{\partial y} + \tau_y, \\ \frac{\partial U}{\partial x} + \frac{\partial V}{\partial y} = 0, (x, y) \in \Omega^0, U \cdot n_x + V \cdot n_y = 0, (x, y) \in \partial\Omega_0. \end{cases} \tag{6}$$

Current function $\Psi(x, y)$ is defined by the formulas:

$$U = \frac{\partial \Psi}{\partial y}, V = -\frac{\partial \Psi}{\partial x}.$$

As a result, we get the following task:

$$\begin{cases} \mu\left(\frac{\partial^2 \Psi}{\partial x^2} + \frac{\partial^2 \Psi}{\partial y^2}\right) + \beta\frac{\partial \Psi}{\partial x} = \frac{\partial \tau_x}{\partial y} - \frac{\partial \tau_y}{\partial x}, (x, y) \in \partial\Omega_0^0, \\ \Psi = 0, (x, y) \in \partial\Omega_0. \end{cases} \tag{7}$$

In (Kochergin et al. 2022), analytical solution to this problem was obtained with a sufficiently general wind stress:

$$\begin{cases} \tau_x = [F_1 \cdot cos(r_l x) + F_2 \cdot sin(r_l x)] \cdot cos(q_m y) \\ \tau_y = [G_1 \cdot cos(r_s x) + G_2 \cdot sin(r_s x)] \cdot sin(q_p y) \end{cases} \tag{8}$$

where:

$$r_l = \frac{\pi l}{r}; r_s = \frac{\pi s}{r}; q_m = \frac{\pi m}{q}; q_p = \frac{\pi p}{q} \tag{9}$$
$$l, s = 0, 1, 2, \ldots; m, p = 1, 2 \ldots.$$

The wind model haves four real: F_1, F_2, G_1, G_2 and four integer: l, s, m, p numerical parameters, the choice of which help to describe a different wind situation. For example, if $F_1 = \frac{F \cdot q}{\pi}$, $F_2 = G_1 = G_2 = 0$; $l = 0$; $m = 1$ we have

$$\tau_x = \frac{F \cdot q}{\pi} cos\left(\frac{\pi y}{q}\right), \tau_y = 0 \tag{10}$$

and when

$$F_1 = \frac{F \cdot q}{\pi}, F_2 = 0, G_1 = -\frac{F \cdot q}{\pi}; G_2 = 0 \cdot l = 0; m = 1 \tag{11}$$

we have a cyclone over the water area.

4   Basic Formulas for the Stationary Model

Let's put all the formulas for stationary model. For the problem (7) we have:

$$\Psi(x, y) = \Psi_1(x, y) + \Psi_2(x, y)$$

where:

$$\Psi_1(x, y) = \left[C_1 e^{Ax} + C_2 e^{Bx} + D_1 \cdot \cos(r_l x) + D_2 \cdot \sin(r_l x)\right] \cdot \sin(q_m y),$$

$$D_1 = \frac{\mu\left(r_l^2 + q_m^2\right) F_1 + \beta r_l F_2}{\mu^2\left(r_l^2 + q_m^2\right)^2 + \beta^2 r_l^2} \cdot q_m, \quad D_2 = \frac{\mu\left(r_l^2 + q_m^2\right) F_2 - \beta r_l F_1}{\mu^2\left(r_l^2 + q_m^2\right)^2 + \beta^2 r_l^2} \cdot q_m,$$

and:

$$C_1 = D_1 \cdot \frac{e^{Br} - (-1)^l}{e^{Ar} - e^{Br}}, \quad C_2 = D_1 \cdot \frac{(-1)^l - e^{Ar}}{e^{Ar} - e^{Br}};$$

$$A = -\frac{\beta}{2\mu} + \sqrt{\left(\frac{\beta}{2\mu}\right)^2 + (q_m)^2}, \quad B = -\frac{\beta}{2\mu} - \sqrt{\left(\frac{\beta}{2\mu}\right)^2 + (q_m)^2}.$$

Similar formulas for the next component of the solution:

$$\Psi_2(x, y) = \left[\overline{C}_1 e^{\overline{A}x} + \overline{C}_2 e^{\overline{B}x} + \overline{D}_1 \cdot \cos(r_s x) + \overline{D}_2 \cdot \sin(r_s x)\right] \cdot \sin(q_p y),$$

where

$$\overline{D}_1 = -\frac{\beta r_s G_1 - \mu\left(r_s^2 + q_p^2\right) G_2}{\mu^2\left(r_s^2 + q_p^2\right)^2 + \beta^2 r_s^2} \cdot r_s, \quad \overline{D}_2 = -\frac{\mu\left(r_s^2 + q_p^2\right) G_1 + \beta r_s G_2}{\mu^2\left(r_s^2 + q_p^2\right)^2 + \beta^2 r_s^2} \cdot r_s;$$

$$\overline{C}_1 = \overline{D}_1 \cdot \frac{e^{\overline{B}r} - (-1)^s}{e^{\overline{A}r} - e^{\overline{B}r}}, \quad \overline{C}_2 = \overline{D}_1 \cdot \frac{(-1)^s - e^{\overline{A}r}}{e^{\overline{A}r} - e^{\overline{B}r}}$$

$$\overline{A} = -\frac{\beta}{2\mu} + \sqrt{\left(\frac{\beta}{2\mu}\right)^2 + (q_m)^2}, \quad \overline{B} = -\frac{\beta}{2\mu} - \sqrt{\left(\frac{\beta}{2\mu}\right)^2 + (q_m)^2}$$

The full streams are determined by the formulas:

$$U(x, y) = \frac{\partial}{\partial y}[\Psi_1(x, y) + \Psi_2(x, y)]$$

$$= q_m \cdot \left[C_1 e^{Ax} + C_2 e^{Bx} + D_1 \cdot \cos(r_l x) + D_2 \cdot \sin(r_l x)\right] \cdot \cos(q_m y)$$

$$+ q_p \cdot \left[\overline{C}_1 e^{\overline{A}x} + \overline{C}_2 e^{\overline{B}x} + \overline{D}_1 \cdot \cos(r_s x) + \overline{D}_2 \cdot \sin(r_s x)\right] \cdot \cos(q_p y)$$

$$V(x, y) = -\frac{\partial}{\partial x}[\Psi_1(x, y) + \Psi_2(x, y)]$$

$$= -\left[AC_1 e^{Ax} + BC_2 e^{Bx} - r_l D_1 \cdot \sin(r_l x) + r_l D_2 \cdot \cos(r_l x)\right] \cdot \sin(q_m y)$$

$$- \left[\bar{A}\bar{C}_1 e^{\bar{A}x} + \bar{B}\bar{C}_2 e^{\bar{B}x} - r_s \bar{D}_1 \cdot \sin(r_s x) + r_s \bar{D}_2 \cdot \sin(r_s x)\right] \cdot \sin(q_p y)$$

5 The Evolutionary Model

Let's consider the problem (1)–(4) in full, without assuming its stationarity. The problem for integral velocities, as before, is obtained if the equations of system (1) are integrated in z from 0 to H, taking into account the boundary conditions (2), (3), (6), and then eliminate the pressure P^5 by cross-differentiation:

$$\begin{cases} \left(\frac{\partial}{\partial t} + \mu\right)\left(\frac{\partial U}{\partial y} - \frac{\partial V}{\partial x}\right) - \frac{\partial(lV)}{\partial y} - \frac{\partial(lU)}{\partial x} = \frac{\partial \tau_x}{\partial y} - \frac{\partial \tau_y}{\partial x}, (x, y) \in \Omega_0^0, t > 0; \\ \frac{\partial U}{\partial x} + \frac{\partial V}{\partial y} = 0, (x, y) \in \Omega_0^0, t > 0; \\ U \cdot n_x + V \cdot n_y = 0, (x, y) \in \partial \Omega_0, t > 0 \end{cases} \quad (12)$$

In (12) and further, there are no initial conditions, since our goal is to find a wide class of solutions to the problem, and the initial conditions will be determined by the choice of the desired solution. As before, we introduce the current function $\Phi(x, y, t)$ according to the formulas (the notation has been changed to take into account the stationary case):

$$U = \frac{\partial \Phi}{\partial y}, V = -\frac{\partial \Phi}{\partial x} \quad (13)$$

Substituting these values into the first equation of (12), we obtain:

$$\begin{cases} \left(\frac{\partial}{\partial t} + \mu\right)\left(\frac{\partial^2 \Phi}{\partial x^2} + \frac{\partial^2 \Phi}{\partial y^2}\right) + \beta \frac{\partial \Phi}{\partial x} = \frac{\partial \tau_x}{\partial y} - \frac{\partial \tau_y}{\partial x}, (x, y) \in \Omega_0^0, t > 0; \\ \Phi = 0, (x, y) \in \partial \Omega_0, t > 0. \end{cases} \quad (14)$$

We will look for solutions to problem (14) in the form:

$$\Phi(x, y, t) = \Psi(x, y) + \Phi_0(x, y, t),$$

where $\Psi(x, y)$ is the solution of a stationary inhomogeneous problem:

$$\begin{cases} \mu\left(\frac{\partial^2 \Psi}{\partial x^2} + \frac{\partial^2 \Psi}{\partial y^2}\right) + \beta \frac{\partial \Phi}{\partial x} = \frac{\partial \tau_x}{\partial y} - \frac{\partial \tau_y}{\partial x}, (x, y) \in \Omega_0^0; \\ \Psi = 0, (x, y) \in \partial \Omega_0; \end{cases} \quad (15)$$

and $\Phi_0(x, y, t)$- solution of a nonstationary homogeneous problem:

$$\begin{cases} \left(\frac{\partial}{\partial t} + \mu\right)\left(\frac{\partial^2 \Phi_0}{\partial x^2} + \frac{\partial^2 \Phi_0}{\partial y^2}\right) + \beta\frac{\partial \Phi_0}{\partial x} = 0, (x, y) \in \Omega_0^0, t > 0; \\ \Phi_0 = 0, (x, y) \in \partial\Omega_0, t > 0. \end{cases}$$

Now it is possible to write out a set of solutions to the problem (14):

$$\Phi(x, y, t) = \Psi(x, y) + \Phi_0(x, y, t)$$

$$= \Psi(x, y) + \sin(r_k \cdot x) \cdot \sin(q_n \cdot y) \cdot e^{-\mu t} \cdot \left[S_1 \cdot \sin\left(\alpha_k^n x + \frac{\beta t}{2\alpha_k^n}\right)\right.$$

$$\left. + S_2 \cdot \cos\left(\alpha_k^n x + \frac{\beta t}{2\alpha_k^n}\right)\right] \tag{16}$$

The choice of a specific solution from the set (16) is determined by the choice of parameters S_1, S_2, k, n in accordance with (11), the integral velocities are found by the formulas:

$$U(x, y, t) = \frac{\partial \Psi}{\partial y} + q_n \sin(r_k \cdot x)\cos(q_n \cdot y)\left[S_1 \cdot \sin\left(\alpha_k^n x + \frac{\beta t}{2\alpha_k^n}\right)\right.$$

$$\left. + S_2 \cdot \cos\left(\alpha_k^n x + \frac{\beta t}{2\alpha_k^n}\right)\right] e^{-\mu t} \tag{17}$$

$$V(x, y, t) = -\frac{\partial \Psi}{\partial y} - r_k \cos(r_k \cdot x)\sin(q_n \cdot y)\left[S_1 \cdot \sin\left(\alpha_k^n x + \frac{\beta t}{2\alpha_k^n}\right)\right.$$

$$\left. + S_2 \cdot \cos\left(\alpha_k^n x + \frac{\beta t}{2\alpha_k^n}\right)\right] e^{-\mu t} - \alpha_k^n \sin(r_k \cdot x)\sin(q_n \cdot y)\left[S_1 \cdot \cos\left(\alpha_k^n x + \frac{\beta t}{2\alpha_k^n}\right)\right.$$

$$\left. - S_2 \cdot \sin\left(\alpha_k^n x + \frac{\beta t}{2\alpha_k^n}\right)\right] e^{-\mu t} \tag{18}$$

Recall that the first terms in formulas (17) and (18) are stationary integral velocities, their values are calculated using the formulas found in the previous section.

6 Conclusions

For the unsteady problem of Ekman-type wind circulation, an analytical solution has been obtained for a space-variable wind effect. The obtained solutions can be used as reference for testing various difference schemes. The results can be used in the construction of numerical models of the dynamics of the ocean and various reservoirs.

Acknowledgements The work was carried out within the framework of the state task on the topic 0555-2021-0005 "Complex interdisciplinary studies of oceanological processes that determine the functioning and evolution of ecosystems of the coastal zones of the Black and Azov Seas" (code "Coastal research").

References

Eremeev VN, Kochergin VP, Kochergin SV, Sklyar SN (2002) Mathematical modeling of hydrodynamics of deep-water basins. ECOSI-Gidrophisika, Sevastopol

Kochergin VP, Dunets TV (2001) Computational algorithm of the evaluations of inclinations of the level in the problems of the dynamics of basins. Phys Oceanogr 3(11):221–232

Kochergin VS, Kochergin SV, Sklyar SN (2020) Analytical test problem of wind currents. In: Olegovna C (eds) Processes in GeoMedia—volume I. Springer Geology. Springer, Cham, pp 17–25

Kochergin VS, Kochergin SV, Sklyar SN (2021) Analytical solution of the test three-dimensional problem of wind flows In: Chaplina T (ed) Processes in GeoMedia, vol. II Springer geology, Springer, pp 65–71

Kochergin VS, Kochergin SV, Sklyar SN (2022) Analytical solution of the equation for the stream function in the model of Ekman-type flows with variable wind stress in space. In: Springer proceedings in earth and environmental sciences. pp 147–158

Kochergin VS, Kochergin SV, Sklyar SN (2023) Testing of numerical methods for solving the stream function problem in a model of stationary fluid motion. In: Chaplina T (ed) Processes in GeoMedia, vol. VI Springer geology. Springer, pp 513–5211

Marchuk GI, Sarkisyan AS (1988) Mathematical modeling of ocean circulation. Nauka, Moscow

Stommel H (1948) The westward intensification of wind-driven ocean currents. Trans Amer Geophys Union 29:202–206

Stommel H (1965b) The GULF STREAM a physical and dynamical description. University of California Press

Stommel H (1965) Gulf stream. Moscow, IL

Assessment of the Wave Asymmetry Contribution to the Dynamics of Sediment Fractions in the Coastal Zone of the Kalamitsky Gulf

K. I. Gurov◉ and V. V. Fomin◉

Abstract The paper describes the results of mathematical modeling of the underwater coastal slope and beach morphodynamics, as well as the dynamics of sand fractions of the granulometric composition under the influence of wind waves, taking into account the influence of wave nonlinearity, for the section of the Lake Sakskoe bay-bar (Western Crimea). It has been established that the maximum erosion of the beach is observed in the northern part of the study area, and the maximum accumulation of gravel-sand material is in the southern part. A significant influence on the results of model calculations of the values of the parameter *facua*, which determines the contribution of nonlinearity (wave asymmetry) to the total transport of sediments near the coast, has been established. It was found that for the study area, the optimal value of *facua* is in the range of 0.3–0.4.

Keywords Sakskoe Lake bay-bar (Western Crimea) · Coastal zone · Bottom sediments · XBeach model · Granulometric composition · Relief storm deformation · Wave nonlineariry

1 Introduction

Coastal regions are zones of active anthropogenic load, as well as intensive hydro- and geodynamic processes. The simultaneous influence of natural and anthropogenic factors determines the high rate of changes in the characteristics of coastal ecosystems. The morphology of the beach and coastline is very dynamic and is largely determined by waves, currents, and regional sediment characteristics (McNinch 2004).

It is known that the change in the higher moments of wave motion (asymmetry, kurtosis, etc.) as the waves approach the coast determines the nature of wave breaking, the generation of infra-gravity waves and wave currents, the processes of sediment

K. I. Gurov (✉) · V. V. Fomin
Marine Hydrophysical Institute of the Russian Academy of Sciences, Sevastopol, Russia
e-mail: gurovki@gmail.com

© The Author(s), under exclusive license to Springer Nature Singapore Pte Ltd. 2023   95
T. Chaplina (ed.), *Processes in GeoMedia—Volume VII*, Springer Geology,
https://doi.org/10.1007/978-981-99-6575-5_9

suspension and transport (Kos'yan et al. 2003). Works (Saprykina et al. 2013, 2017; Saprykina 2020) are devoted to the study of the processes of nonlinear transformation of waves in the coastal zone of the sea. Some of them are devoted to establishing a relationship between wave asymmetry and breaking indices (Saprykina et al. 2013). In others, attention is drawn to the influence of nonlinear wave transformation on the features of sediment movement and the formation of the underwater bottom profile (Saprykina et al. 2017; Saprykina 2020).

The particle size of bottom sediments and their spatial distribution are key parameters in the characterization of sediments and mathematical modeling of the morphodynamics of sandy beaches. Studies, carried out in Prodger et al. (2017); Gallagher et al. 2011; Reniers et al. 2013), show that even small variations in grain size can have a significant impact on the results of beach morphology modeling. According to the results of Prodger et al. (2017), the profiles of the particle size distribution of sediments in the coastal zone are associated with the dissipation of the energy of breaking waves. It has been established that the observed maximum grain sizes and their sorting corresponded to the location of the peak energy dissipation. It is shown that the spatial variations of fractions occur on intratidal time scales. In the wave breaking zone, the transport of suspended sediments prevails; finer sediments are redistributed both towards the foreshore and into the offshore part. According to the authors of Reniers et al. (2013), the key parameter of sediment dispersion and retention of the beach profile is the phase shift between the particle sedimentation rates and turbulence in the water column.

The relevance of studying the dynamics of granulometric fractions of bottom sediments on the coast of Kalamitsky Gulf in the area of the Sakskoye Lake bay-bar is due to a significant increase in the rate of development of the recreational potential of this region. Insufficient knowledge and inaccuracies in the assessments of the characteristics of lithodynamic processes in this area have led to the destruction of coastal infrastructure and disruption of the integrity of the ecosystem of the bay.

The purpose of this work is to study the local features of the redistribution of various sediment fractions in the coastal zones of the Kalamitsky Gulf in the area of the Sakskoye Lake bay-bar under the influence of storm waves, taking into account the influence of wave asymmetry. Previously, studies on the dynamics of bottom sediments were carried out by the authors in relation to sections of the coastal zone of Kalamitsky Gulf in the area of Lake Bogaily bay-bar (Gurov et al. 2019).

2   Characteristics of the Study Area

The bay-bar of Lake Sakskoe is located in the central part of Kalamitsky Gulf (Fig. 1). The length of the bay-bar is about 3300 m, the width is up to 700 m, and the thickness of the sands of the bay-bar is 24 m (Zenkovich 1960). The bay-bar is part of a single accumulative formation that stretches from Cape Karantinny to Lake Kyzyr-Yar (Bratus 1965). The bay-bar is composed of coarse-grained sands with an admixture of gravel and pebbles (Zenkovich 1960; Gurov 2020). The width of the beaches

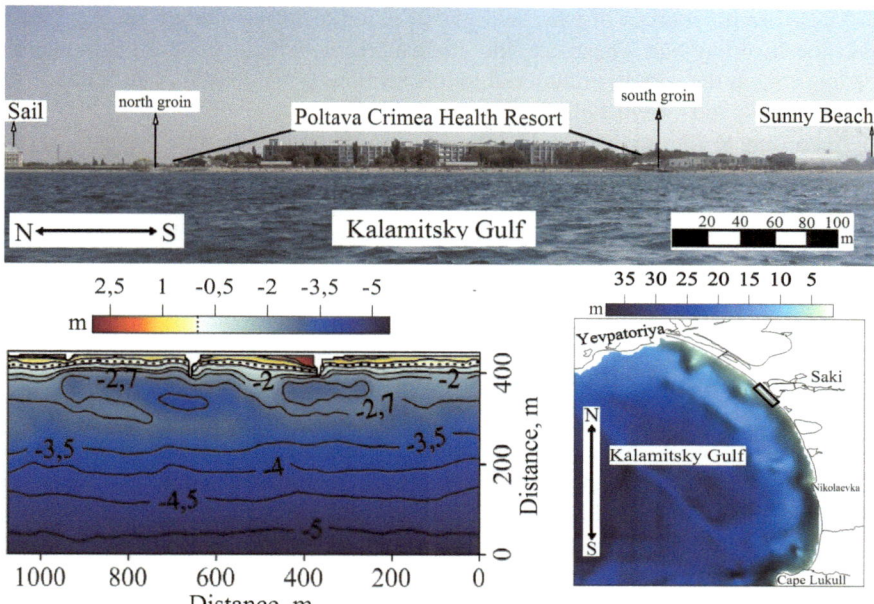

Fig. 1 Schematic location of the study area of the coastal zone

reaches 40 m in the northern, 35 m in the central and 60 m in the southern parts of the bay. The study area is characterized by a relatively straight coastline and isobaths parallel to the coast.

The study area is located in the central part of the bay-bar of Lake Sakskoye between the holiday homes "Sunny Beach" and "Sail" and covers the entire coastal zone area of the Poltava Crimea Heath Resort sanatorium. In the central part of the studied section of the coastline, there are two coastal protection structures in the form of concrete groins 37 m long in the southern part and 54 m in the northern part.

It is noted that in places of the coarse-grained material accumulation, the profile of the nearshore zone has a larger angle of inclination; for areas with a high content of medium and fine-grained sand, there are flatter sections of the profile. It is noted that in the study area, the location of the crest of the bar varies depending on the proximity of coastal structures. So, in the southern part of the studied section of the coastal zone under study, the underwater bar is located at a distance of 25–75 m from the coast, in the central part of the beach of the Poltava Crimea sanatorium it is pulled from the coast to 80–130 m, bypassing the northern groin—again shifts to the coast.

Bottom slopes for sections with and without an underwater bar differ in the area of 2.5–3.4 m. Over the entire area under study, the bottom slopes decrease with depth: near the coast (0–30 m) they are 0.5–0.6; deeper than 3.4 m–0.009. The smallest value of the bottom slope (0.001–0.003) is noted in the upper parts of the crests of the bar.

There are no sources of sand and gravel material in the studied section of the Sakskoe lake bay-bar. Therefore, the alongshore movement of sediments plays a leading role in the supply of material to this section of the coastal zone (Zenkovich 1960; Udovik and Goryachkin 2013).

According to Kharitonova and Fomin (2012), for the region of the Sakskoye Lake bay-bar, the highest frequency of storm wind waves was noted for the southern and southwestern directions, which are close to the normal with respect to the coast-line. This allows one-dimensional models to be used for mathematical modeling of morphodynamics.

3 Mathematical Model and Initial Data

To study the effect of wave asymmetry on the dynamics of the granulometric compo-sition of bottom sediments the XBeach complex numerical model was used (Roelvnik et al. 2009; XBeach Model Description and Manual 2010).

In the model, wave asymmetry is taken into account in the transport-diffusion equation describing the concentration of suspended sediments C:

$$
\frac{\partial hC}{\partial t} + \frac{\partial (u + V_w \cos \theta_m)hC}{\partial x} + \frac{\partial (v + V_w \sin \theta_m)hC}{\partial y} - \frac{\partial}{\partial x}\left[Dh\frac{\partial C}{\partial x} \right]
$$
$$
- \frac{\partial}{\partial y}\left[Dh\frac{\partial C}{\partial y} \right] = \frac{h(C_{eq} - C)}{T_s} \tag{1}
$$

where x, y, t are spatial coordinates and time; u, v are the flow velocity components from the hydrodynamic block; θ_m is the average direction of the waves; h is the depth of the sea; D is the coefficient of horizontal turbulent diffusion; T_s is the time for C to adjust to the equilibrium concentration C_{eq}; $V_w = u_{rms}\, facua\, (S_k - A_s)$ is the amplitude of the velocity of wave currents; u_{rms} is the amplitude of the velocity of near-bottom movements; $0 \le facua \le 1$ is the setting parameter that regulates the contribution of wave asymmetry to sediment transport; S_k, A_s are semi-empirical functions of Ursell number (Ruessink et al. 2012; Rienecker and Fenton 1981):

$$
S_k = B \cos \psi, \; A_s = B \sin \psi
$$

$$
B = 0.79/\left[1 + \exp((0.61 + \log U_r)/0.35)\right], \; \psi = \frac{\pi}{2}\left[\tanh(0.64/U_r^{0.6}) - 1\right] \tag{2}
$$

The XBeach model with a horizontal resolution of 4 m was used. On the seaward boundary of the computational area, the waves were specified in the form of the JONSWAP spectrum, which included the significant wave height (h_s) and the peak wave period (T_p) as parameters. 3 variants of storms were considered: storm 1—h_s = 1.54 m; T_p = 8.6 s; storm 2—h_s = 2.04 m; T_p = 9.2 s; storm 3—h_s = 2.54 m;

$T_p = 9.8$ s. These characteristics of storms were identified based on the ERA wave reanalysis data for 1979–2021. The duration of the storm was 12 h.

In the process of morphodynamic modeling, in each cell of the computational grid (x_i, y_i) the following were determined: $p_{m,n}(x_i, y_j, t)$—volume concentration of fractions; $\Delta z_n(x_i, y_j, t)$—thickness of the layers occupied by sediments. Where m—fraction number; n—number of sediment layer.

Further, for each fraction of bottom sediments, the average volumetric concentration over the layers was calculated ($m = 1, 2, 3, 4$):

$$p_m(x_i, y_j, t) = \frac{\sum_{n=1}^{nd} p_{m,n}(x_i, y_j, t) \cdot \Delta z_n(x_i, y_j, t)}{\sum_{n=1}^{nd} \Delta z_n(x_i, y_j, t)} \tag{3}$$

where $nd = 15$ is the number of bottom layers.

The median diameter of the sediments was presented as:

$$D_{mean}(x_i, y_j, t) = \sum_{m=1}^{ngd} D_{50,m} \cdot p_m(x_i, y_j, t) \tag{4}$$

where $ngd = 4$ is the number of sediment classes; $D_{50,m}$ is the median diameter for m-class.

When carrying out model calculations, data on the relief and granulometric composition of sediments obtained during monitoring observations in this area were used (Gurov 2020).

Based on field observations, four fractions of bottom sediments were identified (Table 1). For each fraction, the values of the median diameter D_{50} and the values of D_{15} and D_{90} are given, obtained by constructing a cumulative curve for each fraction based on the results of a granulometric analysis of sediment samples.

The amount of each fraction in the mixture was set in volume concentrations corresponding to the fractional content of this material per unit volume. At the initial moment of time, the distribution of volumetric concentrations was set based on the results of field observations and is shown in Fig. 2.

The distribution of fine-grained gravel material and coarse-grained sand (fraction 1) is limited by the surf zone between two coastal protection structures. The maximum concentrations of fraction 1 are observed to the north of the southern groin. Thus,

Table 1 Granulometric characteristics of sediment fractions

Fraction (m)	D_{90}, mm	D_{50}, mm	D_{15}, mm
(1) Fine gravel and coarse sand	1.5	0.75	0.5
(2) Medium sand	0.5	0.35	0.25
(3) Fine sand	0.25	0.2	0.13
(4) Aleuritic silt	0.12	0.11	0.1

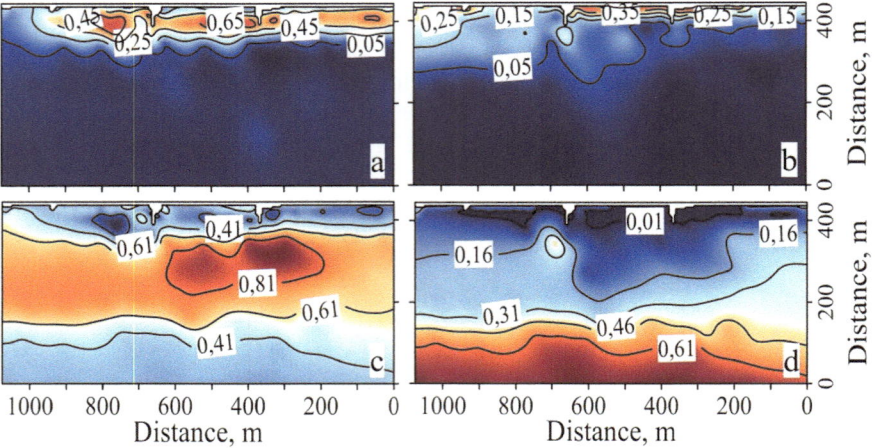

Fig. 2 Initial distribution of granulometric fractions: **a** fine gravel and coarse sand ($m = 1$), **b** medium sand ($m = 2$), **c** fine sand ($m = 3$), **d** aleuritic silt ($m = 4$)

overcoming the coastal protection structure, the sediment flow loses the necessary energy, and large fractions are immediately deposited (Fig. 2a).

The fraction of medium-grained sand (fraction 2) is observed mainly in the sediments of the beach, with maximum values in the central part of the beaches in the interbunal space of the Poltava Crimea sanatorium (Fig. 2b). The proportion of medium-grained sand decreases with depth.

The fraction of fine sand accumulates as the depth increases. Its maximum values are noted in the central part of the study area, which is determined by the location of the shallow slope of the underwater bar directed towards the shore (Fig. 2c). The results obtained showed that the accumulation of the fine sand fraction is noted between the isobaths of 2.5–3.5 m.

The silt fraction accumulates deeper than the 3.4 m isobath. The maximum values of silt material are noted deeper than the 5.5 m isobath (Fig. 2d). The increase in the share of the silty fraction near the northern groin of the Poltava Crimea sanatorium is explained by the features of the bottom topography in this area. According to the bathymetric survey data, in this area there is a depression up to 3.5 m deep, seaward of which there is an underwater bar.

The profiles of calculated characteristics were considered in 4 sections of the study area: in the southern part—$y = 200$ m, in the central part in the interbunal space of the Poltava Crimea sanatorium—$y = 500$ m, to the north of the coast protection structures—$y = 800$ m, in the northern part—$y = 1000$ m. The influence on the results of model calculations of the values of the parameter *facua*, which determines the contribution of nonlinearity (wave asymmetry) to the total transport of sediments in the coastal zone, was estimated.

4   Discussion of Results

The modeling results showed that the evolution of the coastal zone profile for all considered storms leads to the erosional destruction of beaches and the flattening of the underwater slope in the waterfront zone. The eroded material moved in seaward direction and accumulated, forming an underwater bar. It was revealed that for the sites located in the southern ($y = 200$ m) and northern ($y = 1000$ m) parts of the study area, changes in the morphodynamics of the underwater coastal slope occurred within a 100-m zone limited by the 2.5–2.7 m isobath. The bottom topography reshaping band expanded with a change in the value of the *facua* parameter.

It has been established that the maximum erosion of the beach for all selected areas occurs at the value of the parameter *facua* = 0.0. As a result, the width of the beach after 12 h of storm action decreases from 22 to 3 m on the profile $y = 200$ and $y = 1000$ m, to 7 m on the profile $y = 500$ m, and completely disappears on the profile 800 m (Fig. 3).

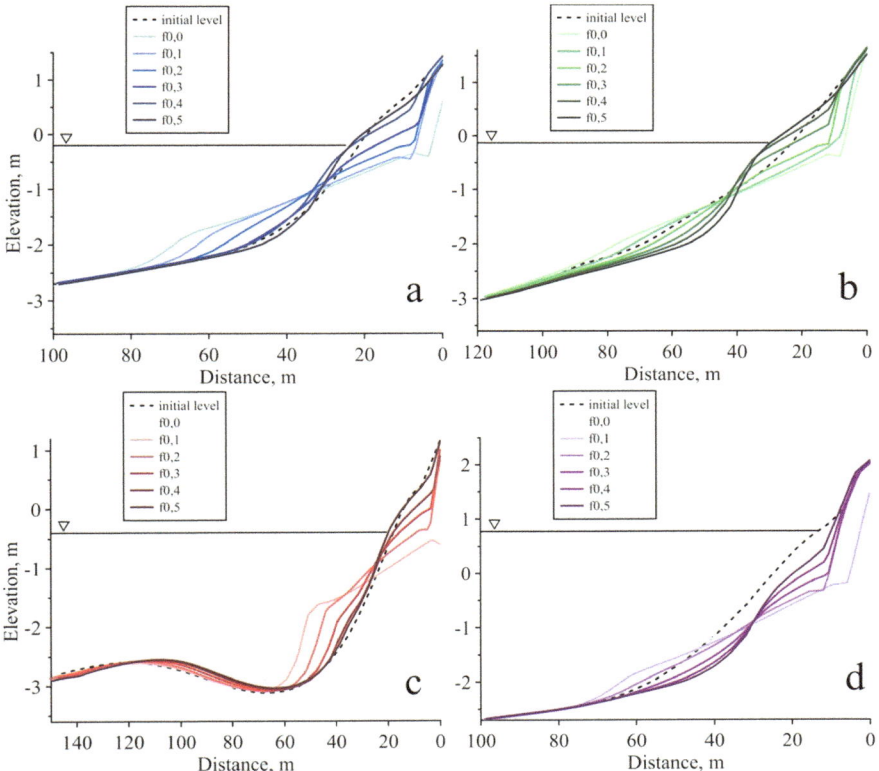

Fig. 3 Changing the profile of the coastal zone in different areas: **a**—200 m; **b**—500 m; **c**—800 m; **d**—1000 m

With an increase in the *facua* parameter, the intensity of the erosion impact on the coastal zone decreases. However, even with an increase in values up to 0.5, the profile deformation is preserved both in the beach area and in the marine part of the underwater coastal slope. As a result, on the profiles $y = 200$ and $y = 800$ m, after 12 h of storm action at *facua* $= 0.5$, erosion occurs in the central part of the beach and redeposition of material in the waterfront zone. For the central part of the coastal zone in the area between two groins of the Poltava Crimea Health Resort ($y = 500$ m), the erosion of the coastal zone profile occurs both in the beach area and in the nearshore zone of the underwater coastal slope. The magnitude of the erosion in the nearshore part with an increase in the value of the parameter *facua*, on the contrary, increases. For the northern section (y $= 1000$), the erosion of the upper part of the coastal zone profile is preserved at large values of *facua*.

Analysis of the obtained results showed that for the fraction of fine-grained gravel and coarse-grained sand after 12 h of storm impact, a decrease in concentrations in the central part of the beach zone in three areas (y $= 200$, y $= 500$ and y $= 800$ m) and redeposition of material in the northern part (y $= 1000$ m) is observed. In the southern part of the study area, coarse-grained material moves to the seaward part, forming an accumulative body. The minimum values of the fraction dynamics (up to 10%) were noted in the intergroin space (y $= 500$), and the maximum (up to 30%) to the north from the northern groin. With an increase in the value of the parameter *facua* in the southern part of the study area, the intensity of the redistribution of coarse-grained material increases, and in the northern part, on the contrary, it decreases.

For the medium-grained sand fraction, an increase in volumetric concentrations is observed in the zone of formation of an accumulative body in the nearshore part of the underwater coastal slope at a distance of 15–60 m from the shoreline. Medium-grained material moves mainly from the seaward part of the underwater coastal slope. The increase in the proportion of the medium-grained fraction does not exceed 15%, and the maximum increase is observed for the area north from the northern groin. It is noted that the maximum values of accumulation of medium-grained sand are observed in the central part of the accumulative body, and with a change in the angle of the bottom, the values decrease sharply. It has been established that with an increase in the value of the *facua* parameter, the width of the formed bar and the volume concentrations of the fraction decrease. At the same time, an increase in the *facua* parameter for the northern part of the study area (y $= 800$, 1000) leads to a significant erosion of the beach and the existing morphodynamic structure near the coast.

The fraction of fine-grained sand in the southern and northern parts of the study area at *facua* values (0.0–0.3) is removed from the surf zone, and at *facua* values (0.4–0.5) it is shifted into the surf zone, as well as into the beach zone. The maximum decrease in the proportion of fine-grained sand (up to 15%) is noted in the southern part of the study area and coincides with the boundaries of the formation of an accumulative body, which indicates the fact that the body itself is formed mainly by coarse and medium-grained sand. To the north of the northern groin, the opposite picture is observed: in the intergroin space, the material is removed from the beach zone and accumulates in the nearshore part of the underwater coastal slope, forming an accumulative body together with the medium-grained fraction. In the northern part of the study area, only the removal of the fine-grained fraction is observed, the scale of which decreases with increasing values of the *facua* parameter.

The fraction of silty silts in the southern part of the study area ($y = 200$, $y = 500$ m) at low values of *facua* (0.0–0.2) is removed to the offshore part of the profile, and at the maximum value (0.5) it accumulates in the foreshore part of the beach. In the northern part, there is a simultaneous removal from the beach zone and the nearshore part of the underwater coastal slope and a slight accumulation in the zone of an accumulative body formation to the north of the northern groin.

An analysis of the resulting parameter, characterizing the granulometric composition of the mixture (D_{mean}), showed that after 12 h of storm impact on the coastal zone, a decrease in the size of the material in the beach zone and in the foreshore zone occurs with the formation of an accumulative body. In addition, an increase in the size of the material in the northern part due to the removal of medium- and fine-grained material from this area and the accumulation of a coarse-grained fraction is observed (Fig. 4).

It has been established that the main redistribution of granulometric fractions occurs in the foreshore and nearshore areas bounded by the 2.5–3 m isobath, and does not exceed 150 m in width, which is consistent with other studies performed in different parts of the coastal zone of the Kalamitsky Gulf (Gurov et al. 2019).

5 Conclusion

Mathematical modeling of the morphodynamics of the underwater coastal slope and the beach, as well as the dynamics of sandy fractions of the particle size distribution under the influence of wind waves, was performed using the example of a 1 km section of the bay-bar of Lake Sakskoe. The calculations were carried out for 3 storm situations, the parameters of which were determined from the ERA wave reanalysis data. The dependence of spatial changes in the granulometric composition on the intensity of the storm was studied. It has been established that the maximum erosion of the beach is observed in the northern part of the study area, and the maximum accumulation of gravel-sand material is in the southern part. It has been established that a decrease in the proportion of fine-grained sand is determined by a change in the angle of the bottom, especially in the northern part of the study area. It was found that in the nearshore zone the main contribution to the variability of the average diameter of bottom sediments (D_{mean}) is made by fine-grained gravel and coarse-grained sand. A significant influence on the results of model calculations of the values of the parameter *facua*, which determines the contribution of nonlinearity (wave asymmetry) to the total transport of sediments near the coast, has been established. It was found that for the study area, the optimal value of *facua* is in the range of 0.3–0.4.

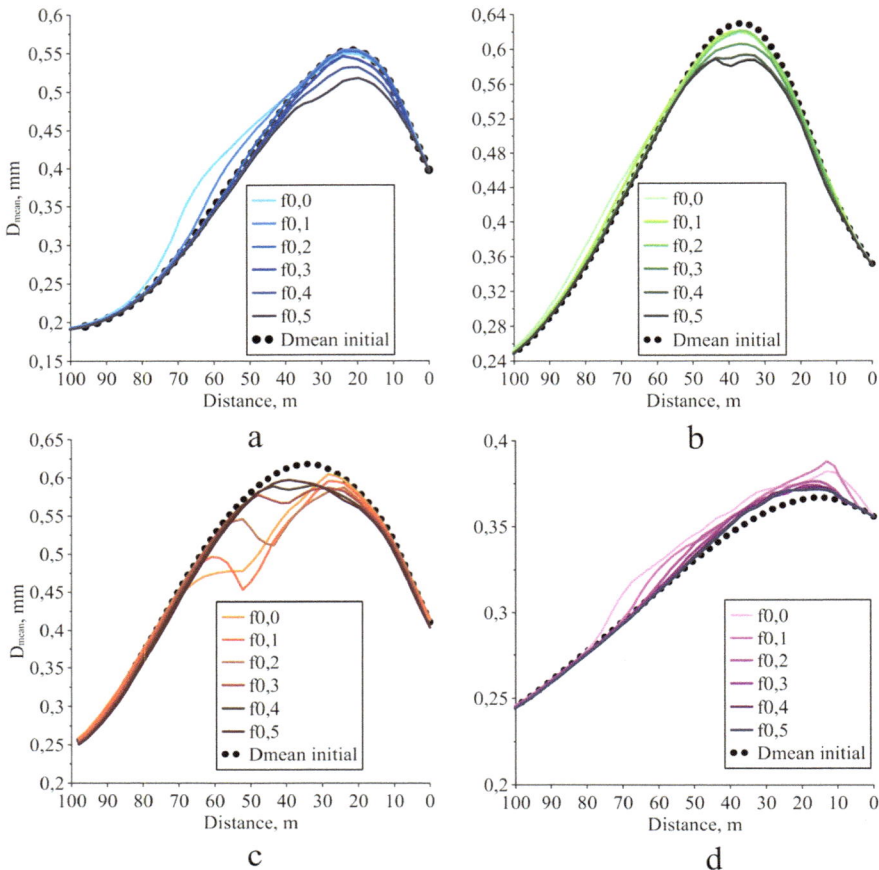

Fig. 4 Change in D_{mean} value in different parts of the coastal zone: **a**—200 m; **b**—500 m; **c**—800 m; **d**—1000

Acknowledgements This research was funded by the framework of the state assignment of the Ministry of Science and Higher Education of the Russian Federation No. FNNN-2021-0005. Model calculations were carried out on the MHI computing cluster (http://www.hpc-mhi.org).

References

Bratus OS (1965) Material composition of the crimean Peninsula beaches. Doklady Earth Sci 163(2):399–402 (in Russ)

Gallagher EL, MacMahan J, Reniers AJHM, Brown J, Thornton EB (2011) Grain size variability on a rip-channeled beach. Mar Geol 287(1–4):43–53. https://doi.org/10.1016/j.margeo.2011.06.010

Gurov KI (2020) Results of coastal zone dynamics and beach sediment granulometric composition monitoring in the central part of the Kalamitsky Gulf. Ecol Saf Coast Shelf Zones Sea 1:36–46. https://doi.org/10.22449/2413-5577-2020-1-36-46 (in Russ)

Gurov KI, Udovik VF, Fomin VV (2019) Modeling of the coastal zone relief and granulometric composition changes of sediments in the region of the Bogaily Lake bay-bar (the Western Crimea) during storm. Phys Oceanogr 26(2):170–180. https://doi.org/10.22449/1573-160X-2019-2-170-180

Kharitonova LV, Fomin VV (2012) Statistical characteristics of wind waves in the coastal area of the western crimea according to retrospective estimation during 1979–2010. Ecol Saf Coast Shelf Zones Integr Use Shelf Resour 26(1):24–33 (in Russ)

Kos'yan RD, Podymov IS, Pykhov NV (2003) Dynamical processes in the sea nearshore zone. Sci World 320 (In Russ.)

McNinch J (2004) Geologic control in the nearshore: shore-oblique sandbars and shoreline erosional hotspots, Mid-Atlantic Bight, USA. Mar Geol 211:121–141. https://doi.org/10.1016/j.margeo.2004.07.006

Prodger S, Russell P, Davidson M (2017) Grain-size distributions on high-energy sandy beaches and their relation to wave dissipation. Sedimentology 64(5):1289–1302. https://doi.org/10.1111/sed.12353

Reniers AJHM, Gallagher EL, MacMahan JH, Brown JA, van Rooijen AA, van Thiel de Vries JSM, van Prooijen BC (2013) Observations and modeling of steep-beach grain-size variability. J Geophys Res Oceans 118(2):577–591. https://doi.org/10.1029/2012JC008073

Rienecker M, Fenton J (1981) A Fourier approximation method for steady water waves. J Fluid Mech 104:119–137. https://doi.org/10.1017/S0022112081002851

Roelvnik D, Reniers A, van Dongeren A, van Thiel de Vries J, VcCall R, Lescinski J (2009) Modelling storm impacts on beaches, dunes and barrier islands. Coast Eng 56(11–12):1133–1152. https://doi.org/10.1016/j.coastaleng.2009.08.006

Ruessink BG, Ramaekers G, van Rijn LC (2012) On the parameterization of the free-stream non-linear wave orbital motion in nearshore morphodynamic models. Coast Eng 65:56–63. https://doi.org/10.1016/j.coastaleng.2012.03.006

Saprykina Y (2020) The influence of wave nonlinearity on cross-shore sediment transport in coastal zone: experimental investigations. Appl Sci 10:4087. https://doi.org/10.3390/app10124087

Saprykina YV, Kuznetsov SY, Shtremel MN, Andreeva NK (2013) Scenarios of nonlinear wave transformation in the coastal zone. Oceanology 53(4):422–431. https://doi.org/10.1134/S0001437013040103

Saprykina YV, Kuznetsov SY, Divinskii BV (2017) Influence of processes of nonlinear transformations of waves in the coastal zone on the height of breaking waves. Oceanology 57(3):383–393. https://doi.org/10.1134/S0001437017020187

Udovik VF, Goryachkin YN (2013) Interannual variability of the along-shore sediment flow in the coastal zone of the Western Crimea. Ecol Saf Coast Shelf Zones Integr Use Shelf Resour 27:363–368 (in Russ)

XBeach Model Description and Manual (2010) UNESCO-IHE Institute for Water Education. Deltares and Delft University of Technology, p 106

Zenkovich VP (1960) Morphology and dynamics of the Soviet Black Sea coast, vol 2. AS USSR Publ., p 216 (in Russ)

Changes in the Repeatability of Dangerous Winds in the Arctic Zone of Siberia for the Summer Months in the Period 1961–2020

A. V. Kholoptsev, S. V. Palaev, and R. G. Shubkin

Abstract The purpose of the article is to identify areas of the territory and water area of the Arctic zone of Siberia, where the values for the modern climatic period of climatic norms for the recurrence of dangerous winds for months of the fire hazardous period (from May to September and winds) are increased, as well as for the period 1961–2020. There were stable and significant trends in their changes towards their increase. Winds of any bearing were considered as dangerous, for which the hourly average values of their velocity modulus at an altitude of 10 m above the earth's surface exceed 7 m/s. Information on reanalysis of ERA-5 on changes of meridional and zonal components of vector of mean hourly velocity of such wind is used as actual material to achieve specified goal. The values of climatic norms for the repeatability of hazardous winds for the months from May to September, as well as the trend of its interannual changes in 1961–2020 in various areas of the territory and water area of the Arctic zone of Siberia. All its areas have been identified, where in the modern climatic period the repeatability of the studied winds is increased and steadily increases. The obtained results suggest that in the coming years in such areas the recurrence of dangerous winds, as well as the risks of emergency situations that are caused by them, will continue to increase. Therefore, the established patterns should be taken into account when planning various types of economic activities in the regions of Russia related to the Arctic zone of Siberia, navigation on the Northern Sea Route, as well as when developing regional divisions of the Ministry of Emergency Situations of Russia.

A. V. Kholoptsev
Sevastopol Branch of the State Oceanographic Institute Named After N.N. Zubov, Sevastopol, Russian Federation

A. V. Kholoptsev · S. V. Palaev
Sevastopol State University, Sevastopol, Russian Federation

R. G. Shubkin (✉)
Siberian Fire and Rescue Academy of GPS EMERCOM of Russia, Zheleznogorsk, Russian Federation
e-mail: rshubkin@yandex.ru

© The Author(s), under exclusive license to Springer Nature Singapore Pte Ltd. 2023
T. Chaplina (ed.), *Processes in GeoMedia—Volume VII*, Springer Geology,
https://doi.org/10.1007/978-981-99-6575-5_10

Keywords Repeatability · Dangerous wind · Warming climate · Arctic zone of
Siberia · Climatic norm · Trend

The impact of wind on the earth's surface can lead to natural and man-made emergencies on it. Therefore, the identification of areas of territories and waters of various regions of the world, where in the modern period the recurrence of such winds was increased and steadily increased, is an urgent problem of physical geography, climatology and safety in emergency situations.

The most interesting solution to this problem is for regions with significant natural resources, in which their wind monitoring systems are not effective enough. In Russia, regions belong to them, as well as the waters of its seas belonging to its Arctic zone (Pavlenko 2013).

The boundaries of the land (continental) part of this zone are determined by Decrees of the President of the Russian Federation of 02.05.2014 No. 296 and 27.06.2017 No. 287. It includes the territories of the Komi Republic, Nenets, Yamalo-Nenets, CNanty-Mansiysk Autonomous Districts, Krasnoyarsk Territory and the Republic of Sacna (Yakutia). Numerous oil fields are located here, where more than 57% of all oil in Russia is produced. Many valuable ores are also mined here, including molybdenum, zinc, silver, gold, etc., as well as diamonds.

The Arctic zone of Russia also includes its waters of the seas of the Arctic Ocean: Barents, Kara, Laptev, East Siberian and Chukotka. They follow the routes of the most important transport communications of the Arctic—the Northern Sea Route (hereinafter referred to as the NSR), which ensures the life of the population and the functioning of the economy of all northern regions of Russia.

Four oil (Prirazlomnoye, Dolginskoye, Varandeyskoye, Medynskoye), three gas (Murmansk, Ludlovskoye, Severo-Kildynskoye), three gas condensate (Pomorskoye, Ledovoye and the world's largest Shtokmanovskoye), as well as the North Gulyaevskoye oil and gas condensate field have been explored on the Barents Sea shelf field.

Numerous gas and gas condensate fields, including unique in their reserves—Leningradskoye and Rusanovskoye, were also discovered on the Kara Sea shelf (Volkov 2019; Pavlenko 2013; Grigoriev 2019).

The predominant part of undeveloped mineral deposits in Russia is located in the territories and water areas of its Arctic zone belonging to Siberia, as well as on the shelf of the Kara, Laptev, and East Siberian seas (Volkov 2019; Melnikov et al. 2019; Prishchepa 2019). Therefore, in the future, accelerated development of their economy is likely, which should be sustainable and environmentally friendly. The latter is possible if this development is carried out comprehensively, as a result of which it is advisable to consider the mentioned territories and water areas as a single object of physical and geographical research—the Arctic zone of Siberia (AZS).

A significant feature of the AZS is the small number of points in which systematic monitoring of its wind regime is carried out. As a result, its features are now studied mainly according to the results of one or another global reanalysis (Pustovalov et al. 2019; Kononova 2019), or according to the results of mathematical modeling (Diansky 2019).

According to the existing ideas on geoecological factors (Aleseev 2014; Report on climate risks in the Russian Federation 2017; Soloviev 2019), it is advisable to consider wind as dangerous if, if it occurs, various kinds of risks for the population, transport, as well as infrastructure facilities increase. The degree of hazard of the wind is characterized by the value of the modulus of its speed and depends on the characteristics of the territories and waters on which it affects.

Geoecological risks for the population of a particular area largely depend on the recurrence of winds on it, considered as dangerous.

The repeatability of the wind of any bearing and force over a certain section of the earth's surface is defined as part of the period of time considered, during which the value of the modulus of their speed exceeded one or another level (BezrukiCN 2005; Gandin and Kagan 1980). In studies assessing the consequences of changes in the repeatability of hazardous wind, this indicator is usually estimated for a given month.

The risks of many hazardous natural phenomena in AZS are determined not by the peak, but by the average intensity of the process of exchange of heat, moisture and momentum between the atmosphere and the earth's surface, which is due to the wind. Their examples are the removal of moisture from combustible material (drying) (Volkov 2019; Vorobyov 2004; Drozdova and Sorokovikova 2021; Nesterov 1979, 1981; Sverlova 2000; Sokolova 2014; Sheshukov et al. 2020; Shubkin 2016), the development of surface disturbance in the seas capable of destroying their shores (Kaplin 2000; Ogorodov, et al. 2020), as well as the drift of the ice cover of the seas (Buzin et al. 2022; Dmitriev 2000; Tsoi 2017).

When studying such phenomena as an indicator essentially similar to repeatability, but independent of the number of days in a particular month, the number of hours related to it (hereinafter, T) during which the average hourly speed of the wind under study exceeded the considered level may be considered.

The threshold level specified in the assessment of T depends on the type of hazardous consequences under study, arising under the influence of the wind in question. One of the lowest values of this level corresponds to such a dangerous effect of the wind action as compression, which threatens ships in the Arctic seas if the average hourly wind speeds exceed 7 m/s (Klyachkin et al. 2010; Mironov and Frolov 2010).

The characteristics of interannual changes T during a certain period of time are the average value of this indicator, as well as the trend defined as the angular coefficient of the linear trend of the time series formed from its values (hereinafter Trend) (Ayvazyan and Mkhitaryan 1998).

As one of the characteristics of the local climate, the climatic norm T (hereinafter referred to as CN), calculated as the average value of this indicator for the month in question for a particular climatic period, can be considered. The trend of CN changes is determined by comparing its values for the modern (1991–2020) and the basic climatic period (1961–1990) (WMO 2017).

The future is not predetermined, but the more likely it is to remain the current trend of changes in the studied indicator, the more stable and significant it is. Therefore, assessment of modern values of CN for different sections of AZN, as well as identification of areas where trends of T changes with specified properties are of theoretical and practical interest.

One of the most advanced sources of information on changes in the characteristics of the wind speed field in the Earth's atmosphere is ERA-5 reanalysis (Hersbach and Dee 2016; Kononova and Lupo 2020), which is supported by the Copernicus service.

Reanalysis contains information on changes of meridional and zonal components of vector of average hourly speed of wind over any point of earth surface, which coincides with one or another node of its coordinate grid with pitch 0.25°.

This information corresponds to different heights above the earth's surface and each hour from the time interval from 00 h. 01.01.1959 to 23 h. 31.12.2021 (Hersbach et al. 1979). Nevertheless, the current values of CN, Trend, as well as trends in CN changes for various months, have not been previously established for any AZS points using them. There are also no sections of AZS for which current trends in interannual T changes can be recognized as stable and significant. The latter does not allow determining the probable trends of further T changes for such areas, as well as taking them into account when planning the main activities of the relevant departments of the Russian Ministry of Emergency Situations.

Given the seasonal changes in the frequency of landscape fires and other hazardous natural events at AZSs (Aleseev et al. 2014; Report on climate risks in the Russian Federation 2017), it is most important to obtain this information for the months of May–September and winds for which the hourly average values of their velocity modulus in the lower troposphere layer exceed 7 m/s.

The purpose of this work is to identify areas of the territory and water area of the AZS, where the current values of CN for the specified months and winds are increased, as well as for the period 1961–2020 stable and significant trends in T changes towards their increase appeared.

When achieving this goal, the following tasks were solved for all points of the AZS corresponding to the nodes of the reanalysis grid and for all months from May to September:

1. Assessment of CN values for the current and basic climatic period, as well as trend values for all time periods of at least 30 years, relating to 1961–2020.
2. Identification of AZS sites where T changes during the period 1961–2020. occurred in the direction of their increase, and the trends manifested at the same time were stable and significant.

When solving these problems, as an actual material, information on reanalysis of the ERA-5 on changes in the meridional and zonal components was used, during the period from 0 h 1.01.1961 to 23 h of 31.12.2020, at an altitude of 10 m above all points of the AZS corresponding to the nodes of its coordinate grid.

Sections of meridians 60° E and 180° E, located between parallels 80° N and 62.25° N, were considered as the western and eastern boundaries of AZS.

It was assumed that the threshold value of the module of the average hourly wind speed is 7 m/s.

The method of solving the first problem included performing the following calculations for each studied station of the AZS corresponding to a particular node of the reanalysis grid and for each month under consideration:

1. For each hour related to the studied time period, calculate the value of the modulus of the hourly mean speed of the wind in question.
2. Determines the values of T corresponding to the selected threshold value of the wind hourly space velocity modulus.
3. Estimate of CN values, for the periods 1961–1990 and 1991–2020.
4. For each period of time of at least 30 years related to the period 1961–2020 calculation of Trend indicator (using the least squares method) (Ayvazian and Mkhitaryan 1998), as well as assessment of its statistical significance.

When assessing the significance of this indicator, the Fisher test was used (Sachs 1975). The decision on the significance of the trend was made if the validity of such a statistical conclusion exceeded 0.95.

When solving the second problem, it was assumed that for a certain point and month the modern trend of the studied process is stable for the entire period 1961–2020, if the condition was met:

$$M = X/445 > 0.95,$$

where X is the number of different time periods of at least 30 years relating to the specified period, for which the sign of the angular coefficient of the linear trend T coincides with the sign of this indicator for the period 1991–2020;

445 is the total number of such segments.

The feasibility of this condition was checked for each point of the AZS and each month under consideration.

Also, for each item and month, in accordance with the recommendation of the WMO, the trend of changes in the CN for the period 1961–2020 was assessed, for which the following was calculated:

$$R = CN (1991 - 2020) - CN (1961 - 1990).$$

Signs R and Trend (1991–2020) are compared to each other. The decision on stability was made when they coincided.

It is not difficult to see that the method used is based on assumptions, the validity of which was not previously evaluated. Therefore, the results that can be obtained with its help can only be of a qualitative nature.

With the use of the considered method, the assigned tasks were solved.

As a result of solving the first problem for all months from May to September and for each section of the AZS, the values of CN (1991–2020), as well as CN (1961–1990), corresponding to the modern and basic climatic period, were calculated. Based

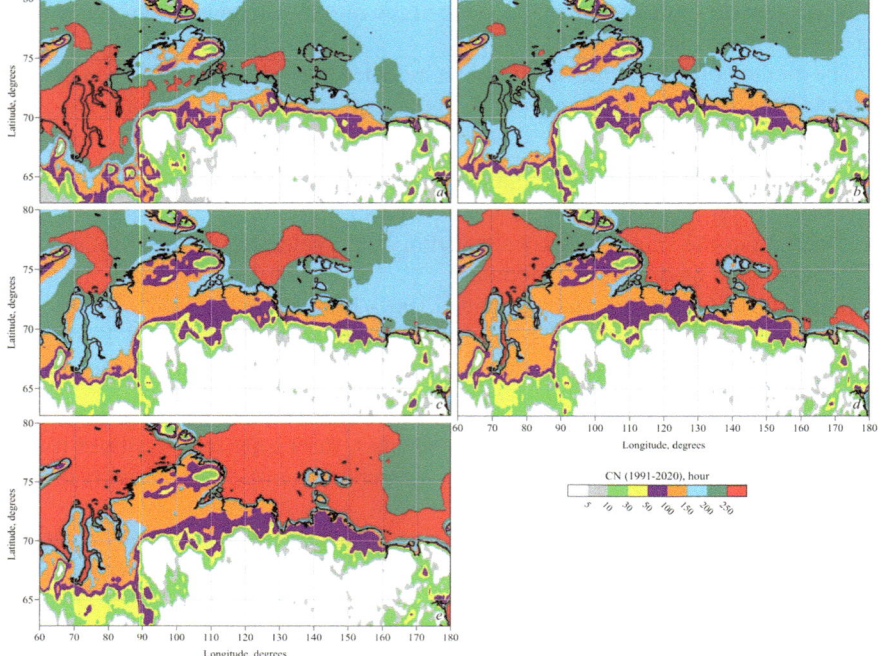

Fig. 1 Locations of AZS sites where CN (1991–2020) took certain values for the months: **a** May; **b** June; **c** July; **d** August; **e** September

on these distribution values within the CN AZS (1991–2020) for the specified months, see Fig. 1.

As can be seen from Fig. 1 a, the values of the CN for May in the modern climatic period exceeded 250 h in the vast area of the southwestern part of the Kara Sea, which also includes the Baydaratskaya, Ob, Gydan and Yenisei bays. All ports of the Kara Sea, as well as all the largest land, sea and coastal hydrocarbon fields of Yamal, fall within the specified area.

The values of the indicator under consideration are equally high for the Kara Sea region, located northeast of Cape Desire (the island of the Northern archipelago Novaya Zemlya), crossed by the routes of the NSR, as well as for the area of the Laptev Sea, located north of the Lena River delta.

At least modern values of the CN also reached for the coastal territories of the Yamalo-Nenets, Nenets Autonomous Districts and the Komi Republic, as well as the Krasnoyarsk Territory.

In the coastal territories of AZS belonging to the Republic of SaCNa (Yakutia), the modern values of the CN for May reached 183 h.

It should be noted that in all the territories of AZS, as they are removed from the sea coasts, modern values of CN decrease to levels of less than 5 h per month.

It follows from Fig. 1b that for June, the current values of the CN exceeded 250 h only over the marine areas of the AZS. In the Kara Sea, the largest of such areas are located northeast of Cape Desire, north of the entrances to the Gulf of Ob, Gydan and Yenisei Bay. In the Laptev Sea, the same area was identified north of the Lena River delta. The total area of such areas for June is significantly less than for May.

Over the continental areas of the AZS in the month under consideration, the modern values of the CN reached 193 h (on the territory of the Yamalo-Nenets Autonomous Okrug).

Figure 1c shows that for July the modern values of the CN exceeded 250 h only for the offshore areas of the AZS, the total area of which is more than for June, but less than for May. In the Kara Sea, such a high average recurrence of dangerous winds is characteristic of its western part (from Cape Desire to the entrances to the bays of the Ob, Gydan and Yenisei Bay). In the Laptev Sea, the same area extends from the delta of the river. Lena to the northeast, to the island of Kotelny and further north of the islands of the Anzhu archipelago.

The largest modern values of CN, lying within 150–200 h, for July are character-istic in the territories of AZS belonging to the Yamalo-Nenets Autonomous Okrug and Krasnoyarsk Territory. It is important to note that for July and August, the winds of the northern rumbas prevail over these territories, bringing relatively dry air, which contributes to an increase in fire hazard on them.

As follows from Fig. 1d, the current values of the CN for August exceeded 250 h throughout the Laptev Sea, as well as over the western parts of the Kara and East Siberian Seas. The total area of such waters exceeds its value for all previous months.

In the coastal territories of the AZS, these indicators reached the highest values (from 150 to 200 h) in the area of the ports of Sabetta and Dudinka. The same levels are characteristic of them in the territories of the northern regions of the Krasnoyarsk Territory remote from the sea.

Figure 1e indicates that for September the total area of the water areas of the AZS, where the current values of CN are at least 250 h, is more than in all the previous months under consideration. Such waters include all areas of the Kara and Laptev seas, as well as the southern regions of the East Siberian Sea.

The locations of the AZS territories, where the modern values of CN T for September reached the highest levels (from 150 to 200 h), practically do not differ from their location for August.

The comparison of Fig. 1a–e allows us to conclude that the wind regime in the AZS for June is the safest. From June to September, the total areas of sea waters, where modern values of CN exceed 250 h, steadily increase.

In the territories of AZS, the winds under consideration arise with the same repeatability only in May (Yamalo-Nenets, Nenets Autonomous Okrugs, Komi Republic and Krasnoyarsk Territory). From May to September, the total area of these regions, where the modern values of CN T ranged from 150 to 200 h, is steadily decreasing.

Also, when solving the first problem for each AZS point, each month and each segment of the time series T with a length of at least 30 members, corresponding to the period 1961–2020 the Trend indicator is calculated.

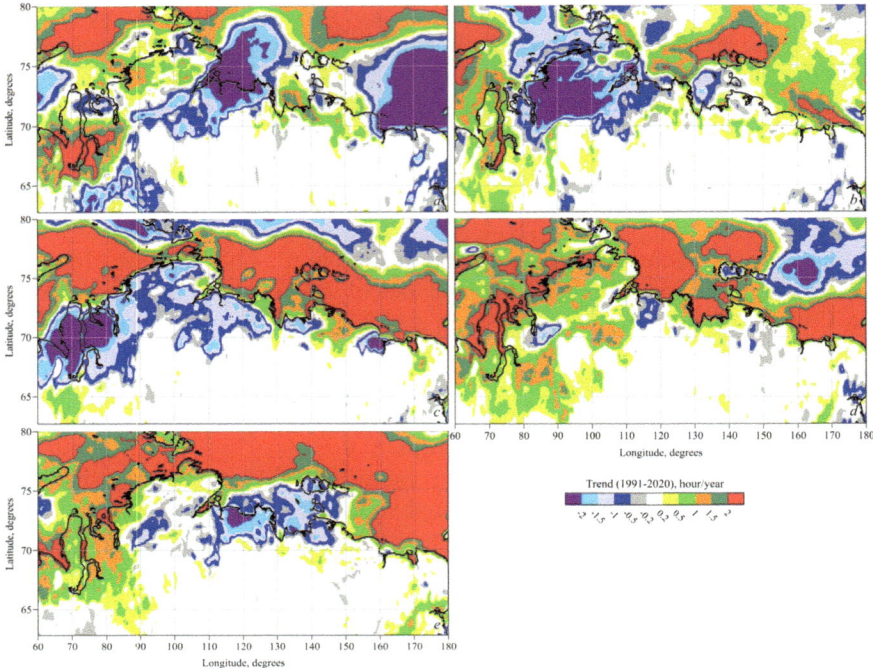

Fig. 2 Locations of AZS sites where the values of the angular coefficient of the linear trend of values of T for the modern climatic period (1991–2020) took certain values for the months: **a** May; **b** June; **c** July; **d** August; **e** September

As an example, Fig. 2 shows the distribution of this indicator on the surface of the AZS for the months of May–September, corresponding to the period of time 1991–2020 (modern climatic period).

Figure 2a indicates that for May there are both sea and continental regions in the AZSs, for which the values of the Trend indicator are positive, and the reliability of the conclusion about their significance exceeds 0.95.

These areas in the Kara Sea are located northeast of Cape Desire and in the Gulf of Ob Bay. Similar areas were also identified in the Laptev seas, the East Siberian and the adjacent part of the Arctic basin of the Arctic Ocean. They are east of the islands of the Severnaya Zemlya archipelago and north of the islands of the Anjou archipelago.

Comparison of Fig. 2 a with Fig. 1a allows us to establish that marine areas, where the trend of the considered indicators for May for 1991–2020 is ascending and significant, are also found in many of the previously established areas of the water area of the same seas. For them, the modern values of CN exceed 250 h.

The continental areas of the AZS, for which for the period 1991–2020 for May there was a significant increase in the frequency of the winds studied, are located in the southern part of the Yamal Peninsula, as well as in the bordering territories of the

Yamalo-Nenets, Nenets Autonomous Okrug and the Komi Republic. It can be seen from Fig. 1a that for these territories the current values of CN also exceed 250 h.

Figure 2b shows that for June in the Kara Sea there are areas where the trend of interannual changes T for 1991–2020 was upward (with a reliability of at least 0.95). These areas are located north of Cape Desire, east of the North Island of the Novaya Zemlya archipelago and in the Gulf of Ob Bay.

The same areas are present in the Laptev Sea: north of the delta of the river Lena and north of the islands of the Anjou archipelago. They were also identified in the East Siberian Sea: off the northern coast of the Kotelny and New Siberia islands, as well as off the mainland coast (on the section from Kolyma Bay and Bear Islands to Cape Yakan).

The continental areas of AZS, in which the trend of interannual changes T for 1991–2020 is recognized as upward (with a reliability of at least 0.9), are located only in the southern and northeastern part of the Yamal Peninsula.

Figure 2c allows us to conclude that the offshore areas of the AZS, where there is an increasing trend in the interannual T changes for July over the 1991–2020 years, include all areas where the modern values of CN exceeded 250 h.

The largest continental area of the AZS, where in 1991–2020 the interannual T changes for July there is an increasing trend, and the modern values of the CN lie within 150–200 h, is located in the eastern part of the Yano-Indigir lowland.

It follows from Fig. 2d that for August, many offshore AZS, where the trend of interannual T changes over the 1991–2020 years is 0.95 recognized as increasing, are located almost in the same place, where, as can be seen from Fig. 1d, the current CN values exceed 250 h.

The continental areas of the AZS, where the Trend indicator for August is recognized as increasing with the same reliability, are located in the northern part of the Yamal Peninsula. In the southern part of the peninsula, the values of this indicator are less (the confidence level does not exceed 0.9). Here, the modern values of CN lie within 200–250 h.

As follows from Fig. 2e, for September, the total area of offshore areas of AZS, in which interannual changes in T for 1991–2020 occurred upward, and the reliability of this conclusion exceeded 0.95, compared to previous months, significantly increased. Such areas occupy almost all the waters of the Laptev and East Siberian seas, and are also very common in the eastern and southern (coastal) parts of the Kara Sea.

A comparison of Figs. 1e and 2e shows that in many of these areas the current CN values exceed 250 h.

The continental region, where the conclusion about the presence of an increasing trend is characterized by a reliability of at least 0.9, but not more than 0.95, is located in the lower reaches of the Yenisei River (it includes the village of Dudinka).

Comparison of Figs. 1a–e and 2a–e allows us to conclude that for the months under consideration in the current climatic period between the locations of areas where the CN exceeds 250 h, and there are increasing trends in T changes with at least 0.95 reliability, there is a correspondence.

When solving the second problem, it was established that a similar correspondence takes place with the locations of the areas where the changes in T with the same

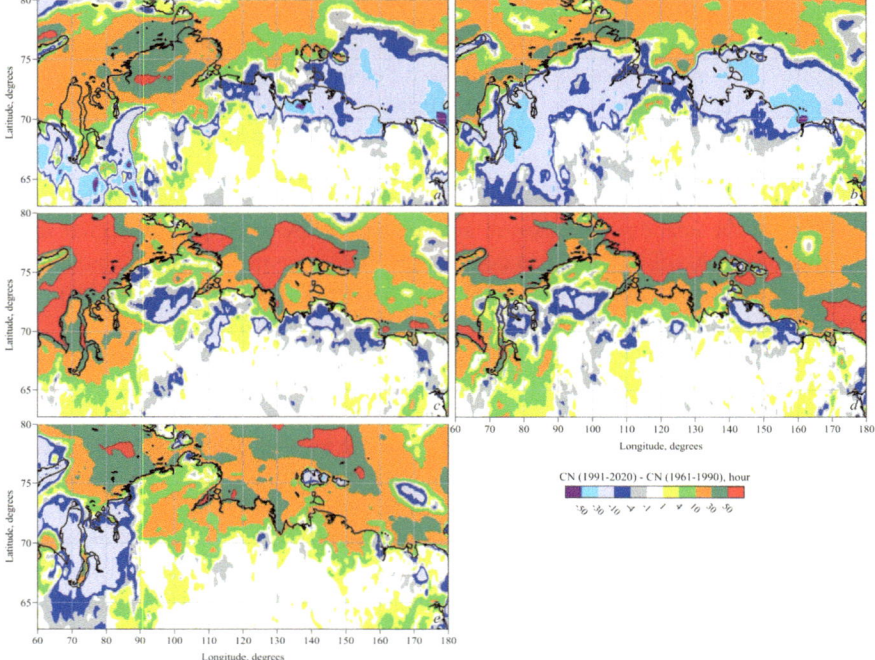

Fig. 3 The locations of the AZS sites, where the values of CN (1991–2020)–CN (1961–1990)) took certain values for the months: **a** May; **b** June; **c** July; **d** August; **e** September

reliability revealed increasing trends for many other periods of time lasting at least 30 years, relating to the period 1961–2020.

For areas where in certain months of the current climatic period CN exceeds 250 h, M values exceed the selected threshold level, which makes it possible to consider the corresponding current trends in T changes as stable.

To confirm the validity of this conclusion, Fig. 3 shows the locations of the AZS areas in which certain levels are exceeded by the difference in the values of the CN for the current climatic period with respect to the baseline period.

It is clear from Fig. 3a that the CN for May for the period 1961–2020 with the highest speed (more than 50 m/year) increased in two areas of the AZS.

The first is located in the Barents Sea, west of Cape Desire, directly on the NSR routes. The second district is located on the territory of the Krasnoyarsk Territory, in the south of the Taimyr Peninsula.

It follows from Figs. 1a and 3a that in many areas where CN for May increased over 1961–2020, compared to similar indicators for the baseline period of 10–30 h per year, the values of these indicators for the modern climate period exceeded 250 h.

Such areas are located both in the identified sea waters and in various territories. The latter include the territories of the Yamal Peninsula, where hydrocarbons are produced, as well as the coasts on which the ports of Harasaway, New Port and Sabetta are located.

Figure 3b allows us to argue that for June in the period 1961–2020 the predominance of trends in the decline of the CN is characteristic in the AZS. Nevertheless, opposite trends have been identified in those marine, as well as its continental areas, where for this month in the modern climatic period the CN exceeded 250 days.

In the Kara Sea, one of the areas where the value of this indicator increased by 30 to 50 h is located north of the entrances to the Gulf of Ob, Gydan and Yenisei Bay.

In the Laptev Sea, the same area where the current value of the CN exceeds the same indicator for the 1961–1990 period by 10 to 30 h is located north of the Lena River delta.

On the coasts of the Kara Sea, as well as on the territory of the Taimyr Peninsula, there are numerous areas where for the modern climatic period the value has increased in relation to the base period by 30–50 h.

As can be seen from Fig. 3c, for July, the fastest increase in CN over the 1961–2020 period (by more than 50 h) occurred precisely in those areas of the Kara and Laptev seas (see Fig. 1c), where in the current climatic period the values of this indicator exceed 250 h. The location of continental regions belonging to the Yamalo-Nenets, Nenets Autonomous Okrug, Krasnoyarsk Territory and the Komi Republic practically coincides, where for July in the modern climatic period the CN exceeds the same indicator for the base period by 10–30 h, as well as the location of areas where modern CN values exceed 150 h.

The trend towards an increase in CN was also revealed in the coastal territories of the Republic of Sacna (Yakutia). The total area of territories and waters, where in 1961–2020 the trends towards an increase in CN prevailed, for July is much more than for June and May.

It follows from Fig. 3d that in August the total area of the offshore areas of the AZS, where in 1961–2020 the trends towards an increase in the CN prevailed, increased markedly in comparison with the same indicator for July.

For continental regions, no significant change in this indicator was revealed. For the territories belonging to the Yamalo-Nenets, Nenets Autonomous Okrug and the Komi Republic, its importance has decreased, and for the territories of the Taimyr Peninsula (Krasnoyarsk Territory) and the coastal regions of the Sacna Republic (Yakutia) it has increased.

Comparison of Figs. 1d and 3d indicates that in areas where the August values of CN for the modern climatic period are maximum, or increased, in 1961–2020 the tendency to increase these indicators at a rate close to maximum prevailed.

From Fig. 3e, it can be concluded that for September the total area of the offshore areas of the AZS, where in 1961–2020 the trends towards an increase in CN prevailed, has significantly decreased compared to the same indicator for August. The total area of the same continental regions has not changed much. Such territories were identified only in the districts of the Krasnoyarsk Territory and the Republic of Sacna (Yakutia), and their area, in comparison with the same indicator for August, increased.

The comparison of Figs. 1a–e and 3a–e suggests that over the 1961–2020 period, trends in the increase in CN prevailed in those areas of the AZS where in the current climatic period the values of these indicators exceeded 150 h.

Comparison of Figs. 2a–e and 3a–e shows that trends towards CN increase have been identified in many of those areas of the AZS where the current trends of T changes for the period 1961–2020 with a reliability of at least 0.95 are recognized as stable (which confirms this).

It should be noted that the continental areas of AZS, in which the trend of inter-annual changes T for 1991–2020 is recognized as upward (with a reliability of at least 0.9) and stable, are located only in its western part. Their territories belong to the Yamalo-Nenets, Nenets Autonomous Okrug and the Komi Republic.

According to existing ideas about the consequences of warming for the summer months of the climate of Siberia (Aleseev et al. 2014; Report on climate risks in the Russian Federation 2017), one of them is an increase in the repeatability and average speed of the winds of the northern rumbas over many of its territories belonging to the AZS.

The latter indicates that with further warming of the AZS climate, the revealed trends in changes in T and the recurrence of dangerous winds in these areas may persist.

Since in a significant number of such areas the current CN values exceed 250 h per month, a further increase in these indicators will lead to an increase in the risks of emergencies in them due to the action of dangerous winds.

In the offshore areas of the AZS, such risks for navigation may be caused by an increase in the recurrence of not only storms, but also unforeseen changes in the state of their ice cover (ice removal to the routes of ships, compression, etc.).

With an increase in the recurrence of dangerous winds in the future, the work of not only water and air transport, but also sea, river ports and airports, where the operation of cranes will be more often suspended, problems arise when mooring ships, taking off and landing aircraft. The risks of surf destruction of the shores of abrasion, thermal abrasion, accumulator and abrasion-accumulator types will also increase.

In the continental areas of the AZS, a further increase in the frequency of hazardous winds will help reduce the average moisture reserves in combustible material, which will increase the likelihood of its ignition. The likelihood of fires in such areas will increase as a result of more frequent breaks in dangerous winds of overhead power lines.

Since the wind delivers oxygen to the combustion area and removes carbon dioxide, sparks and heat from it, in the tundra, a further increase in T will contribute to the faster development of landscape fires and their spread to vaster areas. In forests, it will more often lead to the transformation of grass-roots fires into horse fires, which will significantly increase their danger and complicate their elimination.

The result of an increase in the repeatability of hazardous winds in the future will be an increase in the likelihood of destruction of various engineering structures, as well as man-made accidents. Therefore, the established patterns indicate that in many identified areas of the AZS one should expect a further increase in the risks of occurrence of not only natural, but also man-made emergencies. Their increase is most likely in the territories of the Yamalo-Nenets, Nenets Autonomous Okrug and the Komi Republic, as well as in the sections of the NSR water area along which its Coastal and High Latitude routes pass. Therefore, in these areas, it is necessary to develop local divisions of the Ministry of Emergency Situations of the Russian Federation at a faster pace.

The results obtained generally correspond to the existing ideas about the consequences of modern climate warming in the AZS (Aleseev et al. 2014; Report on climate risks in the Russian Federation 2017; Kotlyakov 2012; Soloviev 2019).

The following facts established above are new:

1. There are numerous areas of the AZS, where in the modern climatic period for the months from May to September, CN values exceed 250 h, and the trends towards an increase in these indicators are statistically significant.
2. For many of these areas, these modern trends between 1961 and 2020 were stable (the reliability of this conclusion is at least 0.95).

The obtained results justify the admissibility of the assumption that in such areas of the AZS, with further warming of its climate, an increase in T is likely, which can pose a danger to the population and transport, as well as economic activity.

Thus, it is established:

1. The climatic norms for the recurrence of dangerous winds in the territories and waters of the Arctic zone of Siberia for the modern climatic period and months from May to September significantly depend on their geographical location. They reach the highest values in the territories of the Yamal Peninsula and in the adjacent areas of the western part of this zone (Yamalo-Nenets, Nenets Autonomous Okrugs, Komi Republic and Krasnoyarsk Territory, as well as the waters of the seas along which the Coastal and High Latitude routes of the Northern Sea Route pass).
2. In many sea and continental regions of the Arctic zone of Siberia, where monthly climatic norms of recurrence of hazardous winds for the modern climatic period are close to maximum levels (they exceed 250 h), in the modern period significant trends towards an increase in this indicator prevail, which were stable throughout the entire interval of 1961–2020.
3. With further warming of the climate of the Arctic zone of Siberia, in such areas, the risks of natural and man-made emergencies caused by the action of dangerous winds may increase, which is advisable to take into account when planning the implementation of any types of economic activities in them, as well as the development of the regional grouping of the RF Ministry of Emergency Situations.

References

Aleseev GV, Ananicheva MD, Anisimov OA, Ashik IM, Bardin MYu, Bogdanova EM, Bulygina ON, Georgievsky VYu, et al (2014) The second assessment report of Roshydromet on climate changes and their consequences in the Russian Federation. M.: Publishing House of Roshydromet, p 1009

Ayvazyan SA, Mkhitaryan VS (1998) Applied statistics and fundamentals of econometrics. M.: Unity, p 1022

BezrukiCN PP (2005) The concept of using wind energy in Russia. M.: Book–Penta, p 128

Buzin IV, Klyachkin SV, Frolov SV, Smirnov KG, MiCNaltseva SV, Sokolova YV, Gudoshnikov YP, Voinov GN, Grigoriev MN (2022) Compression of the ice cover in the Pechora Sea: natural phenomenon and its influence on marine operations. Ecol Econ 12(4):500–512. https://doi.org/10.25283/2223-4594-2022-4-500-512

Diansky NA (2019) Prognostic estimates of climatic changes in the Arctic based on the combined scenario [Electronic resource]. In: Diansky NA, Solomonova IV, Gusev AV (eds) Russian arctic, vol 4. pp 24–33

Dmitriev AA (2000) Dynamics of atmospheric processes over the seas of the Russian Arctic. Hydrometeoizdat, St. Petersburg, p 233

Drozdova TI, Sorokovikova EV (2021) Analysis of forest fires in the Irkutsk region in 2010–2019//XXI century. Technosphere Saf 6(1):29–41

Gandin LS, Kagan RL (1980) Application of statistical methods in meteorology. In: Voeikov AI (ed) Proceedings of the order of the red banner of labor of the main geophysical observatory named after. Leningrad, Hydrometeoizdat, p 146

Grigoriev GA (2019) Prospects for developing the hydrocarbon potential of the russian arctic shelf from petroleum, technological, and financial and economic positions [Electronic Resource]. In: Modern challenges of petroleum geology. Alternatives and development prospects: a collection of reports of the Anniversary Conference (St. Petersburg, November 6–8, 2019. VNIGRI,). St. Petersburg

Hersbach H, Dee D (2016) ERA5 reanalysis is in production. ECMWF Newsletter 147:7

Hersbach H, Bell B, Berrisford P, Biavati G, Horányi A, Muñoz Sabater J, Nicolas J, Peubey C, Radu R, Rozum I, Scheper D, Simmons A, Soci C, Dee D, Thépaut J-N (1979) Database ERA5 hourly data on pressure levels from 1979 to present. [Electronic resource]. Режим доступа. https://cds.climate.copernicus.eu/cdsapp#!/dataset/reanalysis-era5-pressure-levels?tab=form. https://doi.org/10.24381/cds.bd0915c6

Hoffmann L, Günther G, Li D, Stein O et al (2019) From ERA-Interim to ERA5: the considerable impact of ECMWFs next-generation reanalysis on Lagrangian transport simulations. Atm Than Phys 19: 3097–3124

Hólm E, Janiskova M, Keeley S, Laloyaux P, Lopez P, Lupu C, Radnoti G, de Rosnay P, Rosum IV, Vamborg F, Vili laume S, Thépaut J-N (2020) The ERA5 global reanalysis. Quart J R Metorol Soc 146, 1999–2049. https://doi.org/10.1002/qj.3803

Kaplin PA (2000) Forecast of the development of the coastal zone of the seas of Russia [including the Arctic ones] in conditions of increasing their level and warming climate. In: Kaplin PA, Pavlidis YA, Selivanov AO (eds) Humanity and the coastal zone of the World Ocean in the XXI century: XX International. conf., 90th anniversary of V.P. Zenkovich (February 4–5, 2000). Moscow, pp 16–28

Klyachkin SV, Gudkovich ZM, May RI, Frolov SV (2010) Ice compression. In: Mironova EU (ed) Dangerous ice phenomena for shipping in the Arctic. AANII, St. Petersburg, pp 33–91

Kononova NK (2019) Study of circulating factors of climate change in the territory of the Siberian sector according to the typization data of B. L. Dzerdzeevsky [Electronic resource]. In: Kononov NK, Latyshev IV (eds) Modern trends and prospects for the development of hydrometeorology in Russia: materials of the II All-Russian scientific and practical conference dedicated to the 55th anniversary of the Department of Hydrology and Nature Management of ISU (Irkutsk, June 5–7, 2019). ISU Publishing House, Irkutsk, pp 578–586

Kononova NK, Lupo AR (2020) Changes in the dynamics of the northern hemisphere atmospheric circulation and the relationship to surface temperature in the 20th and 21st Centuries. Atmosphere 11(3):255. https://doi.org/10.3390/atmos11030255

Kotlyakov V (2012) On the causes and consequences of modern climate change. Solar-Terrest Phys 21:110–114

Melnikov PN, Skvortsov MB, Kravchenko MN, et al (2019) Results of geological exploration on the arctic shelf of Russia in 2014–2019 and prospects for work in the near future. Oil Gas Geol 6:5–18. https://doi.org/10.31087/0016-7894-2019-6-5-18

Mironov EU, Frolov SV (2010) Effect of sea ice on shipping and classification of hazardous ice phenomena. In: Mironov EU (ed.) Dangerous ice phenomena for shipping in the Arctic. AANII, St. Petersburg, pp 12–32

Nesterov VG (1979) Forest combustibility and methods of its determination. M.: Goslesbumizdat, p 76

Nesterov VG (1981) Nature management issues. M.: Lesn. prom, p 263

Ogorodov S, et al (2020) Coastal erosion of the Russian Arctic: an overview. J Coast Res 95(1):599–604

Pavlenko VI (2013) Arctic zone of the Russian Federation in the system of ensuring the national interests of the country. Ecol Econ 4(12):16–25

Prishchepa OM (2019) Prospects for further study and development of the hydrocarbon potential of the arctic shelf of the russian federation. In: Prishchepa OM, Nefedov YV, Grigoriev GA (eds) Scientific Journal of the Russian Gas Society. 3(4):5–20

Pustovalov KN, Cnaryutkin EV, Korolkov VA, Nagorsky PM (2019) Variability of solar and wind energy resources in the Russian sector of the Arctic. Optics Atmos Ocean 32(11):908–914. https://doi.org/10.15372/AOO20191105

Report on climate risks in the Russian Federation (2017) St. Petersburg. p 106

Sachs ST (1975) Statistical inference theory. M.: Mir., p 776

Sheshukov MA, Kovalev AP, Orlov AM, Pozdnyakova VV (2020) Problems and prospects for protecting forests from fires. Siberian For J 2:14–20. https://doi.org/10.15372/SJFS20200202

Shubkin RG (2016) Results of long-term forecasting of large-scale forest fires in the Baikal region. In: Shubkin RG, Shirinkin PV (eds) Scientific and analytical journal Siberian Fire and Rescue Bulletin, vol 3, pp 35–38

Sokolova GV (2014) Method of long-term prediction of fire hazard indicators in the forests of the amur region based on atmospheric circulation parameters accounting. Forest Log Office 5

Soloviev DA (2019) Modern climate changes in the Arctic: causes and environmental consequences [Electronic resource]. In: Problems of ensuring environmental safety and sustainable development of Arctic territories: II Yudakhin readings readings: a collection of materials from the All-Russian Conference with international participation (June 24–28, 2019). Printing house, Arkhangelsk pp 152–156

Sverlova LI (2000) Method of assessing fire hazard in forests according to weather conditions, taking into account the zones of atmospheric aridity and seasons of the year. Khabarovs, p 46

Tsoi LG (2017) To assess the risk when sailing ships along the Northern Sea Route. In: Study of ice qualities and substantiation of rational parameters of ice navigation ships: Collection of works. Nestor-History, St. Petersburg, pp 386–400

Volkov AV (2019) Strategic metals deposits of the Arctic region. In: Volkov AV, Bortnikov NS (eds) Scientific support of implementation of priorities of scientific and technological development of the Russian Federation. T. 2: Scientific sessions of the general meetings of the branches of the Russian Academy of Sciences. Russian Academy of Sciences, Moscow, pp 104–114

Vorobyov YL (2004) Forest fires in Russia: State and problems. In: Vorobyov YL, Akimov VA, Sokolov YI (eds) Under the general. EMERCOM of Russia.–M.: DEX-PRESS, p 312. ISBN 5–9517–0008–6

WMO (2017) WMO guidelines for calculation of climatic standards, vol 1203. p 21

Repeatability of Dangerous Winds Over the Sea of Azov and Black Sea and Their Coasts

A. V. Kholoptsev and G. F. Batrakov

Abstract Values of climatic norms of the repeatability of dangerous winds for all the months from November to April over various areas of the Sea of Azov and the Black Sea, as well as over their coasts, were estimated. The trends in their interannual changes in 1961–2020 were determined. All the areas where the repeatability of the winds under consideration is higher and steadily increases in the modern climatic period were identified. The results obtained suggest that in the coming years, the repeatability of dangerous winds in such areas, as well as the risks of emergencies caused by them, will continue to increase. Therefore, it is advisable to take into account the identified patterns both when planning various economic activities in the Russian regions washed by these seas and when developing navigation on their water routes.

Keywords Repeatability · Dangerous wind · Climate warming · Azov-Black Sea region · Climate norm · Trend

1 Introduction

The wind with the strength exceeding a certain limit is one of the most powerful environmental factors creating sea storms that complicate navigation and destruct numerous natural and technogenic objects on the sea coasts. Therefore, the identification of the areas in various regions of the world, where the repeatability of such wind was elevated and has steadily increased in the current period, is an urgent problem.

The problem is of the greatest interest for densely populated and economically developed regions with significant recreational potential. The Sea of Azov and the

---

A. V. Kholoptsev
Sevastopol Branch of FSB Institution "N. N. Zubov State, Oceanographic Institute". 299011, Sovetskaya St. 61, Sevastopol, Ukraine

G. F. Batrakov (✉)
FRC "Marine, Hydrophysical Institute of RAS". 299011, Kapitanskaya St. 4, Sevastopol, Ukraine
e-mail: batrakovgf@gmail.com

T. Chaplina (ed.), *Processes in GeoMedia—Volume VII*, Springer Geology,
https://doi.org/10.1007/978-981-99-6575-5_11

Black Sea together with the adjacent territories of Russia and other states (hereinafter referred to as the ABSR) are among them (Hydrometeorology 1991).

An essential feature of the ABSR is a significant number of points where systematic monitoring of the wind regime is carried out (Hydrometeorological 2009; Hydrometeorology 1991*) as well as the availability of numerous reanalyses (Huler 2004; Database) obtained using mathematical models verified by the results of such observations.

The environmental risks existing in a certain region for its population are largely determined by such a characteristic of its wind regime as the repeatability of winds considered as dangerous.

It is advisable to consider wind as dangerous if various risks for the population, natural ecosystems, transport, as well as infrastructure facilities, increase as a result of its action. The degree of danger of a particular wind is characterized by the value of its speed modulus and duration of action, depending also on the peculiarities of the territories and water areas affected by the wind.

The repeatability of the wind of certain direction and speed over some area of the earth's surface is defined as part of the period under consideration, during which the value of the speed modulus exceeded one or another level. Since the wind regime in the ABSR changes significantly within a year, studies devoted to assessing the consequences of changes in the repeatability of dangerous wind, usually estimate this indicator for a particular month.

The greatest danger everywhere is the short-term strengthening of the wind, at which its speed reaches peak values within seconds (gusts). They can cause the most serious mechanical destruction of both natural components of landscapes and engineering structures, and complicate the work of all types of transport. As a rule, these winds occur unexpectedly, but the probability of such events significantly increases during periods when the hourly average wind speed exceeds a certain threshold value (Application 1980).

During the periods of such winds, the intensity of many other dangerous natural processes in the ABSR also increases. They are caused by the average intensity of the exchange of heat, moisture and impulse between the atmosphere and the earth's surface. Examples of such processes include the removal of moisture from combustible material (drying), which increases the fire hazard for natural landscapes (Berdnikov et al. 2019) as well as the grows of surface waves in the seas, which can destroy their shores (Huler 2004) and complicate the work of ship crews.

When assessing the average repeatability of the studied winds, the number of hours related to the period under consideration (hereinafter referred to as T) during which they were observed can also be used as an indicator.

The threshold level indicated when assessing T depends on the type of the studied dangerous effect resulting from the wind under consideration. In this paper, the value of the modulus of the hourly average wind speed, which is 10.8 m/s, is chosen as the threshold (which, according to the Beaufort scale, allows us to consider the wind as strong).

The characteristics of interannual changes in T over a period of time are the average value of this parameter, as well as the trend, defined as the angular coefficient of the linear trend of a time series formed from its values (hereinafter Trend).

The climatic norm of T (hereinafter CN) can be considered as one of the of the local climate characteristics. It is calculated as the average value of this parameter for the month in question for a particular climatic period. The trend of CN changes is identified by comparing its values for the modern (1991–2020) and the base climatic periods (1961–1990).

One of the most advanced sources of information about changes in the characteristics of the wind speed field in the earth's atmosphere is the ERA-5 reanalysis (Ayvasyan and Mkhitaryan 1998), which is supported by the Copernicus service.

The reanalysis contains information about changes in the meridional and zonal components of the vector of the hourly average wind speed over any point on the earth's surface that coincides with one or another node of its 0.25° coordinate grid.

This information corresponds to different heights above the earth's surface and to each hour from the time interval from 0 a.m. 01.01.1959 to 11 p.m. 31.12.2021 (Report 2017). However, the data was not used previously to determine the current values of CN, Trend and CN trends for different months for any points of the ABSR.

Neither areas of ABSR for which the current trends of interannual changes in T can be recognized as stable and significant, were discovered. This fact does not allow for such areas to determine the likely trends of further changes in T, as well as to take them into account when planning navigation in the seas and carrying out economic activities in coastal territories.

Taking into account seasonal changes in the ABSR wind regime (Hydrometeorological 2009; Hydrometeorology 1991*), the development of existing ideas about the repeatability of dangerous winds in it seems to be most important for the months from November to April.

The purpose of this work is to identify the ABSR areas where the current CN values for winds are elevated and where stable and significant trends of increase in their repeatability T for the period 1961–2020 occurred.

When this goal was achieved, the following tasks were solved for all the ABSR points corresponding to the nodes of the reanalysis coordinate grid and for all the months from November to April:

1. Obtaining estimates of CN values for the current and basic climatic period, as well as Trend values for all periods of at least 30 years, related to 1961–2020;
2. Identification of the ABSR areas where changes in T for the period 1961–2020 occurred in the direction of their increase, and the trends manifested at the same time were stable and significant.

2   Materials and Methods

In solving the tasks, information from the ERA-5 reanalysis on changes in the meridional and zonal components of the hourly average wind speed, in the period from 0 a.m. 1.01.1961 to 1 p.m. 31.12.2020, at a height of 10 m above all the ABSR points corresponding to the nodes of its coordinate grid, was used.

Meridian segments of 26° E and 42° E between 48° N and 40° N were taken as the western and eastern boundaries of the ABSR.

The threshold value of the module of the hourly average wind speed was supposed to be 10 m/s.

The reanalysis results were tested by their selective comparison with the results of monitoring the same real wind characteristics carried out in Yevpatoria, Sevastopol, Yalta, Feodosia, Zavetnoye, Kerch, Taganrog, Yeysk, Primorsko-Akhtarsk, Temryuk, Novorossiysk, Gelendzhik, and Sochi. The data was obtained from the archive of the Sevastopol branch of the FSBI "N. N. Zubov State Oceanographic Institute".

The test showed that the results of actual observations satisfactorily correspond to the reanalysis results that allows us applying the latter to identify qualitative patterns.

The method of solving the first task included the following calculations for each studied ABSR point corresponding to one or another node of the ERA-5 reanalysis coordinate grid and for each month under consideration:

1. Calculation of the module of the hourly average speed of the wind in question for each hour related to the studied period.
2. Determination of T values corresponding to the selected threshold value of the module of the hourly average wind speed.
3. Estimation of CN values for the periods 1961–1990 and 1991–2020.
4. Calculation of Trend indicator (using the least squares method), as well as an assessment of its statistical significance for each period of at least 30 years between 1961 and 2020.

In assessing the significance of this indicator, the Fisher criterion was applied. The decision on the trend significance was made if the reliability of such a statistical conclusion exceeded 0.95.

When solving the second problem, it is assumed that for a certain point and month, the current trend of the process under study is stable for the entire period of 1961–2020, if the following condition was met:

$$M = X/445 > 0.95,$$

where X is the number of different periods of at least 30 years, related to the specified period, for which the sign of the angular coefficient of the linear trend T coincides with the sign of this indicator for the period of 1991–2020; 445 is the total number of such periods.

Verification of the feasibility of this condition was carried out for each ABSR point and each month under consideration.

For each ABSR point and the month from November to April, in accordance with the WMO recommendation, the trend of CN changes for the period of 1961–2020 was estimated as well, and $R = CN(1991 - 2020) - CN(1961 - 1990)$ was calculated for that purpose.

The signs of R and Trend (1991–2020) were compared with each other. The decision on sustainability was made when they coincided.

3 Results and Discussion

As a result of the first problem solution, for all the months from November to April and for each ABSR area, the CN values (1991–2020), as well as CN (1961–1990) corresponding to the current and basic climatic periods were calculated. The CN (1991–2020) distributions within the ABSR for the above-mentioned months constructed using these values are shown in Fig. 1.

As can be seen from Fig. 1a, the CN values for November in the current climatic period reached levels of 150–200 h in the central part of the Sea of Azov. They reached levels of 100–150 h in many open areas of the Northwestern and Western Black Sea, in areas located near the Kerch Strait, in the Strait itself, as well as in the Sea of Azov. In all coastal areas of the Sea of Azov, as well as similar areas of the Black Sea located near its northern and western shores, the CN values lie in the range of 50–100 h. On the coasts, as well as in the coastal areas of the Eastern Black Sea, their values do not exceed 50 h (the repeatability of dangerous winds is minimal).

It should be noted that in all the ABSR territories, as they move away from the seacoasts, the current CN values for November decrease to levels of less than 5 h per month.

From Fig. 1b, it follows that for December, the CN values for the current period reached the highest levels of 150–200 h over the open areas of the Sea of Azov, the Northwestern and Western Black Sea, as well as over the Kerch Strait and the Taman Peninsula. The average repeatability of dangerous winds decreased in 1991–2020 in the Southeastern Black Sea, as well as near its southern shores (Turkey and Georgia).

Figure 1c, shows that for January, the current CN values for the marine ABSR areas exceeded 150 h only in certain open areas of the Azov Sea, the Northwestern and Western Black Sea, as well as near the Kerch Strait and on the Taman Peninsula. In most of the ABSR water area (except for coastal areas, as well as for the Southeastern Black Sea), the CN values lie in the range of 100–150 h.

As it follows from Fig. 1d, in the current climatic period, the CN values for February exceed 150 h only on the Taman Peninsula and in some adjacent areas of the Black Sea. In many open ABSR areas (except for both coastal areas and the Southeastern Black Sea), the CN values for February, as well as for January, lie within 100–150 h.

Figure 1e, indicates that for March, the total ABSR water area where the current CN values are at least 100 h is significantly less than for the winter months. Such CN

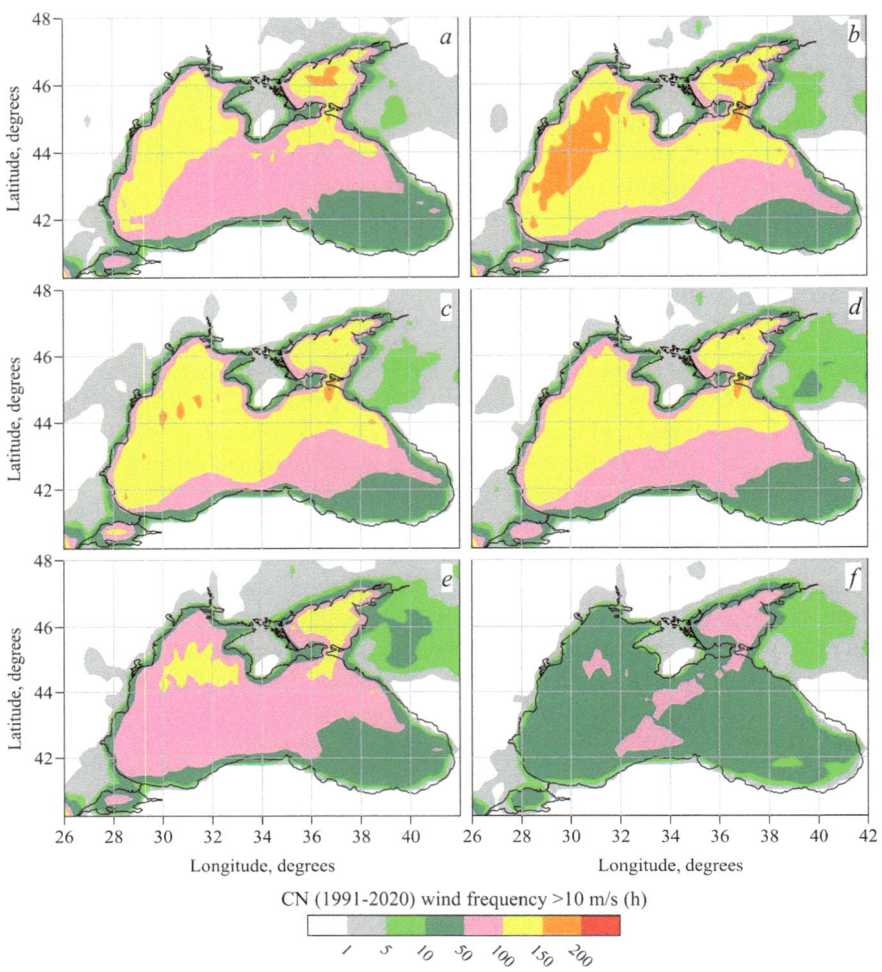

Fig.1 The locations of the ABSR areas, where CN (1991–2020) assumed certain values for **a** November; **b** December; **c** January; **d** February; **e** March; **f** April

values correspond to the open areas of the Sea of Azov and the Northwestern Black Sea.

Unlike the previous months, March is characterized by the maximum ABSR area where the CN exceeded 50 h. Such regions occupy a significant part of Krasnodar Krai.

Comparison of Fig. 1a–f, allows us to conclude that for the months under consideration, the ABSR wind regime typical for the current period, is generally the same as for the second half of the twentieth century (Hydrometeorology 1991*; Hydrometeorological 2009).

When solving the first problem, the Trend indicator was calculated for each ABSR point, each month and each segment of the time series T with a length of at least 30 years related to 1961–2020. The threshold values of the module of this indicator, exceeding which allows us to consider the results obtained as significant (with a confidence of at least 0.95) are 1 h/year.

As an example, the distributions of this indicator over the ABSR surface for November–April, related to 1991–2020 (current climatic period) are shown in Fig. 2a–f.

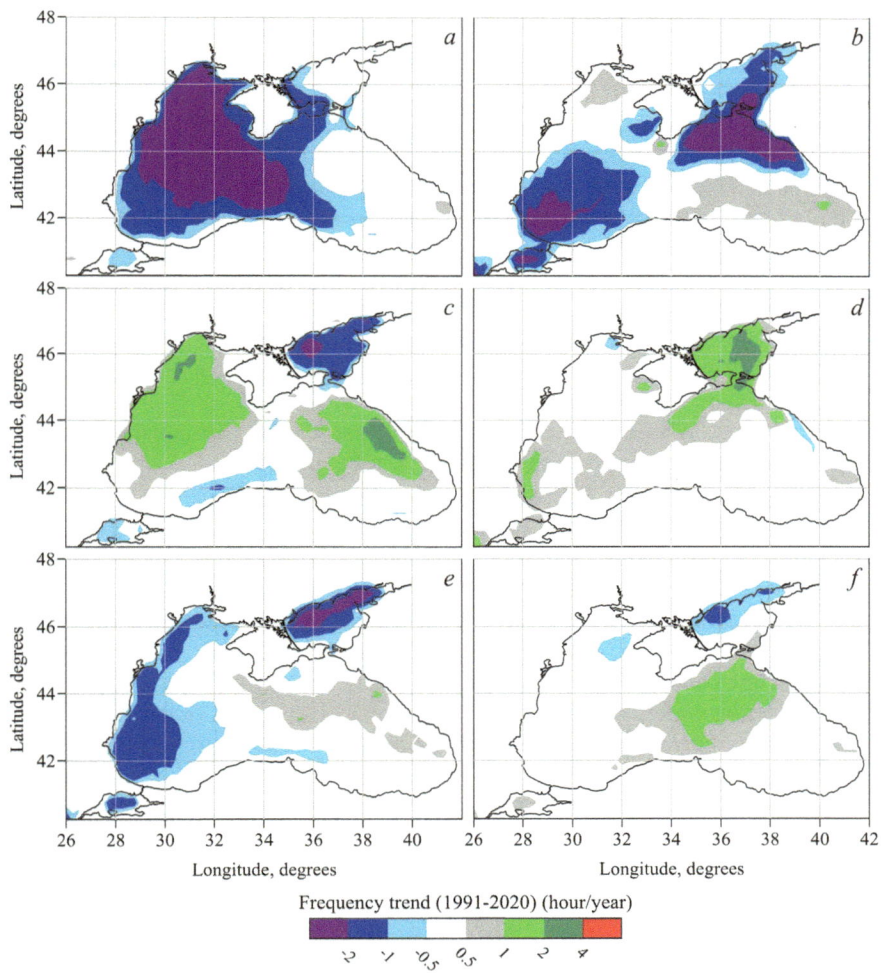

Fig. 2 Locations of the ABSR areas, where the angular coefficient of the linear trend of T values for the current climatic period (1991–2020) took certain values for **a** November; **b** December; **c** January; **d** February; **e** March; **f** April

Figure 2a, shows that, for November, numerous ABSR areas have been identified for which the Trend indicator is significant and negative. Such areas predominate in the Northwestern and Western Black Sea, as well as in the Eastern part of the Gulf of Sivash and adjacent areas of the Sea of Azov.

The continental ABSR areas, for which there was a significant decrease in the repeatability of the studied winds for November during 1991–2020, are located on the Kerch Peninsula.

Figure 2b, shows that for December, the ABSR areas where a significant trend of interannual T changes for 1991–2020 decreased, were mainly found. In the Black Sea, they are located in its West (near the Bosporus Strait), as well as in the North-East (at the entrance to the Kerch Strait). In the Sea of Azov, the areas under consideration are located in the East (from the Kerch Strait to the mouth of the Don River in Taganrog Bay).

The ABSR continental regions, where a significant trend of interannual T changes for 1991–2020 decreases, are located on the Kerch and Taman peninsulas.

The parts of the ABSR water area, where a significant upward trend T for December was detected, are located only to the south of Cape Sarych (the Crimean Peninsula), as well as to the west of Batumi (Georgia).

Figure 2c, allows us to conclude that for January, in contrast to the previous months, there are numerous parts in the ABSR water area, where T values increased significantly for 1991–2020. Such areas prevail in the North-Western Black Sea, as well as in its West (in the economic zones of Romania and Bulgaria) and in the East (in the economic zones of Russia, Georgia and Turkey).

There are also ABSR areas where the significant T trend for January decreased. They occupy almost the entire water area of the Sea of Azov.

The largest continental ABSR area, where in 1991–2020 there is a significant downward trend in interannual T changes for January, is located on the Taman Peninsula.

It follows from Fig. 2d, that the ABSR areas, where a significant trend of interannual T changes for February in 1991–2020 is recognized to be increased, are located in the Western Black Sea (near the coast of Bulgaria), in the East (near the southern coast of Crimea and the coast of Krasnodar Krai). In the Sea of Azov, such areas occupy its entire water area, except for Taganrog Bay. No parts of the ABSR water area with a significant decreasing trend of interannual T changes for 1991–2020 were revealed.

The continental region, where the trend of T variations for the same period for February is significant and increasing, is located on the Taman and Kerch peninsulas.

As it follows from Fig. 2e, only two small ABSR areas, where the interannual T changes for 1991–2020 increased and the reliability of this conclusion exceeded 0.95, were identified for March. One of them is located off the Black Sea coast of the Caucasus (in the economic zone of Russia), and the second one is in the center of the Eastern Black Sea.

Over the entire Sea of Azov water area, as well as in the Northwestern and Western Black Sea (in the economic zones of Ukraine, Romania, Bulgaria and Turkey), downward T trends prevailed for the same period.

Figure 2f, indicates that for April in the period of 1991–2020 significant trends towards an increase in T were characteristic of many areas of the Eastern Black Sea mainly in the economic zone of Russia. Opposite trends took place in the open areas of the Sea of Azov.

Comparison of Fig. 1a–e, and Fig. 2a–e, allows us to conclude that for the months under consideration in the current climatic period, there is a correspondence between the locations of a number of areas where CNs exceed 250 h, and there are increasing trends in T changes with a confidence of at least 0.95. The wind regime there, which is currently the most dangerous for navigation, is changing towards a further increase in storm risks.

When solving the second problem, it was found that a similar correspondence also takes place with the location of areas where the same trends in T changes were revealed with the same reliability for many other periods of at least 30 years, related to 1961–2020.

For areas where, in certain months of the current climatic period, CNs exceed 250 h, the M values exceed the selected threshold level, which makes it possible to consider the current trends in T changes corresponding to them as stable.

To confirm the validity of this conclusion, Fig. 3 shows the locations of the ABSR areas, where certain levels are exceeded by the difference in the CN values for the current climatic period in relation to the base period.

As it follows from Fig. 3a, the upward CN trend for November in 1961–2020 prevailed mainly in the eastern part of the ABSR (the Sea of Azov with coasts, the Eastern Black Sea, the Eastern Crimean Peninsula, the Kerch and Taman Peninsulas). In the Western and Northwestern Black Sea, opposite trends were revealed. From the comparison of Fig. 3a, and Fig. 2a, it can be seen that for November, the signs of the trends of the processes under consideration coincide in the Western and Northwestern Black Sea.

As it can be seen from Fig. 3b, December is characterized by the predominance of the ABSR areas where the CNs in 1961–2020 decreased. Significant upward CN trends have been identified in the Northern and Eastern Crimea adjacent to Sivash Bay, on the northern and eastern coasts of the Sea of Azov, on the Black Sea coast at the entrance to the Bosporus Strait (Turkey), as well as in the open sea to the South of the Crimean Peninsula. As it is easy to see from Fig. 3b, and Fig. 2b, the signs of the trends of the same processes coincided in many areas of the Western and Eastern Black Sea, as well as of the Sea of Azov.

Figure 3c, indicates that for January, significant upward CN trends were revealed only in the Southeastern Black Sea near the Caucasus, as well as in the territory of Turkey adjacent to the entrance to the Bosporus Strait. In other ABSR areas (except for its Southeastern part), significant downward CN trends prevail. The coincidence of the trends under consideration takes place only in the of the Sea of Azov water area, as well as in the Black Sea off the Caucasian coast (in the economic zone of Russia).

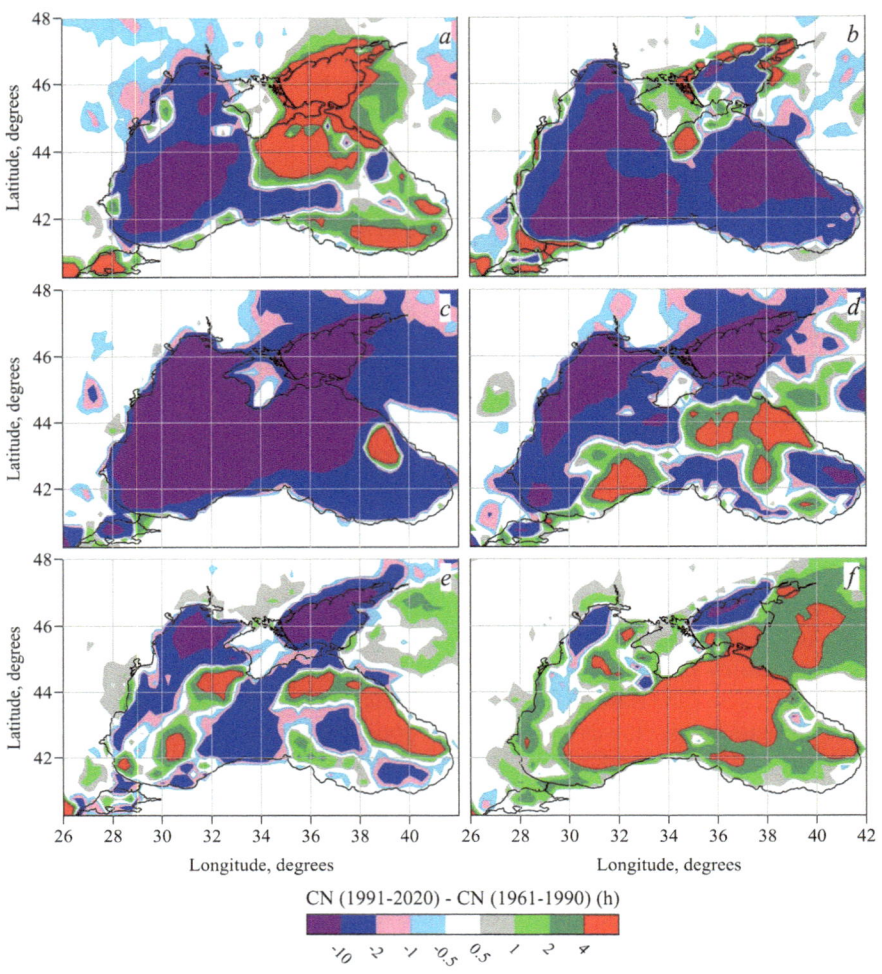

Fig. 3 Locations of the ABSR areas, where in the period of 1961–2020 certain trends in CN changes prevailed for **a** November; **b** December; **c** January; **d** February; **e** March; **f** April

It can be concluded from Fig. 3d, that for February, the ABSR sea areas with significant downward CN trends were found in the Northwestern, Western and South-eastern Black Sea, as well as in the Sea of Azov. Opposite trends were recorded in the Eastern Black Sea and off the Anatolian coast of Turkey.

It should be noted that no areas where the studied trends coincide (see Fig. 3d, and Fig. 2d) were identified for February. This example refers to those 5% in which the correspondence under consideration is violated. The reasons for this violation need further study.

It is clear from Fig. 3e, that, for March in the period of 1961–2020, the CNs increased with the highest rate (more than 50 m/year) in some areas of the Eastern and Western Black Sea. Throughout the Sea of Azov water area and its coasts, as well as in many areas of the North-Western and Western Black Sea, the values of these indicators significantly decreased (as it follows from the comparison of Fig. 3e, and Fig. 2e, the trends of the studied processes coincide here confirming their stability).

From Fig. 3f, it follows that in many ABSR areas occupying almost the entire Black Sea and the Southern Sea of Azov, the CNs have increased significantly for April for the period of 1961–2020. Opposite trends were revealed only in the North-western Black Sea and the Northern Sea of Azov. The ANSR areas where the trends under consideration coincide (see Fig. 3f, and Fig. 2f) are located in the Sea of Azov water area and in the Eastern Black Sea.

Comparison of Figs. 2 and 3 indicates that for the months from November to January, as well as for March and April, there are numerous ABSR areas, where the trends in both CN and T are significant, and their signs coincide. The latter indicates that with further ABSR climate warming, the identified trends in T changes and repeatability of dangerous winds in these areas may persist. Since in a significant number of such areas the current CN values are elevated, their further increase can lead to an increase in the risks of emergencies in them, which are caused by the action of dangerous winds. In the identified ABSR water areas, such risks for navigation may be caused by an increase in the repeatability of storms. In the corresponding coasts, their erosion and abrasion may become more active.

A likely increase in the repeatability of dangerous winds in the identified ABSR areas can complicate the work of water and air transport, as well as of sea, river and airports, where cranes will oftener stop working and some problems may arise during mooring of ships and takeoff and landing of aircraft.

In the identified continental ABSR areas, a further increase in the repeatability of dangerous winds will contribute to accidents on overhead power lines, in housing and communal services and at other technosphere facilities. It will also lead to deterioration in wintering conditions for plants and thus increase the risks for agriculture and horticulture. In such areas, the likelihood of wildfires will also increase.

Therefore, the identified patterns indicate that in many areas of the Kerch and Taman Peninsulas, a further increase in the risks of both natural and technogenic emergencies should be expected.

For February, in the example under consideration, there were no ABSR areas where the trends in changes in CN and T coincide in sign. Nevertheless, in many other examples (with other durations of the studied periods exceeding 30 years), they are found for this month as well.

As it follows from the compared figures, the coincidence of significant trends in interannual changes in the repeatability of winds considered as dangerous was neither found in many ABSR areas in other months. The change in the trends in the processes under consideration in such areas confirms the validity of existing ideas about the climate warming effects in the ABSR (Dotsenko and Ivanov 2010), probably being one of them.

4 Conclusion

1. There are numerous marine and continental areas in the ABSR, where in the current climatic period for the months from May to September, the CN values of the repeatability of dangerous winds increase and the upward trends for these indicators are statistically significant.
2. For many of these areas, the indicated current trends from 1961 to 2020 were stable (the reliability of this conclusion is not less than 0.95). The results obtained substantiate the admissibility of the assumption that in such areas, with further climate warming, an increase in T is likely and can pose a danger to the population and transport, as well as to economic activity.
3. Climatic norms of the repeatability of dangerous winds in the water area and territory of the Azov-Black Sea Region for the current climatic period significantly depend on the month and their geographical location. They reach the highest values in the Eastern Crimea and the Taman Peninsula, as well as in the Sea of Azov and Black Sea water areas.
4. In many marine and continental areas of the Azov-Black Sea Region, where the monthly climatic norms for the repeatability of dangerous winds for the current climatic period are close to maximum levels, significant upward trends of this indicator prevail in the current period and they were stable over the entire interval of 1961–2020.
5. With further warming of the climate in the Azov-Black Sea Region, the risks of natural and technogenic extreme disasters caused by dangerous winds may increase in such water areas and territories. This fact should be taken into account when planning any types of economic activity in them.

References

Ayvazyan SA, Mkhitaryan VS (1998) Applied statistics and fundamentals of econometrics. Unity, M., p 1022

Berdnikov SV, Dashkevich LV, Kulygin VV (2019) Climatic conditions and hydrological regime of the Sea of Azov in the 20th–early 21st centuries. Aquatic Bioresour Habitat. 2(2):7–19

Dotsenko SF, Ivanov VA (2010) Natural disasters of the Azov-Black Sea Region. SPC "EKOSI-Gidrofizika", Sevastopol, p 174

Gandin LS, Kagan RL (eds) (1980) Application of statistical methods in meteorology. In: Proceedings of Voeikov Main Geophysical Observatory. Hydrometeoizdat, L, p 146

Huler S (2004) Defining the wind: the beaufort scale, and how a 19th-century admiral turned science into poetry. Crown Publishers, N-Y., p 290

Hydrometeorological conditions of the seas of Ukraine. The Sea of Azov (2009) Sevastopol: URNMI MB, p 400

Hydrometeorology and hydrochemistry of the seas of the USSR. The Black Sea (1991) Iss. 1: Hydrometeorological conditions. SPb: Hydrometeoizdat, p 430

Hydrometeorology and hydrochemistry of the seas of the USSR. The Sea of Azov (1991*) SPb: Hydrometeoizdat, p 236

Investigation of Evolution and Thermal State of the Snow Cover Thickness at the Observation Site of the Lomonosov MSU in Winter of 2022/2023

D. M. Frolov, Yu. G. Seliverstov, A. V. Koshurnikov, V. E. Gagarin, and E. S. Nikolaeva

Abstract Some results of field observations conducted at the Lomonosov Moscow State University meteorological observation site during the winter period 2022/2023 are presented in the paper. The aim of these observations is to investigate the evolution of the snow thickness column, its thermal state and its stratigraphy alteration during this winter period. The calculations of the snow cover thickness thermal state are conducted via finite difference numerical method on rectangular grid within MATLAB software computer program. The obtained results are presented on the diagrams.

Keywords Snow cover · Thermal state · Spatial and temporal heterogeneities

1 Introduction

In June 2023, the average surface temperature of the world ocean was $+18.354°C$ according to the reanalysis data of ERA5, which is a record high. This value is $0.418°C$ higher than the climatic norm for the period 1991–2020. The average air temperature in the Northern Hemisphere in June 2023 was $+20.715°C$, which is the highest indicator in the entire history of observations. This value is $0.572°C$ higher than the climatic norm for the period 1991–2020, and the temperature on the planet temporarily exceeded 1.5 °C, according to data provided by the European Earth Observation Program Copernicus. So since the beginning of the industrial era, the average global temperature has risen by 1.1 degrees Celsius. This has led to a series of climatic events including warmer winters in the cold and temperate zone, and changes in the structure and properties of snow cover.

Since snow cover is important indicator of climate change and plays an important role in the balance of surface energy (Flanner et al. 2011), interaction of the surface with the atmosphere (You et al. 2020), and hydrological processes (Beniston

D. M. Frolov (✉) · Yu. G. Seliverstov · A. V. Koshurnikov · V. E. Gagarin · E. S. Nikolaeva
Lomonosov Moscow State University, Moscow, Russia
e-mail: denisfrolovm@mail.ru

© The Author(s), under exclusive license to Springer Nature Singapore Pte Ltd. 2023   135
T. Chaplina (ed.), *Processes in GeoMedia—Volume VII*, Springer Geology,
https://doi.org/10.1007/978-981-99-6575-5_12

et al. 2018). At the meteorological observation site of the faculty of geography of Lomonosov Moscow State University there are annual winter observations of the snow cover including internal physical parameters and processes of snow cover formation (snow cover granularity, snow cover density, snow cover hardness, snow cover height, snow cover water content, snow cover duration, snow temperature, etc.) like in Rikhter (1954), Gold and Williams (1957), Benson (1962), Irwin (1979). The study of the properties of snow cover is used primarily for the needs of the educational process and scientific research at the department of Cryolithology and Glaciology and the laboratory of Snow Avalanches and Mudflows of the Faculty of Geography of Lomonosov Moscow State University.

2 Materials and Methods and Results

The results of field observations conducted at the meteorological observation site of Lomonosov Moscow State University for the winter period 2022/2023 as well as meteorological data processing are presented in the paper. The purpose of the observations was to study the development of the snow mass and its thermal state and variations over this winter period. Field observations consisted of analyzing the stratigraphic layers of the snow mass and measuring their density like in Golubev and Frolov (2015), Golubev et al. 2010). The calculations of the snow cover thickness thermal state are conducted via finite difference numerical method on rectangular grid within MATLAB software computer program.

The winter of 2022–2023 turned out to be heterogeneous in terms of temperature regime, with the average monthly temperature in December relatively close to the norm. The variations of the air temperature, precipitation and snow cover thickness for the winter period 2022/23 as well as calculated snow thickness thermal state displayed in the Fig. 1.

The calculated snow thickness thermal state is obtained numerically via finite difference method. It consists in the fact that the calculated area is discretized in time and space, that is, it is covered by a rectangular grid, in the nodes of which the thermophysical characteristics are set and the desired temperatures are determined. In this case, all derivatives are replaced by finite differences, and the original differential equation is replaced by a finite-difference analogue of the grid equation.

To discretize the computational domain, points are selected on each of the coordinate axes. All the initial physical characteristics and all the desired functions are determined and taken into account only in the nodes of the grid. Time is also discretized: the whole system is considered only at certain moments. So here the time step is constant as well as the step of a uniform spatial grid.

The investigation of the snow thickness stratigraphy at the Lomonosov Moscow State University meteorological observation site during the winter period 2022–2023 were made on December 22, January 12 and 17, February 1 and 21, and March 2 and 14. Stratigraphic columns for December 22, 2022 and February 21, 2023 are displayed at Fig. 2.

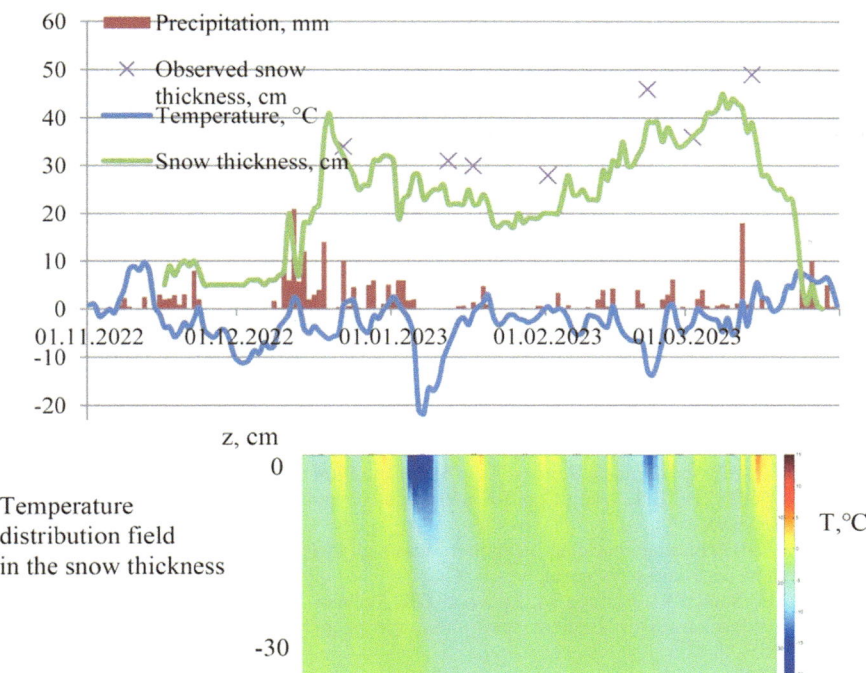

Fig. 1 Variations of air temperature, precipitation and snow cover thickness as well as calculated snow thickness thermal state for the winter period 2022/23

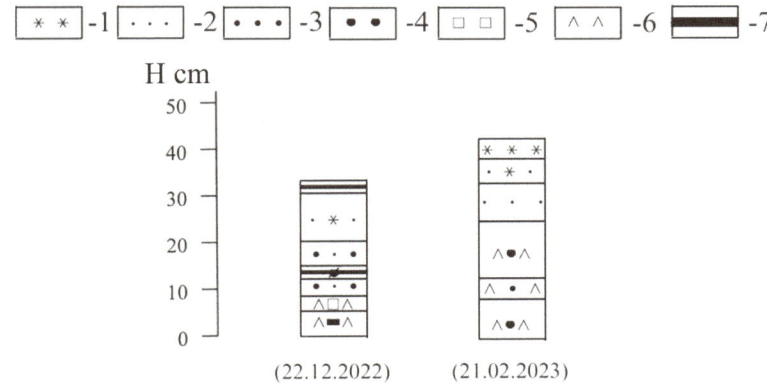

Fig. 2 The obtained snow cover thickness stratigraphic column at the meteorological site of Lomonosov MSU on 22 December and 21 February of winter period 2022/2023 (1—fresh snow, 2—small size grained snow (up to 0,5 mm), 3—medium size grained snow(0,5–1 mm), 4—coarse-grained snow (1–3.5 mm); 5—faceted crystals; 6—crystals of deep frost; 7—ice crust) (According to the international classification (Firtz et al. 2012))

The obtained data makes it possible to characterize and evaluate variations and evolution in snow cover thickness, its structure, density and thermal state during the winter period 2022/2023 and to add the necessary data to the interannual snow cover variation investigation.

The work was carried out in accordance with the state budget theme "Evolution of the cryosphere under climate change and anthropogenic impact" (121,051,100,164–0), "Danger and risk of natural processes and phenomena" (121,051,300,175–4).

Literature/References

Beniston M, Farinotti D, Stoffel M, Andreassen LM, Coppola E, Eckert N, Fantini A, Giacona F, Hauck C, Huss M, et al (2018) The European mountain cryosphere: a review of its current state, trends, and future challenges. Cryosphere 12:759–794 [Green Version]

Benson CS (1962) Stratigraphic Studies in the Snow and Firm of the Greenland Ice Sheet; RR 70, NTIS AD 288219; U.S. Army Cold Regions Research and Engineering Laboratory, Hanover, NH, USA, p 93

Firtz S et al (2012) International classification for seasonally falling snow (a guide to the description of the snow mass and snow cover

Flanner, MG, Shell KM, Barlage M, Perovich DK, Tschudi MA (2011) Radiative forcing and albedo feedback from the Northern Hemisphere cryosphere between 1979 and 2008. Nat Geosci 4:151–155

Gold LW, Williams GP (1957) Some results of the snow survey of Canada; National Research Council of Canada-Division of Building Research, Ottawa, ON, Canada, p 8

Golubev VN, Frolov DM (2015) Peculiarities of water vapor migration at the atmosphere–snow cover and snow cover–underlying soil interfaces. Cryosphere Earth V 19(1):22–29

Golubev VN, Petrushina MN, Frolov DM (2010) Patterns of formation of snow cover stratigraphy. Ice Snow (1):58–72

Irwin GJ (1979) Snow classification in support of off-road vehicle technology; Report 801; defense research establishment of Canada, Ottawa, ON, Canada, p 22

Rikhter GD (1954) Snow cover, its formation and properties; U.S. army cold regions research and engineering laboratory, Hanover, NH, USA, p 66

You Q, Wu T, Shen L, Pepin N, Zhang L, Jiang Z, Wu Z, Kang S, AghaKouchak A (2020) Review of snow cover variation over the Tibetan Plateau and its influence on the broad climate system. Earth Sci Rev 201:103043

Experimental Estimates of the Rate of Deepening of Individual Thermals

A. E. Kupriyanova and V. A. Gritsenko

Abstract A series of laboratory experiments was carried out as part of the study of the process of cooling water from the surface. Video recording of the deepening of individual thermals (small volumes of water with negative buoyancy) using high-speed video recording (up to 100 Hz) and serial photography (up to 8 Hz) made it possible to obtain quantitative estimates of the sinking speed of thermals (about 1 cm/s) and compare them with previous indicators. Model calculations made it possible to explain the features of the dynamics of thermal subsidence observed in the flume.

Keywords Quantitative estimates · The sinking speed of thermals · Sea surface cooling · Negative buoyancy · Laboratory experiment · Numerical model

1 Introduction

It is known that the thermal boundary layer that occurs when water is cooled from the surface includes a thin upper layer (up to 5–8 mm in free convection and 1–4 mm in forced convection) with a quasi-linear temperature gradient determined by molecular mixing mechanisms (Fedorov 1981). Below it is a transition layer with a blurred lower boundary, in which individual thermals are formed randomly (in time and space).

The appearance of local regions with a hydrostatically unstable density distribution in the transition layer is a prerequisite for the development of convective motions in the surface water layer (Ginzburg and Fedorov 1978). The resulting thermals limit the thickness of the cold transition layer to a few millimeters. These thermals are the subjects of cold transfer to the underlying waters. Many works are devoted to various

A. E. Kupriyanova (✉)
Immanuel Kant Baltic Federal University, 14, St. A. Nevskogo, Kaliningrad 236016, Russia
e-mail: kupriyanova_ae@mail.ru

A. E. Kupriyanova · V. A. Gritsenko
Shirshov Institute of Oceanology, Russian Academy of Sciences, 36, Nakhimovsky Prosp., Moscow 117997, Russia

© The Author(s), under exclusive license to Springer Nature Singapore Pte Ltd. 2023
T. Chaplina (ed.), *Processes in GeoMedia—Volume VII*, Springer Geology,
https://doi.org/10.1007/978-981-99-6575-5_13

aspects of the study of convective processes in the surface layer of water, including the dynamics and transformation of the shape of thermals—a fungus, a vortex ring, etc.—in a homogeneous and stratified liquid (Turner 1973; Scorer 1978; Album and of Fluid Motion, edited by Milton Van Dyke. 1982; Romanov 2009; Deremble 2016; Sin'kova et al. 2017; Ingel 2019). At the same time, the authors note the lack of experimental estimates of the rate of sinking from the surface of thermals (volumes of water with negative buoyancy) (Ginzburg et al. 1981; Dikarev and Zatsepin 1983; Bune et al. 1985).

The purpose of this work was to obtain quantitative estimates of the rate of penetration of individual thermals in a homogeneous medium based on the data of laboratory experiments.

2   Materials and Methods

2.1   Laboratory Setup

Laboratory experiments were carried out in the hydrotray of the Laboratory of Marine Physics (LPhM) of the Atlantic Branch of the Institute of Oceanology. P.P. Shirshov Academy of Sciences (AO IO RAS, Kaliningrad), a general view of which is shown in Fig. 1. Preparation of the tray for experiments included slow filling (~1 h through a hole in the bottom of the tray) of its working space (W × H × D, 800 × 600 × 125 mm) with fresh water (T ~ 14° C) and placement of special input devices on its surface. Work with the installation began approximately 3–4 h after the attenuation of the induced perturbations in the water.

To form a density heterogeneity in the surface layer of water, a special "funnel" type inlet device was used, the general scheme of which is shown in Fig. 2. The base of the device is a polystyrene foam tray (with dimensions of 200 × 100 × 15 mm and a thickness of 3.5 mm). A funnel cut into its bottom (upper edge diameter ~30 mm, nozzle inner diameter 6.5 mm). The lower section of the funnel was deepened under the water surface by 22 mm; the inlet and outlet of the "funnel" were covered with a sieve cloth (333 microns), and the inner space between the two grids (see Fig. 2b) was filled with sand (1–2 mm); the gaps between the plate and the funnel were rubbed with sealant (Fig. 2).

To prepare the salt solution of thermals, before the start of the experiment, 200 ml of water was taken from the lower part of the tray into a measuring cup. The required amount of salt (mg, NaCl, chemically pure) and food coloring were added to the sampled water to obtain the required negative buoyancy of the water of thermal.

At the beginning of the experiment, a small volume of salt solution was slowly fed into the "funnel", which entered the surface water layer of the tray. Video recording of successive stages dive of thermal in fresh water was performed using a Baumer high-speed video camera (up to 100 Hz), as well as serial shooting on a Sony-58α digital camera with a frame rate of up to 8 Hz. In the course of the experiments, the

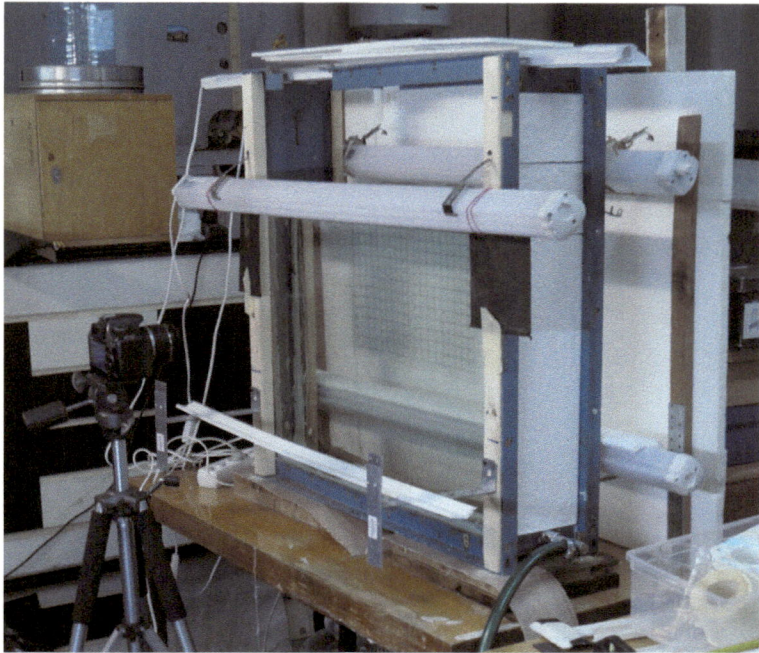

Fig. 1 General view of the hydrotray LPhM, filled with fresh water, with a lighting system. The camera is mounted on a tripod and allows you to fix the working surface of the tray in plan. The side walls of the metal frame of the tray and the cover of the working space are insulated with polystyrene foam plates. A 20 × 20 mm calibration grid is fixed on the back wall, metal rulers were placed vertically on the bottom of the tray

following changes were made: the magnitude of the difference in negative buoyancy between fresh water in the flume and the saline solution (10^{-4} or 2×10^{-4} g/cm^3), the volume of the future thermal (2 or 3 ml), the parameters of the survey of the process with a frequency of 2 Hz (Sony-58α) or 10 Hz (Baumer). Recording times on the Baumer camcorder were fixed at 60 and 40 s for volumes of 2 and 3 ml, respectively; recording on a digital camera lasted up to 85 s.

2.2 Experimental Estimates of the Penetration Rate of the Density Front of Thermal

As part of the study, 27 experiments were performed (9 series of 3 launches). Figure 3 shows two pictures of one of the experiments of the sinking of thermal ($\Delta\rho = 0.0001$ g/cm^3, V = 2 ml). It turned out that the transition from the initial form of the thermal to the mushroom shape (Fig. 3b) and the instability of the vortex ring usually occurred at a depth of 7–10 cm. At the phase of the onset of instability of the

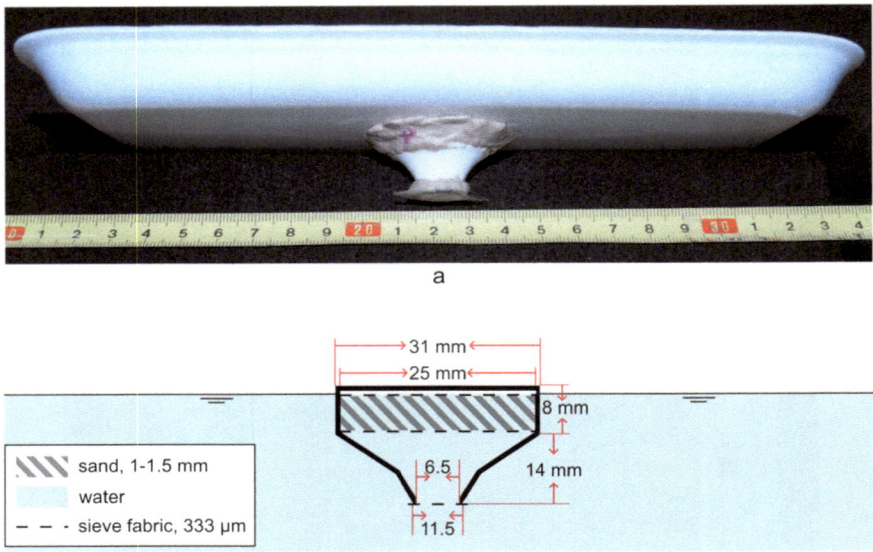

Fig. 2 General view of the "funnel" type inlet device (**a**) and a schematic representation of the cross section of the funnel (**b**)

vortex ring, filming with the camera was stopped. Further, the results of calculations of the sinking rates of individual thermals based on the generated data array will be presented.

After previewing the results of video and photography, it turned out that the time for the formation of developed convection is on average ~50 s (at 2 ml of volume) and ~35 s (at 3 ml of volume). Note that in previous works (Ginzburg et al. 1981; Bune et al. 1985), the time of convection development in similar laboratory experiments was estimated at 83 and 45 s, respectively.

To calculate the desired quantitative estimates of the rate of thermal sinking, the obtained image arrays were converted into data arrays using the ImageJ program. The analysis of the extracted data made it possible to obtain the dependence of the vertical displacement of the density front of thermal on time. The corresponding graph for one of the experiments is shown in Fig. 4b.

The movement of the thermal density front on the pixel matrix of the obtained images was noted by the operator's hand, which explains the appearance of small jumps on the curve. Accuracy of measuring the position of the thermal was ~ 1–2 pixels per frame. To get rid of the "noise" in determining the position of the front by the least squares method, various analytical dependencies were built. Graphs of the variability of the vertical immersion velocity (mm/s) of thermals for the series under consideration are shown in Fig. 5.

The finite-difference calculation of the sinking speed of thermal for this series of images (65 frames) showed that the maximum and minimum values of the vertical

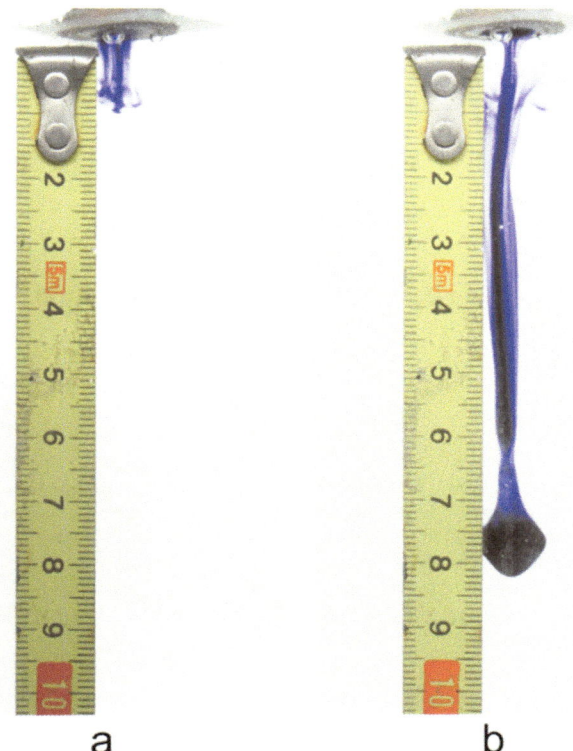

Fig. 3 The photographs show two phases of the sinking of thermal (plan view): entry through the inlet nozzle (**a**), mushroom-shaped transformation of the negative buoyancy volume (**b**)

velocity are, respectively, ~6.5 mm/s and ~0.54 mm/s, and the average velocity of the density front was ~1.8 mm/s. The average quadratic approximation of empirical data by polynomials of various degrees (up to 5) showed slight differences in the estimates obtained with their help—about 0.1 mm/s—from those previously obtained in discrete calculations.

One of the features of the sinking was the increase in the rate of deepening of their density front at the initial stage. In Fig. 5, the local maximum growth rate is marked with a red line through the points of the initial data. A jump in the speed of this was observed in each of the laboratory experiments within 5–10 s from the start of the experiment. After a short growth phase, a rapid, then slow, quasi-asymptotic decrease in the penetration rate was observed.

The obtained estimates of the dynamics of thermals from laboratory experiments were summarized in Table 1. In the left column of the table there are values of density differences of 10^{-4} g/cm^3 and $2 \cdot 10^{-4}$ g/cm^3, the right column of the table is classified by the volume of the thermal (Volume, mL). The two rows located to the right of the density values (Density, g/cm^3) represent discrete velocity data (upper row) and

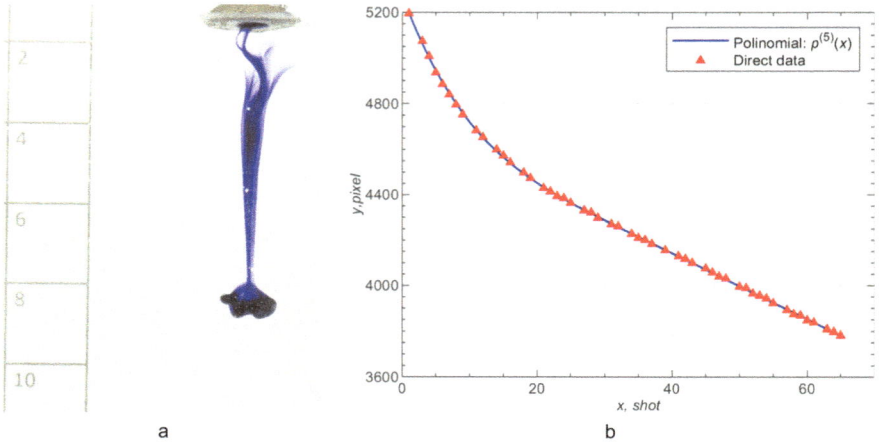

a   b

Fig. 4 The phase of the beginning of the decay of the vortex ring (**a**), grid on the left 2 × 2 cm and the graph (**b**) of the movement of the density front (vertical axis, pixels) of thermal during shooting (horizontal axis, frames). In the illustrated case, the graph shows 65 frames and the movement of pixels along the matrix from top to bottom (from 5195 to 3782 pixels), the maximum pixel value (5195 pixels) corresponds to the starting point of the density front position. Experiment characteristics: volume of saline solution 2 ml, excess density 10^{-4} g/cm^3, Sony camera (2 fps), 0.051 mm/pcs, experiment time 36 s

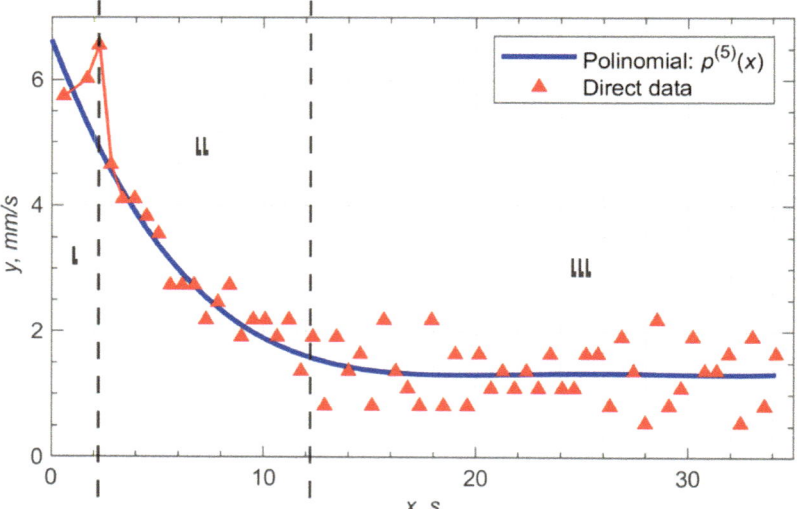

Fig. 5 Graph of the dependence of the vertical velocity (m/s) on the immersion time for one of the thermals with a density difference between fresh and salt water of 10^{-4} g/cm^3. The speed jump (0–4 s) is marked with a thin red line. Empirical speed estimates are red dots, approximation data is a blue line. Roman numerals characterize the modes of the speed of sinking

Table 1 Quantitative estimates of the rate of penetration of the density front of thermals obtained during laboratory experiments. The values of the average depth rate (mm/s) are highlighted in bold: the upper line is empirical data, the lower one is approximate

Volume, mL	2									
	#	time	#	time	#	time	#	time	#	time
	1 / 40 s		2 / 46 s		3 / 36 s		4 / 77 s		5 / 33 s	
	Velocity, mm/s									
Density, 10^{-4} g/cm^3	**2.2**		**3.3**		**1.8**		**1.8**		**3.4**	
	2.2		**3.17**		**2**		**1.76**		**3**	
	#	time	#	time	#	time	#	time	#	time
	1 / 78 s		2 / 77 s		3 / 40 s		4 / 37 s		5 / 43 s	
	Velocity, mm/s									
Density, $2 \cdot 10^{-4}$ g/cm^3	**2.4**		**2.1**		**3.06**		**2.4**		**3.2**	
	2.4		**2.03**		**3.0**		**2.3**		**3.2**	
Volume, mL	3									
	#	time	#	time	#	time	#	time	#	time
	1 / 37 s		2 / 35 s		3 / 24 s		4 / 55 s		5 / 38 s	
	Velocity, mm/s									
Density, 10^{-4} g/cm^3	**4.1**		**4.68**		**4.6**		**3.1**		**3.6**	
	3.97		**4.6**		**3.9**		**3.1**		**3.54**	
	#	time	#	time	#	time	#	time	#	time
	1 / 30 s		2 / 26 s		3 / 40 s		4 / 36 s		5 / 30 s	
	Velocity, mm/s									
Density, $2 \cdot 10^{-4}$ g/cm^3	**5.8**		**7**		**4.7**		**5**		**5.3**	
	5.68		**6.6**		**4.7**		**4.76**		**4.6**	

approximation data (lower row). Above these lines is the number (#) and the duration of the experiment (time). In each cell with the value of the sinking velocity for the time interval before the collapse of the vortex ring of the thermal, the average sinking velocity (Velocity, mm/s) is indicated.

In total, the table shows the parameters of the 20 most representative experiments. As can be seen from Table 1, a small change in density differences with the same volume has little effect on the change in velocity and ranges from 0.1–0.2 mm/s.

A more significant factor was the change in the volume of the thermal (in fact, the total reserve of negative buoyancy). From the table values, it can be seen that for a thermal of any density, when the volume changes by 50%, the average speed increases by about 2 times.

Note that all the obtained values of the immersion rate of thermals are in good agreement with the results of other authors (Ginzburg et al. 1981; Dikarev and Zatsepin 1983; Bune et al. 1985; Zatsepin et al. 2005). From a qualitative comparison of the graphs of the obtained analytical dependence and experimental data, it follows that the nature of the change in empirical data at the stage of velocity drop at different parameters of thermals has been preserved.

In the course of the study, the estimates of the immersion rate of laboratory and calculated thermals were compared with the velocity scales of T. Benjamin (1968), and R. Scorer (1957). The velocity scales proposed for near-bottom gravitational currents and thermals in the atmosphere turned out to be in satisfactory agreement with the values of the sinking velocity obtained in the experiments. Moreover, the tendency to increase the rate of deepening of the density front with the variability of the magnitude of negative buoyancy turned out to be within the calculation error (0.1–0.2 mm/s). In the calculations for T.Benjamin, taking into account the value of the constant $C = 0.44$ led to an underestimation of the speed indicators, and the correlation with the data in Table 1 is ~1.05. The estimate of the thermal immersion rate by R. Scorer at $C \approx 1.2$ greatly overestimated the values compared to Table 1. The correlation of experimental estimates and estimates by R. Scorer turned out to be ~2.31.

2.3   Analysis of Calculated Flows

Model calculations were carried out on a nonlinear 2d model of the dynamics of a fluid inhomogeneous in density (Kupriyanova and Gritsenko 2022). The main equations of the model in the variables vorticity—stream function—density are given below:

$$\frac{D\omega}{Dt} = \frac{g}{\rho_0} \frac{\partial \sigma}{\partial x} + v_0 \Delta \omega, \tag{1}$$

$$\Delta \psi = \omega, \tag{2}$$

$$\frac{D\sigma}{Dt} = D_0 \Delta \sigma, \tag{3}$$

where $\omega = \partial u/\partial z - \partial w/\partial x$—vorticity, ψ—stream function, $u = \partial/\partial z$ и $w = -\partial/\partial x$—horizontal and vertical speed components, g—gravitational acceleration, ρ_0 и $\rho = \rho_0 + \sigma$—densities of fresh and salt water, v_0 and D_0—molecular viscosity of water and salt diffusion, D/Dt и Δ—full derivative and Laplace operators. The horizontal (Ox) and vertical (Oz) coordinate axes coincide with the bottom and side wall of the model space. Two computational grids 601×801 and 1001×601 with a spatial sampling value of 0.05 were used. The thermals were set on the grid areas of 31×11 or 31×31 calculation nodes and with characteristic scales close to the flows in the tray. The advective terms of the finite-difference equations of the model were constructed using the second scheme with differences against the flow (Rouch 1980). The calculation of the current function was carried out iteratively.

Figure 6 shows a typical view of a calculated thermal that has almost completed active immersion and is close to the beginning of transformation into a vortex ring.

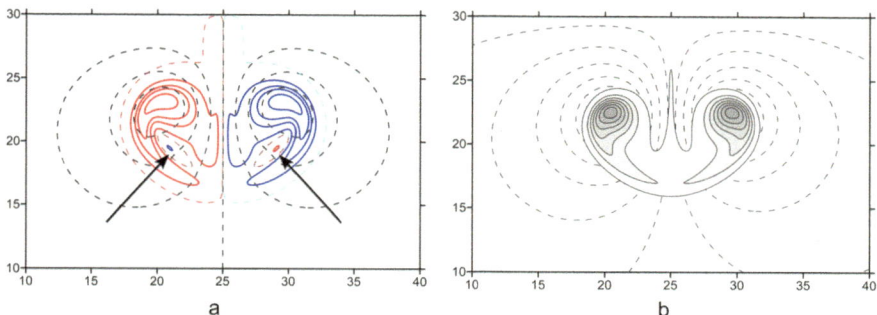

Fig. 6 Typical mushroom-like appearance of the formed thermal. The characteristic scales were as follows: h = 1 cm, $\Delta\rho = 0.0002$ g/cm^3, u ≈ 0.63 cm/s. **a**—values of vorticity isolines $\omega \in [-0.65, 0.65, 0.15]$. The blue color of the isolines corresponds to negative values of vorticity, red—positive. The zero values of vorticity are highlighted with thin dashed lines (± 0.001). The isolines of the stream function are indicated by gray dashed lines and are drawn from -1.6 to 1.6 in increments of 0.4. **b**—density isolines $\sum = \sigma/\Delta\rho$ with gray fill are carried out in the range [0.05, 0.4, 0.1]. Dashed lines are drawn $\psi \in [-1.6, 1.6, 0.2]$

The thermal is represented by the distributions of the main model fields (density, vorticity, stream function).

The vortex nature of the movement of water inside and in the vicinity of the thermal is clearly visible, as well as the appearance of secondary vorticity of a viscous nature. Local maxima of secondary vorticity are marked with arrows. Here and further on, the figures show only part of the model space for a better demonstration of the details of the studied flows.

To obtain quantitative estimates sinking speed of thermal, the position of its density front was determined by the depth of the isopycn $\sum \approx 0.05$ using the capabilities of the Surfer package. The finite-difference calculation of the diving speed showed a result close to the estimates of laboratory experiments, both in terms of the nature of the variability of the speed and in quantitative indicators. Figure 7 shows the results of the change sinking speed of thermal from time to time for one of the model flows.

The physical meaning of the appearance of a jump of the sinking speed of thermal with a maximum of 0.43 cm/s at a time of $\approx$5.0 s turned out to be quite understandable. Figure 8 shows three phases of the initial stage of sinking of thermal, which allow us to explain the observed effect. Immediately after the start of the dive for a short period of time (up to 5 s), the lower boundary of the thermal is almost horizontal (see Fig. 8a on the right). The generation of vorticity of baroclinic nature, in accordance with Eq. 1, occurs at the lateral boundaries of the thermal (Turner 1964). The vortices on the left and right have a linear scale of the order of the thickness of the thermal (see Fig. 8 a on the left) and are at an early stage of their development. The involvement of external water in the thermal is minimal. At this phase evolution of the thermal, the maximum rate of its immersion is achieved. Starting from this moment, the sinking speed of thermal only decreases.

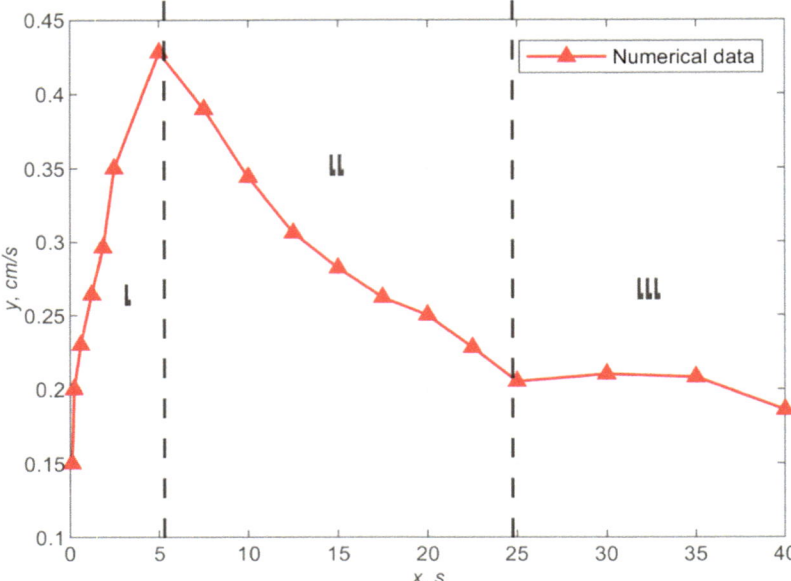

Fig. 7 Graph of changes in the values of the sinking speed of thermal from time to time. The jump and quasi-asymptotic attenuation of the velocity are observed in the same way as in laboratory experiments. Roman numerals characterize the speed modes. The flow characteristics are the same (Fig. 6)

After acquiring a developed mushroom shape and involving all the water of the termal in the vortex motion (see Fig. 8b), its inhibition occurs due to an increase in the horizontal dimensions of the density front and the involvement of the surrounding water. On the determination of the vorticity field, the emergence of a secondary vorticity of a viscous nature is noted (arrows are marked in Fig. 8b). In the next phase (see Fig. 8c) the reasons for the drop in speed are clearly visible—a significant increase in the involvement of water in the movement, both inside the thermal and outside it, generates the consumption of negative buoyancy for all volumes of water involved in the movement. It is this circumstance, according to Scorer (1957), Turner (1964), that determines the drop of the sinking speed of thermal.

R. Scorer has previously considered a similar problem within the framework of several propositions (1957): on the shape of a thermal—an oblate spheroid, on the similarity of the movement of water inside the thermal at all horizons. Following by R. Scorer, the main inhibitory factor is the involvement of stationary water in the vicinity of the thermal. The results of our model calculations have shown that the currents arising from the immersion of the thermal from the very beginning have a vortex structure and do not fit into the framework of R. Scorer's assumptions. This allows us to explain the differences in speed scales. Finally, in the analyses of the experiments of R. Scorer's a short time interval of growth in the rate of thermal sinking was not noticed and taken into account. The technique used by us to register

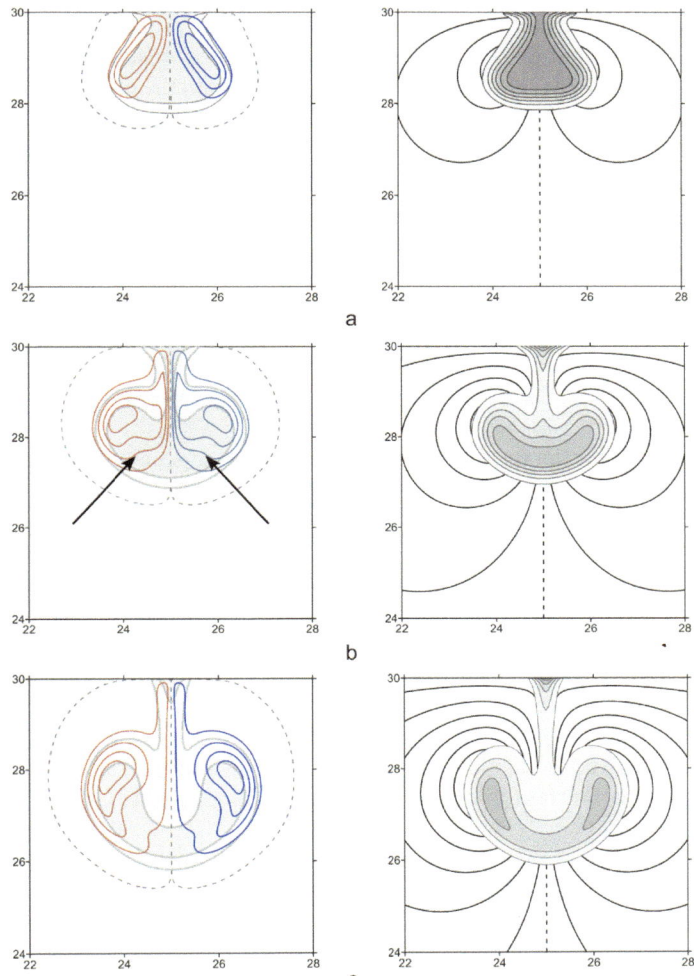

Fig. 8 Three successive phases of sinking of thermal, illustrating the stages of achieving the maximum sinking speed of thermal. **a**—sinking speed of thermal it is still growing, **b**—the maximum speed has been reached, **c**—the braking of the thermal immersion has already occurred. The characteristic scales of the calculated cross-section are the same as in Fig. 6. Values of the isolines of the main flow parameters: for the left row of graphs—density isolines ($\sum = 0.05$ and 0.3, gray fill), vortices $\omega = [-2; +2; 0.4]$); for the right graphs—density isolines $\sum = [0.1; 0.85; 0.15]$, current isolines ψ for **a** $[-0.35; 0.35; 0.1]$, for **b** $[-0.55; 0.55; 0.1]$; for **c** $[-0.75; 0.75; 0.1]$

the position of the thermal made it possible to clarify the processes occurring at the same time and to highlight a previously unobserved growth stage. Further analysis of the rate of immersion of the thermals showed that, with the exception of the short initial stage described above, of the sinking speed of thermals during this rapid fall decreases inversely proportional to the square of time.

3 Discussion and Results

The analysis of the data of all 27 flows allowed us to distinguish, as in the case of the collapse of the spots (Barenblatt 1978; Zatsepin et al. 1978; Wu 1969; Abramyan and Kudin 1983), three different stages by the nature of the variability the sinking speed of thermal.

At the first stage, after the formation of density inhomogeneity in the surface layer of water and the beginning of immersion of the thermal, there is a rapid increase in the rate of its loading. The thermal lacks a characteristic mushroom-shaped "hat". The vortex structure is in the of the sinking of thermal phase of its formation. These factors determine the minimality of drag and the involvement of surrounding waters.

At the second stage, there is a rapid drop in the rate of immersion and the formation of the vortex structure of the submerging thermal (see Fig. 3). When submerged to a depth of 7–10 cm, the thermal acquires a mushroom shape with the vortex nature of the movement of its waters clearly visible in the images. This circumstance generates an increase in the volume of both the waters of the thermal itself and the surrounding fresh waters involved in the movement. An increase in drug also slows down the rate of sinking. At the finish of this stage, the thermic completes its transformation into a vortex ring.

The third (viscous) mode was characterized by a slow decrease in the rate of immersion and the decay of the vortex ring into a set of mini-fungi. This stage is the longest in terms of time in the series of experiments under consideration, it begins at ~12–15 s from the beginning of the sinking of thermal.

Model calculations with a linear scale of thermals from 1 to 10 cm and density differences from $2 \cdot 10^{-4}$ to 10^{-3} g/cm^3 allowed us to obtain quantitative estimates of the rate of thermal sinking.

The ranges of variability of the rate of sinking of thermal were small (from 2–3 m/s to 1–2 cm/s), and were dependent on the total reserve of negative buoyancy. In particular, a threefold reduction in volume (from 31 × 31 knots to 31 × 10 knots) led to a 1.5–1.7-fold reduction in speed. This detail of the behavior of thermals was previously noted by other authors (Scorer 1957; Turner 1964).

An analysis of quantitative estimates of the sinking speed of thermals for all the experiments considered with different initial conditions showed that the change in the sink rate at the stage of rapid deceleration is inversely proportional to the square of time and depends on the value of the total reserve of negative buoyancy of thermals.

Acknowledgements The investigations are supported by Russian Science Foundation via grant number 23-27-00150, https://rscf.ru/en/project/23-27-00150/.

References

Abramyan TO, Kudin AM (1983) Laboratory study of the interaction of mixed liquid spots during their spreading in a stratified medium. Izvestiya AN SSSR. Series Phys Atmos Ocean 19(8):888–891

An Album of Fluid Motion, edited by Milton Van Dyke. The Parabolic Press. 1982, p 176

Barenblatt GI (1978) Dynamics of turbulent patches and intrusions in a stably stratified fluid. Izvestiya AN SSSR. Series Phys Atmos Ocean 14(2):195–206

Bune AV, Ginzburg AI, Polezhaev VI, Fedorov KN (1985) Numerical and laboratory modeling of the development of convection in a layer of water cooling from the surface. Izvestiya AN SSSR. Series Phys Atmos Ocean 21(9):956–963

Benjamin T (1968) Gravity currents and related phenomena. J Fluid Mech 31(2):209–248. https://doi.org/10.1017/S0022112068000133

Deremble B (2016) Convective plumes in rotating systems. J Fluid Mech 799:27–55

Dikarev SN, Zatsepin AG (1983) Development of convection in a two-layer unstable stratified fluid. Okeanologiya 23(6):950–953

Fedorov KN (1981) On the physical structure of the near-surface oceanic layer. Meteorol Hydrol 10:58–66

Ginzburg AI, Fedorov KN (1978) Cooling of water from the surface under free and forced convection. Izvestiya AN SSSR. Series Phys Atmos Ocean 14(1):79–87

Ginzburg AI, Dikarev SN, Zatsepin AG, Fedorov KN (1981) Phenomenological features of convection in a fluid with a free surface. Izvestiya AN SSSR. Series Phys Atmos Ocean 17(4):400–407

Ingel LK (2019) On the limiting laws of buoyant convective jets and thermals from local sources of a heat releasing impurity. J Eng Phys Thermophys 92(6):1481–1488.https://doi.org/10.1007/s10891-019-02067-6

Kupriyanova AE, Gritsenko VA (2022) Sinking patches of salt water on a slope of bottom surrounded by fresh water: dynamics and structural features of density front propagation up the slope. J Oceanol Res 50(2):106–124, https://doi.org/10.29006/1564-2291.JOR-2022.50(2).5

Romanov NP (2009) Experimental investigation of free (downward) convection near horizontal cold spots. Izvestiya RAS. Series Phys Atmos Ocean 45(3):356–370

Rouch P (1980) Computational hydrodynamics. Mir, Moscow, p 618

Scorer R (1957) Experiments on convection of isolated masses of buoyant fluid. J Fluid Mech 2(6):583–594. https://doi.org/10.1017/S0022112057000397

Scorer R (1978) Environmental aerodynamics. Wiley, p 488

Sin'kova OG, Statsenko VP, Yanilkin YuV (2017) Analytical and numerical study of the formation of a vortex ring in the event of a thermal in the atmosphere. Izvestiya RAS, Mekhanika Zhidkosti i Gaza 4:39–51. https://doi.org/10.7868/S0568528117040041

Turner JS (1973) Buoyancy effects in fluids. New York, Cambridge Univ, Press, p 367

Turner J (1964) The flow into an expanding spherical vortex. J Fluid Mech 18(2):195–208. https://doi.org/10.1017/S0022112064000155

Wu J (1969) Mixed region collapse with internal ware generation in a density stratified medium. J Fluid Mech 35(3):531–544

Zatsepin AG, Fedorov KN, Voropayev SN, Pavlov AM (1978) Experimental study of the spreading of a mixed spot in a stratified fluid. Izvestiya AN SSSR. Series Phys Atmos Ocean 14(2):234–237

Zatsepin AG, Gritsenko VA, Kremenetsky VV, Poyarkov SG, Stroganov OY (2005) Laboratory and numerical studies of the dense water spreading along the sloping bottom. Okeanologiya 45(1):5–15

Interaction of Nonlinear Waves Taking into Account Planetary Scale

S. I. Peregudin and S. E. Kholodova

Abstract Nonlinear flows and waves are considered in a thin layer of an ideal incompressible fluid rotating with angular velocity between concentric hemispheres. Along with the theoretical study, an important applied aspect is the result of numerical simulation with possible further implementation through modern integrated software development environments.

Keywords Waves of finite amplitude · Nonlinear flows and waves · Nonlinear wave dynamics · Mathematical modeling · computational experiment

Let us consider nonlinear flows and waves in a thin layer of an ideal incompressible fluid rotating with angular velocity ω between concentric hemispheres.

The study of nonlinear fluid flows in rotating spherical layers is necessary to understand many dynamic processes of a global scale in the ocean. In this case, the form of motion essentially depends on two important factors: the spherical geometry of the volume and rotation. The application and necessity of this kind of research for the problems of geophysics is beyond doubt (Roesner 1978; Kholodova 2008, 2009; Peregudin and Kholodova 2011; Peregudin et al. 2019, 2021; Shamin 2019).

Assuming that the pressure is distributed according to the hydrostatic law, the basic equations of motion of the theory of shallow water (Pedloski 1984) in a spherical coordinate system associated with the rotating Earth can be reduced to a single nonlinear partial differential equation in terms of a stream function $\psi(t, \theta, \lambda)$ of the form

S. I. Peregudin (✉)
St. Petersburg State University, Saint Petersburg, Russia
e-mail: s.peregudin@spbu.ru

S. E. Kholodova
ITMO University, Saint Petersburg, Russia

$$\frac{\partial \Delta \psi}{\partial t} + \frac{1}{r_0^2 \sin \theta} \frac{D(\Delta \psi, \psi)}{D(\theta, \lambda)} + 2\omega \frac{\partial \psi}{\partial \lambda}$$
$$= \frac{\Delta \psi + 2r_0^2 \omega \cos \theta}{h} \left[\frac{\partial h}{\partial t} + \frac{1}{r_0^2 \sin \theta} \frac{D(h, \psi)}{D(\theta, \lambda)} \right], \tag{1}$$

where r_0—Earth radius, h—fluid depth, Δ—the Laplace operator in spherical coordinates:

$$\Delta = \frac{1}{\sin \theta} \frac{\partial}{\partial \theta} \left(\sin \theta \frac{\partial}{\partial \theta} \right) + \frac{1}{\sin^2 \theta} \frac{\partial^2}{\partial \lambda^2}.$$

The horizontal velocity components v_θ, v_λ are related to the stream function ψ by the relations:

$$v_\theta = -\frac{1}{r_0 \sin \theta} \frac{\partial \psi}{\partial \lambda}, \qquad v_\lambda = \frac{1}{r_0} \frac{\partial \psi}{\partial \theta}. \tag{2}$$

The vertical projection of the motion velocity vortex ζ, taking into account relation (2), is written as

$$\zeta = \frac{\Delta \psi}{r_0^2}.$$

For a liquid layer located between two concentric hemispheres, the depth function $h(t, \theta, \lambda)$ is a constant and Eq. (1) can be written as

$$\frac{\partial \Delta \psi}{\partial t} + \frac{1}{r_0^2 \sin \theta} \frac{D(\Delta \psi, \psi)}{D(\theta, \lambda)} + 2\omega \frac{\partial \psi}{\partial \lambda} = 0. \tag{3}$$

The solution to this equation will be sought in the form

$$\psi(t, \theta, \lambda) = -r_0^2 \alpha \cos \theta + J_n(\theta, \lambda - \sigma_n t), \tag{4}$$

where $J_n(\theta, \lambda)$—spherical function of n-order; α—some constant that determines the angular velocity of the zonal flow; σ_n—the frequency of the wave.

Substituting expression (4) into Eq. (3), we obtain the relation for the wave frequency σ_n:

$$\sigma_n = \alpha - \frac{2(\lambda + \omega)}{n(n+1)} = 0. \tag{5}$$

Thus, wave (4) is an exact solution of the nonlinear Eq. (3) when the relation for wave frequency (5) is fulfilled.

The frequency of the wave σ_n characterizes the rotation of the wave relative to the earth's surface. To move to the linear velocity $v_{\lambda n}$ of the wave at an arbitrary latitude θ, it is necessary to multiply the angular velocity σ_n by the radius $r_0 \sin \theta$ of the circle of latitude

$$v_{\lambda n} = \alpha \, r_0 \, \sin \theta \left[1 - \frac{2\left(1 + \frac{\omega}{\lambda}\right)}{n(n+1)} \right], \tag{6}$$

where $v_{\lambda n} = \alpha \, r_0 \, \sin \theta$—zonal fluid flow velocity at latitude θ. According to formula (6), the waves corresponding to small values of the meridional wave number n, and hence the zonal number m, since the summation extends m, $(m \le n)$, to propagate from east to west.

For such waves, the speed $v_{\lambda n} < 0$. Waves corresponding to large values of m and n propagate from west to east. In this case $v_{\lambda n} > 0$.

Solution (4) is a wave superimposed on the west–east current, the angular velocity of which is α. A solution similar in appearance is contained in the works of Blinova in the problem of waves in the atmosphere (Blinova 1943).

In solution (4), in connection with the boundary condition $\psi = 0$, $\theta = \frac{\pi}{2}$, it is assumed $n - m = 2k + 1$.

The stationary solution of Eq. (3) is determined by the relation

$$\frac{\Delta \psi}{r_0^2} + 2\omega \cos \theta = \tilde{h}(\psi).$$

The same ratio can be obtained in another way, similar to that used by Chia-Shun (1977).

In this case, the shallow water theory equations reduce to the equation

$$\left(\frac{\Delta \psi}{r_0^2} + 2\omega \cos \theta \right) d\psi = dF, \tag{7}$$

where $F = \frac{1}{2r_0^2} \left(\frac{1}{\sin^2 \theta} \left(\frac{\partial \psi}{\partial \lambda} \right)^2 + \left(\frac{\partial \psi}{\partial \theta} \right)^2 \right)$ is a Bernoulli function depending only on ψ.

The study of wave motions, representing perturbations propagating parallel to the ocean surface, allows us to restrict ourselves to the study of two-dimensional fluid motions that depend only on time and the angular coordinates of a point on the surface of a spherical Earth. Usually, to simplify the analysis, it is customary to consider such (planetary) motions in a limited area of the sphere surface—in a certain β-plane in which the sphere is locally replaced by a plane, but the change of the Coriolis parameter in the north direction is taken into account (Pedloski 1984).

Consider a plane-parallel motion of a rotating incompressible fluid occurring in a β—plane on which a system of Cartesian coordinates (x, y) is introduced, so that the axis Ox is directed to the east, and Oy—to the north.

These movements are described by the following system of equations:

$$\frac{\partial \mathbf{v}}{\partial t} + \nabla\left(\frac{\mathbf{v}^2}{2}\right) - [\mathbf{v},\, rot\, \mathbf{v}] + [\boldsymbol{\alpha},\, \mathbf{v}] = -\nabla p, \tag{8}$$

$$\mathrm{div}\, \mathbf{v} = 0, \tag{9}$$

where $\mathbf{v} = (v_x,\, v_y)$—the velocity vector of fluid particles, $\boldsymbol{\alpha} = (0,\, 0,\, \alpha(y))$—the Coriolis parameter depending on y, that is, on the latitude of the β-plane, and the vector $\boldsymbol{\alpha}$ is perpendicular to the β-plane, is the dynamic pressure, the density of the fluid is assumed to be equal to unity for ease of recording. Equation (9) allows us to introduce the stream function $\psi(x,\, y)$:

$$v_x = \frac{\partial \psi}{\partial y}, \qquad v_y = \frac{\partial \psi}{\partial x}.$$

Let us write Eq. (8) component by component:

$$\frac{\partial v_x}{\partial t} + \frac{\partial}{\partial x}\left(\frac{\mathbf{v}^2}{2}\right) - v_y \Omega - \alpha(y)\, v_y = -\frac{\partial p}{\partial x},$$

$$\frac{\partial v_y}{\partial t} + \frac{\partial}{\partial y}\left(\frac{\mathbf{v}^2}{2}\right) + v_x \Omega + \alpha(y)\, v_x = -\frac{\partial p}{\partial y},$$

where $\Omega = \frac{\partial v_y}{\partial x} - \frac{\partial v_x}{\partial y}$. Differentiating the first by, and the second by, and subtracting one from the other, we get:

$$\frac{\partial}{\partial t}\left(\frac{\partial v_y}{\partial x} - \frac{\partial v_x}{\partial y}\right) + v_x \frac{\partial \Omega}{\partial x} + v_y \frac{\partial \Omega}{\partial y} + \alpha'(y)\, v_y = 0,$$

or in terms of the stream function:

$$\frac{\partial}{\partial t}(\Delta \psi) + \frac{\partial \psi}{\partial y}\frac{\partial}{\partial x}(\Delta \psi) - \frac{\partial \psi}{\partial x}\frac{\partial}{\partial y}(\Delta \psi) + \alpha'(y)\frac{\partial \psi}{\partial x} = 0,$$

or

$$\frac{\partial \Delta \psi}{\partial t} + \frac{D(\Delta \psi,\, \psi)}{D(x,\, y)} + \alpha'(y)\frac{\partial \psi}{\partial x} = 0. \tag{10}$$

Here $\Omega = -\Delta \psi$. Equation (10) can be written as

$$\left(\frac{\partial}{\partial t} - \frac{\partial \psi}{\partial x}\frac{\partial}{\partial y} + \frac{\partial \psi}{\partial y}\frac{\partial}{\partial x}\right)(\Delta \psi - \alpha(y)) = 0. \tag{11}$$

For a liquid infinitely extended horizontally, we have the exact solution of the nonlinear Eq. (11)

$$\psi = A \exp{(i(kx + ly - \sigma t))}, \qquad \sigma = -\frac{k\beta}{k^2 + l^2},$$

where $\beta = \alpha'(y)$. The parameter β will be considered constant.

Let us consider the problem of reflection of non-stationary planetary waves with a finite amplitude from a solid wall oriented in the latitudinal direction.

Let the wall be along the axis x. The no-flow condition in terms of the stream function ψ requires that

$$v_y = \frac{\partial \psi}{\partial x} = 0, \qquad y = 0. \tag{12}$$

Consider the solution corresponding to the linear superposition of the incident and reflected waves:

$$\psi = \psi_i + \psi_r,$$

$$\psi_i = A_i \exp{(i(k_i x + l_i y - \sigma_i t))}, \qquad \psi_r = A_r \exp{(i(k_r x + l_r y - \sigma_r t))}.$$

Condition (12) on the wall $y = 0$ takes the form

$$k_i A_i \exp(i(k_i x - \sigma_i t)) + k_r A_r \exp(i(k_r x - \sigma_r t)) = 0.$$

This implies the equalities

$$\sigma_i = \sigma_r = \sigma, \qquad k_i = k_r = k, \qquad A_i = -A_r.$$

Thus, the expected solution will take the form

$$\psi = A \exp(i(kx + l_i y - \sigma t)) - A \exp(i(kx + l_r y - \sigma t)),$$

and l_i, l_r are the roots of the equation

$$l^2 + k^2 \sigma + k\beta = 0. \tag{13}$$

Equation roots (13):

$$l_r = \sqrt{-\frac{k^2 + k\beta}{\sigma}}, \qquad l_i = -\sqrt{-\frac{k^2 + k\beta}{\sigma}}.$$

A linear superposition of partial solutions in the form of incident and reflected waves may not be a solution to Eq. (11) for ψ.

Therefore, we substitute the function $\psi = \psi_i + \psi_r$ into Eq. (11). The condition for the left side of Eq. (11) to turning to zero for ψ will be the zero product of multiplication $(l_r - l_i)^2 (l_r + l_i)$. When $l_r \neq l_i$ we get $l_r + l_i = 0$, that is $l_r = -l_i$.

Thus, the solution of the corresponding nonlinear equation can be represented as a linear superposition of the incident and reflected waves.

References

Blinova EN (1943) Hydrodynamic theory of pressure waves, temperature waves and atmospheric action centers. Doklady AN SSSR 39(7):284–287

Chia-Shun Y (1977) Wave motions in layered liquids. Nonlinear Waves. M.: Mir, pp 271–296

Kholodova SE (2008) Dynamics of a rotating layer of an ideal electrically conductive incompressible fluid. Comput Math Math Phys 48(5):882–898

Kholodova SE (2009) Wave motions in a stratified electrically conductive rotating fluid. J Comput Math Math Phys 49(5):916–922

Pedloski J (1984) Geophysical hydrodynamics. In 2 vols. M.: Mir,. T. 1 - 400 p., T. 2 - 411 p

Peregudin SI, Kholodova SE (2011) On the features of the propagation of non-stationary waves in a rotating spherical layer of an ideal incompressible stratified electrically conductive fluid in the equatorial latitudinal belt. Appl Mech Tech Phys 52(2,306):44–51

Peregudin S, Peregudina E, Kholodova S (2019) The influence of dissipative effects on dynamic processes in a rotating electrically conductive liquid medium. J Phys: Conf Ser 1359(1):012118

Peregudin S, Peregudina E, Kholodova S (2021) Mathematical modeling of the dynamics of a rotating layer of an electrically conducting fluid with magnetic field diffusion effects. Springer Geology, pp 277–287

Roesner KG (1978) Numerical calculation of hydrodynamic stability problems with time dependent boundary conditions. In: 6th international conference on numerical methods in hydrodynamics, Tbilisi

Shamin RV (2019) Machine learning in economic problems. M: Green Print, p 140

Hydrodynamic Interaction Between Two Systems of Intense Intra-Mantle Vortices as a Possible Cause of the Connection Between Distant Earthquakes

S. Y. Kasyanov

Abstract During the tangential intrusion of a large temporary satellite, two systems of intense hot quasi–stationary vortices were formed in the Earth's mantle, previously considered as independent—a system of three energy-carrying vortices formed during the decay of a vortex ring that arose in the mantle during the complete immersion of the satellite, and a system of parts of vortex rings formed during the gliding of a partially submerged satellite on the surface of the lithosphere and mantle. The paper proposes a kind of hydrodynamic connection between two systems of vortices in the mantle, which explains the spatiotemporal structure of recent earthquakes in the world.

Keywords Earthquake · Gliding · Vortices · Satellite

1 Introduction

Earlier, we proposed a hypothesis about the tangential fall and intrusion into the Earth's mantle of a large temporary satellite with a radius of about 930 km and described some natural geophysical consequences of this event (Kasyanov 2012, 2021; Kasyanov and Samsonov 2017, 2018, 2021). According to this hypothesis, when the satellite invaded the Earth's mantle, hot intense two-phase vortices formed in the mantle thickness. The process of establishing a quasi-stationary state in the vortex system has led to the fact that by now two systems of intense hot quasi-stationary vortices have been preserved in the Earth's mantle, which were formed at different stages of the evolution of the vortex system. These vortex systems are schematically presented in Figs. 1 and 2. The spatiotemporal structure of earthquakes in the world requires a hydrodynamic connection between these vortex systems in the mantle, previously considered as independent.

S. Y. Kasyanov (✉)
N.N.Zubov's State Oceanographic Institute, Roshydromet, Moscow, Russia
e-mail: skas53@yandex.ru

© The Author(s), under exclusive license to Springer Nature Singapore Pte Ltd. 2023   159
T. Chaplina (ed.), *Processes in GeoMedia—Volume VII*, Springer Geology,
https://doi.org/10.1007/978-981-99-6575-5_15

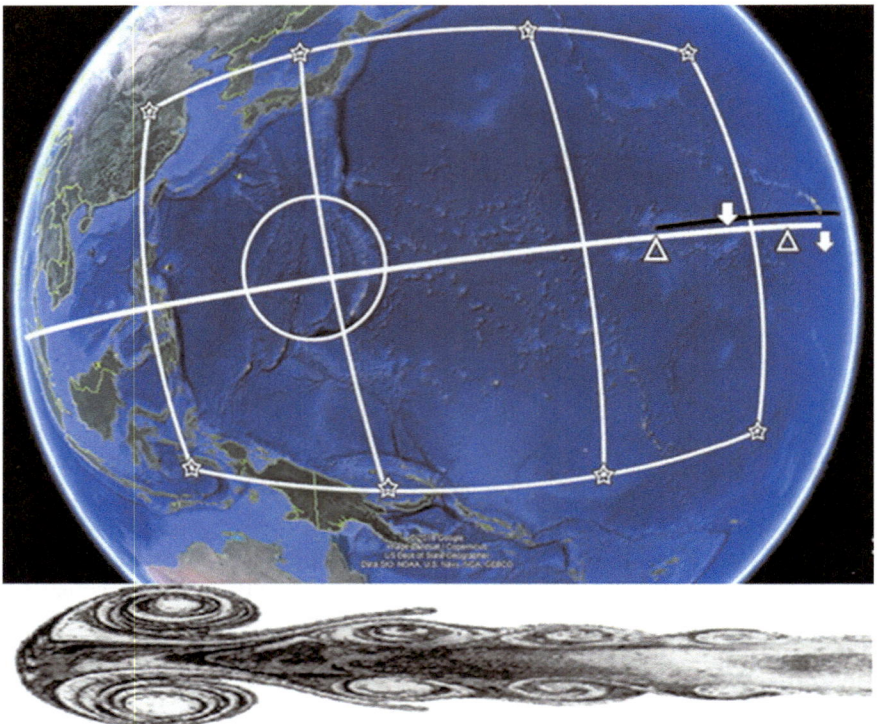

Fig. 1 The trajectory of the body's sinking in the area from Hawaii to the Philippine Islands (Kasyanov 2021). Triangles are the areas of ascent, and arrows are the areas of descent of the trajectory. The black line is a seismic section (Cao et al. 2011). The stars show the preserved southern exits of the "vortex half-rings" (that is, parts of the vortex rings in the mantle ending in two exits of the vortex to the daytime surface) and the positions of the northern exits symmetrical to them relative to the trajectory of the non-preserved northern exits. The area of the Mariana Basin is highlighted in a circle. Below, for comparison, a chain of vortices around a jet invading a liquid is shown, which occurs in a laboratory experiment (Gharib et al. 1998)

2 The Formation of Parts of Intense Intra-Mantle Vortex Rings ("Vortex Half-Rings") and Traces of a Shock Wave During the Partial Penetration of the Satellite into the Mantle

The trajectory of the satellite during its immersion into the mantle passed from the Hawaiian Islands to the Philippine Basin (Kasyanov 2021). In Fig. 1, the trajectory of the satellite corresponds to the central white line. The complete immersion of the satellite into the mantle occurred near the Philippine Islands.

When the satellite was partially submerged along its trajectory, a sequence of parts of vortex rings (that is, "vortex half-rings") formed in the mantle, similar to the

Fig. 2 A modern system of three energy-carrying vortices in the Earth's mantle. Vortices: (1) Himalayas + Tien Shan–Kolyma, (2) Kolyma–Yellowstone and (3) Yellowstone–Hudson Bay—the eastern part of the Mediterranean Sea (Kasyanov and Samsonov 2021)

sequence of vortex rings around the jet invading the liquid (see Fig. 1) (Kasyanov 2021). The parts of the vortex rings end with paired exits of the vortices to the daytime surface to the north and south of the trajectory of the satellite, indicated by the stars in Fig. 1.

The velocity of the satellite during immersion into the mantle turns out to be greater than the velocity of P-waves in the lithosphere and upper layers of the mantle (Kasyanov 2012; Kasyanov and Samsonov 2017, 2018). Therefore, the movement of the satellite in this part of the trajectory turns out to be supersonic and a shock wave occurs in the lithosphere and mantle (Kasyanov and Samsonov 2018). Two branches of the shock wave, symmetrical with respect to the trajectory of the body, appear on the daytime surface, manifested in the structure of the Pacific Ocean floor (see Figs. 3 and 4). The branches of the shock wave stretch from the Philippine Islands: the northern branch to Kamchatka, the southern branch to the Fiji Islands. On the southern branch of the shock wave, manifestations of corrugation instability are noticeable, leading to the formation of local vorticity regions.

At the front of the shock wave, the mantle substance is destroyed, accompanied by significant heat release and melting of the mantle substance (Kasyanov and Samsonov 2018). At present, the boundaries of lithospheric plates partially extend

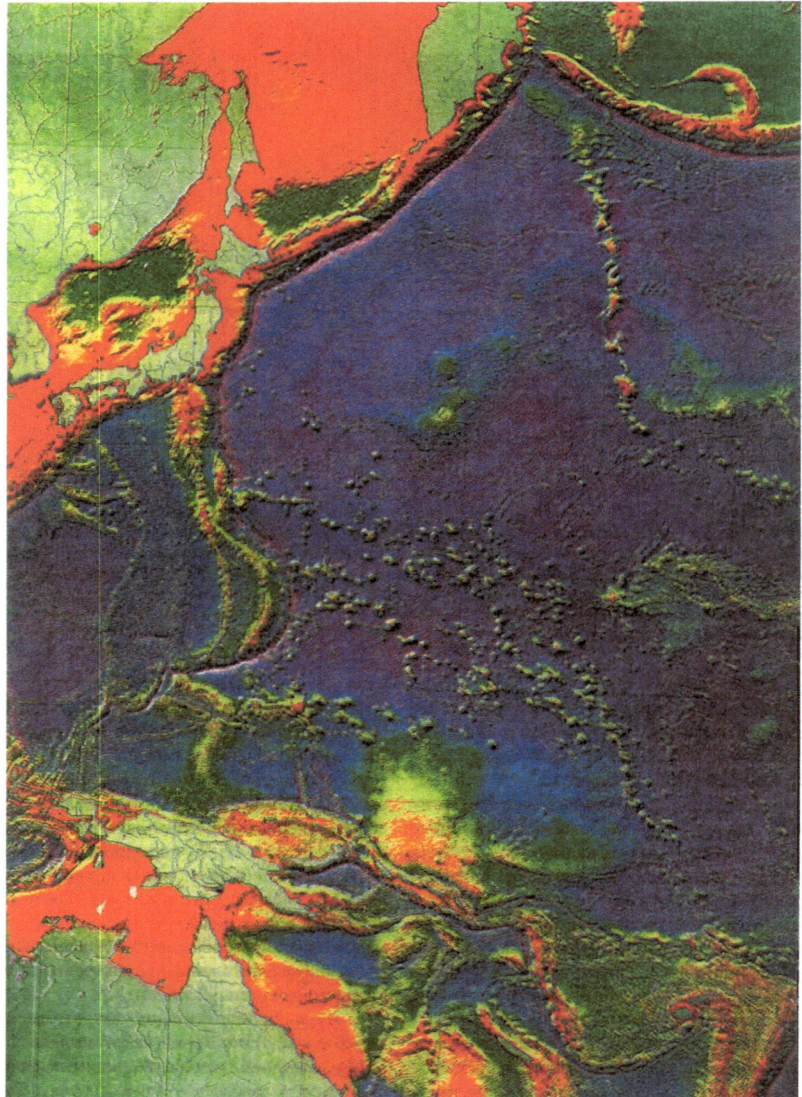

Fig. 3 The western part of the Pacific Ocean. Images from space (Dobretsov et al. 2001)

along the areas of the shock wave front passage. Zones of increased seismic activity are associated with these areas (see Fig. 4). Earthquake epicenters along the line from the Philippine Islands to the Fiji Islands and further to New Zealand in Fig. 4 trace the position of the southern branch of the shock wave formed when the satellite was partially buried in the mantle.

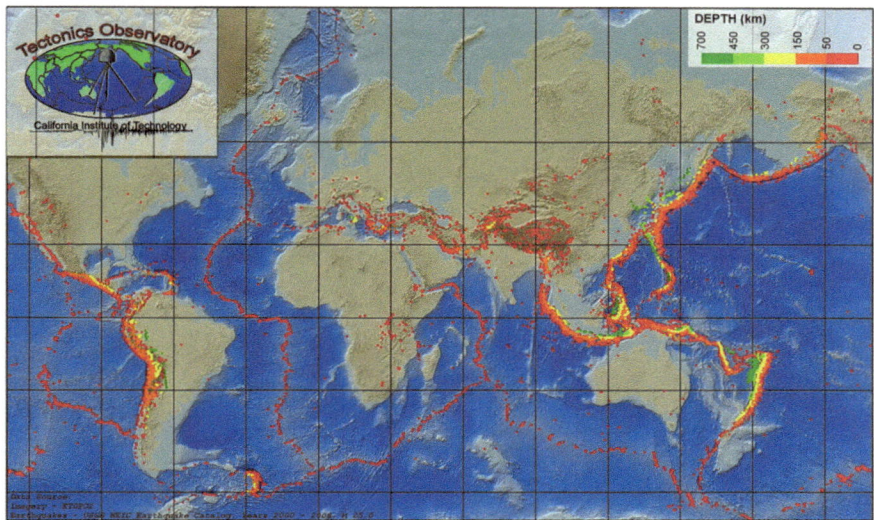

Fig. 4 Location of epicenters and depth of hypocenters of earthquakes of 2000–2008 with magnitude M ≥ 5.0 (http://www.geo.uu.nl/~paulssen/ISS/iss11-2020.pdf)

And earthquake epicenters along the line from the Philippine Islands to Kamchatka trace the position of the northern branch of the shock wave. The branches of the shock wave encircle the paired exits to the daytime surface of the sequence of parts of the vortex rings (indicated by the stars in Fig. 1) and come into contact with "vortex half-rings". It should be noted that the foci of the most deep-focus earthquakes with depths of more than 300 km are located in the places where parts of the vortex rings exit to the daytime surface and in the areas of vorticity resulting from the corrugation instability of the shock wave.

The seismic zone of the Sunda volcanic arc is a trace of the front of the head shock wave that occurs when the satellite plunges into the mantle. Seismic zones located along the Izu-Bonin trench and the Philippine Trench have traces of a shock wave at Mach number M = 1.

In addition, the wake of the shock wave at Mach number M = 1 is the chain of the Imperial seamounts in Fig. 5 and the intramantic vortices of Kamchatka (see Fig. 5) and the Fiji Islands (see Fig. 6), clearly distinguished in the field of gravitational anomalies. The arc of the great circle connecting the intramantile vortices of Kamchatka and the Fiji Islands runs along the chain of the Imperial Seamounts.

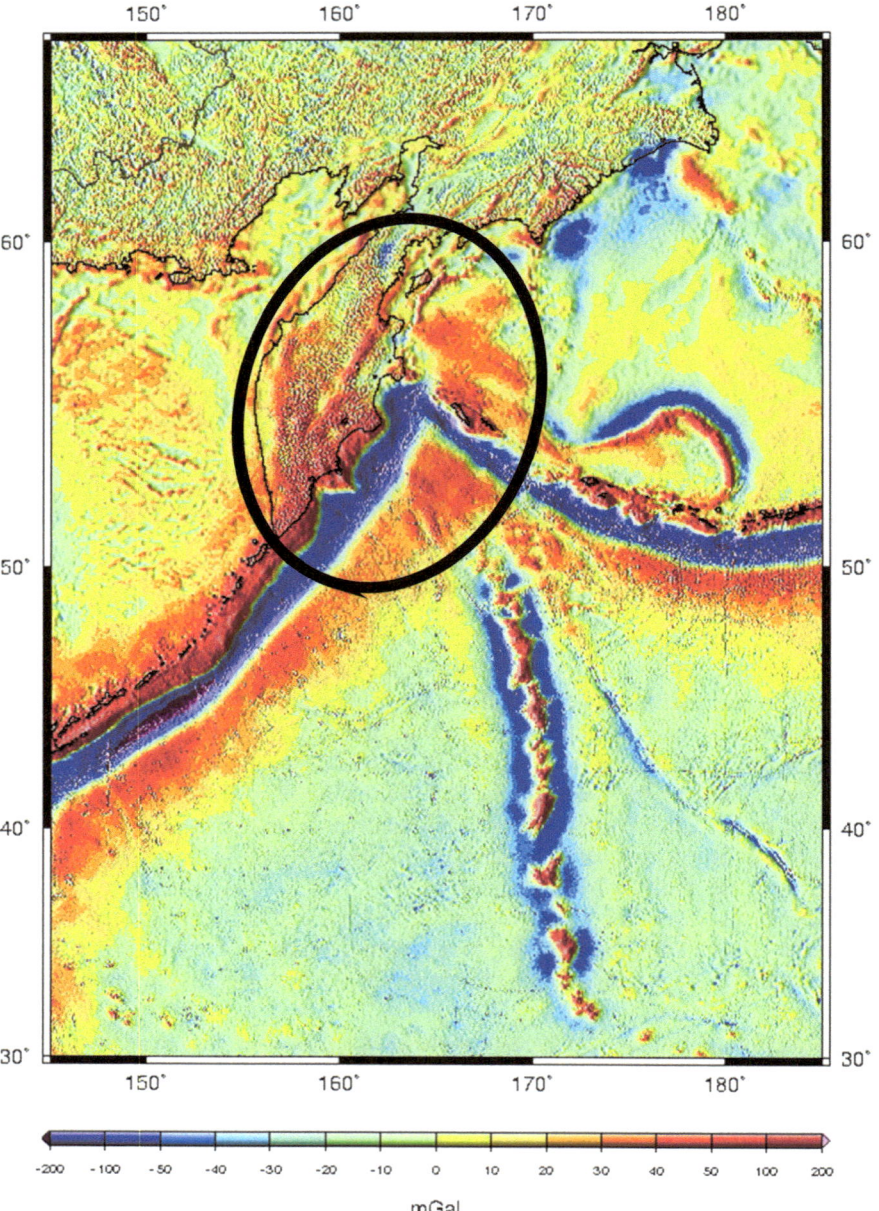

Fig. 5 The intramantium vortex of Kamchatka (thick black line) and the shock wave traces that match it: from the southwest—the northern branch of the shock wave, from the south—the chain of the Imperial Seamounts, which is the wake of the shock wave at Mach M = 1. WGM2012— Complete spherical isostatic gravity anomaly (Airy-Heiskanen, Tc = 30 km) (Bonvalot et al. 2012; https://bgi.sedoo.fr/bgi_affiche_image.php?file=plot_isostatic_20230319041945.png)

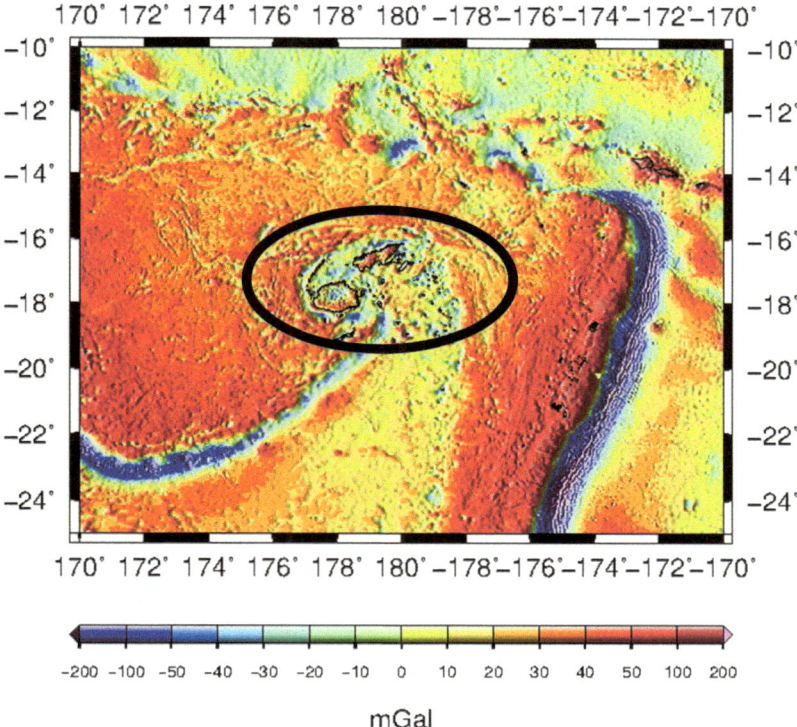

Fig. 6 The intra-mantle vortex of the Fiji Islands (thick black line). WGM2012—Complete spherical isostatic gravity anomaly (Airy-Heiskanen, Tc = 30 km) (Bonvalot et al. 2012; https://bgi.sedoo.fr/bgi_affiche_image.php?file=plot_isostatic_20230319044325.png)

3 Interaction of a System of Parts of Intra-Mantle Vortex Rings with a System of Three Energy-Carrying Vortices in the Earth's Mantle

According to the previously proposed hypothesis (Kasyanov 2021; Kasyanov and Samsonov 2021), a vortex ring formed in the mantle as a result of the satellite intrusion. The collapse of the ring led to the formation of a quasi-stationary system of hot two-phase liquid–gas vortices in the mantle (Kasyanov and Samsonov 2021). The central role in it is played by a system of three intense large-scale vortices, comprising the bulk of the kinetic energy of the entire system of intra-mantle vortices of the Earth. This system of vortices is schematically presented in Fig. 2.

The spatio-temporal analysis of the earthquakes that occurred in February–March 2023 shows the presence of a connection in time of earthquakes in three zones that are far from each other. Namely: earthquakes in the zone of the exits of vortices (1) and (3) belonging to the system of three energy-carrying vortices, with earthquakes in the zone of the sequence of "vortex half-rings" and earthquakes along the strike of the northern and southern branches of the shock wave. This fact speaks in favor of the presence of a hydrodynamic connection between these areas. Therefore, it can be assumed that at least one vortex from the system of the sequence of "vortex half-rings" shown in Fig. 1 is in contact with at least one of the vortices (1), (2), (3) shown in Fig. 2 in the system of three vortices and exchanges matter with it.

Figure 7 shows a diagram of the hydrodynamic interaction of the Himalayan-Tien Shan tip of the vortex (1) with the northern outlet of a part of the vortex ring ("vortex semicircle") under the southeast of China, which is located symmetrically relative to the trajectory of the satellite's sinking relative to the southern outlet of a part of the vortex ring in the Banda Sea in Fig. 1. In the outputs of both of these interacting vortices to the daytime surface, the motion of the medium is anticyclonic. Therefore, a secondary cyclonic vortex is formed between them, through which two systems of intense vortices in the mantle exchange hot moving matter. This vortex is clearly distinguished on the maps of the lithosphere relief in Figs. 8 and 9 and on the map of isostatic gravitational anomalies shown in Fig. 9.

The system of three intense large-scale vortices in the Earth's mantle is controlled by the location of the planets of the Solar System through the effect of gravitational anomalies on the substance of the vortices and the liquid core of the Earth (Kasyanov and Samsonov 2021). As a result of the influence of gravitational anomalies, hydrodynamic disturbances arise in the vortex system.

Fig. 7 Diagram of the hydrodynamic interaction of the Himalayan-Tien Shan tip of the vortex (1) (thin black line) with the northern outlet of the part of the vortex ring ("vortex semicircle") under the southeast of China (thick white line). An intermediate secondary cyclonic vortex ring (a thick black line) formed between them. The cartographic basis of Google Earth Pro (https://www.goo gle.com/earth/about/versions/#download-pro)

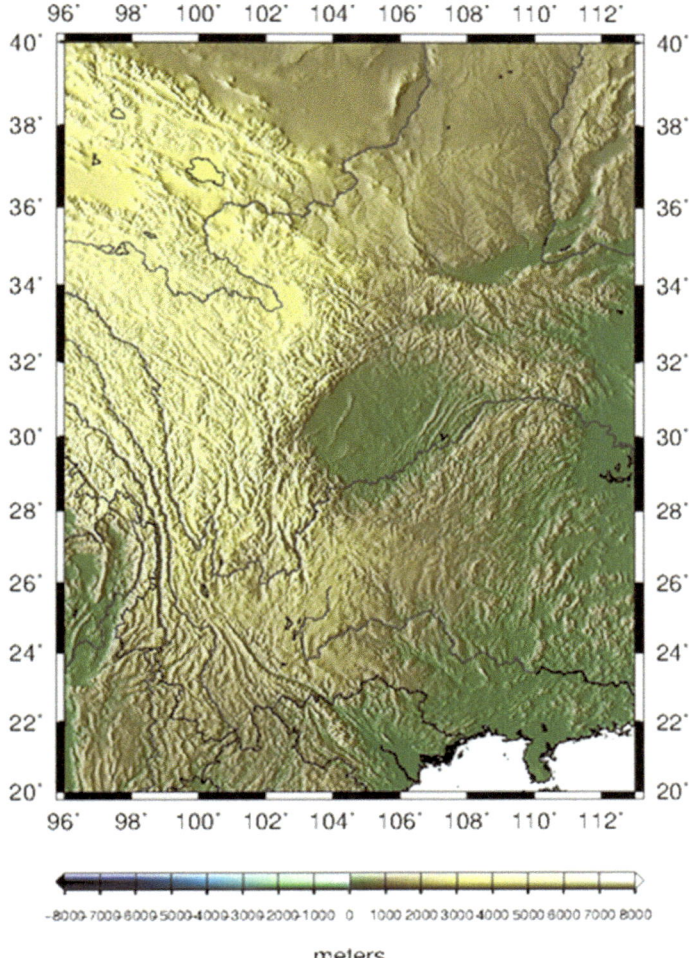

Fig. 8 Intermediate secondary cyclonic vortex ring between the Himalayan-Tien Shan tip of the vortex (1) and the northern outlet of the part of the vortex ring ("vortex semicircle") under the southeast of China. ETOPO1 topography (Bonvalot et al. 2012; https://bgi.sedoo.fr/bgi_affiche_i mage.php?file=plot_topography_20230317050440.png)

During hydrodynamic disturbances through a secondary cyclonic vortex, a system of three intense large-scale vortices exchanges hot moving matter with a system of vortices from a sequence of "vortex half-rings". The parts of vortex rings belonging to the sequence of "vortex half-rings", coming into contact with molten matter in the zones of shock wave traces, exchange matter with it. As a result of the presence of hydrodynamic connections between two systems of intense intra-mantle vortices and the shock wave zone, earthquakes in areas ranging from the Philippine Islands up to Kamchatka and from Sumatra and the Philippine Islands up to the islands of

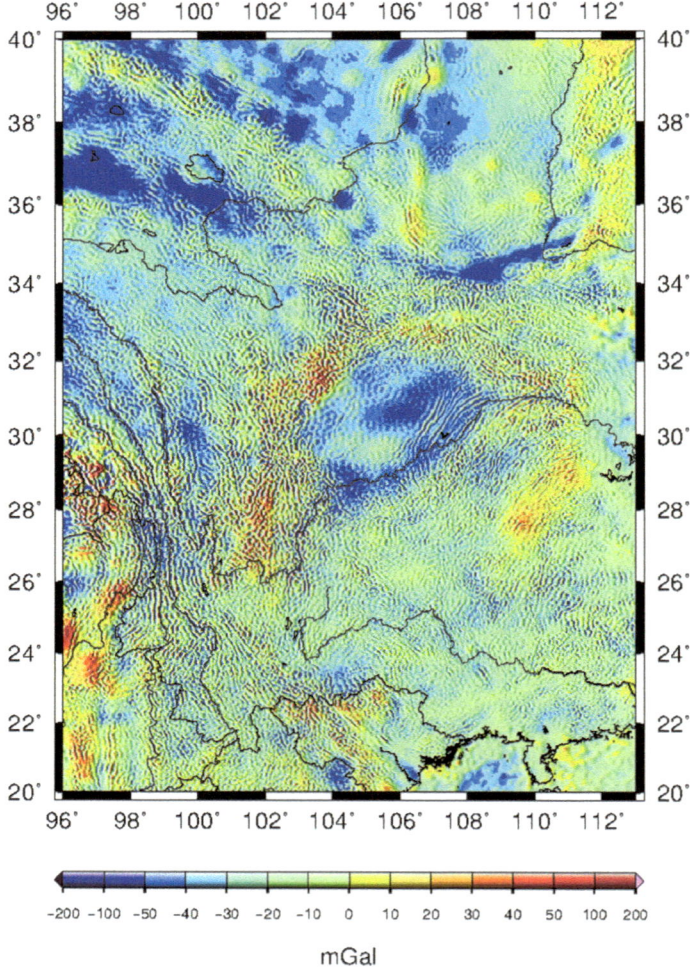

Fig. 9 Intermediate secondary cyclonic vortex ring between the Himalayan-Tien Shan tip of the vortex (1) and the northern outlet of the part of the vortex ring ("vortex semicircle"). WGM2012—Complete spherical isostatic gravity anomaly (Airy-Heiskanen, Tc = 30 km) (Bonvalot et al. 2012; https://bgi.sedoo.fr/bgi_affiche_image.php?file=plot_isostatic_20230317051111.png)

Fiji and New Zealand are associated with earthquakes in the Mediterranean region, Central Asia and Iceland.

4 Conclusions

The system of three intense large-scale vortices in the mantle is hydrodynamically connected with the system of a sequence of "vortex half-rings" and with the melt region in the shock wave front zone through an intermediate secondary cyclonic vortex ring and, possibly, also through the liquid core of the Earth. Consequently, there may be a hydrodynamic connection between recent earthquakes in remote regions: in the Mediterranean region, in Central Asia, in the area of the Philippine Islands and Sumatra, in Kamchatka, on the islands of Fiji and New Zealand, on Lake Baikal, in Iceland and North America.

References

Bonvalot S, Balmino G, Briais A, Kuhn M, Peyrefitte A, Vales N, Biancale R, Gabalda G, Reinquin F, Sarrailh M (2012) World gravity map. Commission for the geological map of the world. Eds. BGI-CGMW-CNES-IRD, Paris

Cao Q, Van der Hilst RD, de Hoop MV, Shim S-H (2011) Seismic imaging of transition zone discontinuities suggests hot mantle west of Hawaii. Science 332(6033):1068–1071. https://doi.org/10.1126/science.1202731

Dobretsov NL, Kirdyashkin AG, Kirdyashkin AA (2001) Deep geodynamics. Publishing House of SB RAS, Novosibirsk, p 247 (In Russian)

Gharib M, Rambod E, Shariff K (1998) A universal time scale for vortex ring formation. J Fluid Mech 360:121–140

Google Earth Pro. https://www.google.com/earth/about/versions/#download-pro

Kasyanov SY (2012) Modeling of the orbital motion of a large asteroid during a descent with a deepening into the lithosphere and mantle of the Earth. Phys Problems Ecol (Ecol Phys) (18):151–164 (in Russian). http://ocean.phys.msu.ru/pubs/ecophys/ecophys-18_pp150-246.pdf

Kasyanov SY (2021) Formation of system of intense vortices in the mantle when a large temporary earth satellite is immersed. In: Chaplina T (eds) Processes in GeoMedia—Volume III. Springer geology. Springer, Cham, pp 217–232. https://doi.org/10.1007/978-3-030-69040-3_21

Kasyanov SY, Samsonov VA (2017) Estimation of tidal wave heights during planing of a large satellite on the surface of a liquid attracting spherical layer with a solid core. Proc MIPT 9(3):14–20 (in Russian). https://mipt.ru/upload/medialibrary/d1a/2_kas_yanov_14_20.pdf

Kasyanov S, Samsonov VA (2018) Effect of a tidal wave caused by large gliding satellite on formation of 220 km seismic boundary and split of the mantle into blocks. In: Karev V, Klimov D, Pokazeev K (eds) Physical and mathematical modeling of earth and environment processes. PMMEEP 2017. Springer geology. Springer, Cham, pp 360–370. First Online 25 March 2018. https://doi.org/10.1007/978-3-319-77788-7_37, Print ISBN 978-3-319-77787-0, Online ISBN 978-3-319-77788-7

Kasyanov SY, Samsonov VA (2021) About the modern system of three energy-carrying intensive vortices in the earth's mantle. In: Chaplina T (eds) Processes in GeoMedia—Volume II. Springer geology. Springer, Cham, pp 273–291. https://doi.org/10.1007/978-3-030-53521-6_31

http://www.geo.uu.nl/~paulssen/ISS/iss11-2020.pdf

https://bgi.sedoo.fr/bgi_affiche_image.php?file=plot_isostatic_20230319041945.png

https://bgi.sedoo.fr/bgi_affiche_image.php?file=plot_isostatic_20230319044325.png

https://bgi.sedoo.fr/bgi_affiche_image.php?file=plot_topography_20230317050440.png

https://bgi.sedoo.fr/bgi_affiche_image.php?file=plot_isostatic_20230317051111.png

Wave Motions of a Liquid in a Rotating Annular Layer

S. I. Peregudin and S. E. Kholodova

Abstract Wave motions are considered in an annular layer of an ideal incompress-ible homogeneous rotating fluid formed by two cylinders with an inner radius, an outer radius. The vessel rotates around its vertical axis with an angular velocity. In the case of shallow water, the assumption of a linear distribution of pressure with depth is acceptable. The corresponding hydrodynamic problem for small amplitude waves is reduced to a boundary value problem for a linear partial differential equa-tion with variable coefficients depending on the depth of the fluid. Along with the theoretical study, an important applied aspect is the result of numerical simulation with possible further implementation using modern integrated software development environments.

Keywords Waves of finite amplitude · Nonlinear flows and waves · Rotating liquid · Nonlinear wave dynamics · Mathematical modeling · Computational experiment

Let us consider wave motions in an annular layer of an ideal incompressible homo-geneous rotating fluid formed by two cylinders with inner radius r_1, outer radius r_2.

The vessel rotates around its vertical axis with an angular velocity ϖ. In the case of shallow water, the assumption of a linear distribution of pressure with depth is acceptable. The corresponding hydrodynamic problem for small amplitude waves is reduced to a boundary value problem for a linear partial differential equation with variable coefficients depending on the depth of the fluid. This equation has the form

$$\frac{\partial}{\partial t}\left[\left(\frac{\partial^2}{\partial t^2} + f^2\right)\eta - g\,\nabla\cdot(H\,\nabla\eta)\right] - gf\,J(H,\,\eta) = 0, \tag{1}$$

S. I. Peregudin (✉)
St. Petersburg State University, Saint Petersburg, Russia
e-mail: s.peregudin@spbu.ru

S. E. Kholodova
ITMO University, Saint Petersburg, Russia

where $\eta(x, y, t)$—deviation of a free surface from its position at rest, g—gravitational acceleration, f—the Coriolis parameter, in this case equal to 2ω, $H(x, y)$—the thickness of the liquid layer at rest. The notation is introduced for the Jacobian of two functions $J(A, B) = \frac{\partial A}{\partial x}\frac{\partial B}{\partial y} - \frac{\partial A}{\partial y}\frac{\partial B}{\partial x}$.

On the side walls of the vessel, the condition of equality to zero of the normal component of the velocity must be satisfied, which in terms of the elevation of the free surface η is equivalent to the following equation:

$$\frac{\partial}{\partial t}\left(\frac{\partial \eta}{\partial n}\right) + f\frac{\partial \eta}{\partial s} = 0, \qquad r = r_1, r_2. \tag{2}$$

Equations (1), (2) in a cylindrical coordinate system will be written as follows:

$$\frac{\partial}{\partial t}\left[\left(\frac{\partial^2}{\partial t^2} + f^2\right)\eta - gH\left(\frac{\partial^2 \eta}{\partial r^2} + \frac{1}{r}\frac{\partial \eta}{\partial r} + \frac{1}{r^2}\frac{\partial^2 \eta}{\partial \phi^2}\right) - g\left(\frac{\partial H}{\partial r}\frac{\partial \eta}{\partial r} + \frac{1}{r^2}\frac{\partial H}{\partial \phi}\frac{\partial \eta}{\partial \phi}\right)\right]$$
$$- \frac{gf}{r}\left(\frac{\partial H}{\partial r}\frac{\partial \eta}{\partial \phi} - \frac{\partial H}{\partial \phi}\frac{\partial \eta}{\partial r}\right) = 0, \tag{3}$$

$$\frac{\partial^2 \eta}{\partial t \partial r} + \frac{f}{r}\frac{\partial \eta}{\partial \phi} = 0, \qquad r = r_1, r_2. \tag{4}$$

We will look for periodic and wave solutions of the form

$$\eta(r, \phi, t) = \text{Re } R(r)\, e^{i(k\phi + \sigma t)}, \tag{5}$$

where k—integer non-negative number. Substituting solution (5) into Eq. (3) and boundary condition (4), we obtain the eigenvalue problem for $R(r)$:

$$\left(R'' + \frac{1}{r}R' - \frac{k^2}{r^2}R\right) + \left(\frac{\partial \ln H}{\partial r} - \frac{f}{i\sigma r}\frac{\partial \ln H}{\partial \phi}\right)R' +$$
$$+ \left(\frac{\sigma^2 - f^2}{gH} + \frac{ik}{r^2}\frac{\partial \ln H}{\partial \phi} + \frac{fk}{\sigma r}\frac{\partial \ln H}{\partial r}\right)R = 0, \tag{6}$$

$$R' + \frac{kf}{\sigma r}R = 0, \qquad r = r_1, r_2. \tag{7}$$

Consider next the following tasks.

1. Let the Depth H—Constant. Then Eq. (6) Takes the Form

$$r^2 R'' + r R' + \left(\lambda^2 r^2 - k^2\right)R = 0, \tag{8}$$

where

$$\lambda^2 = \frac{\sigma^2 - f^2}{gH}. \tag{9}$$

Equation (8) is the Bessel equation, the general solution of which is

$$R(r) = A\, J_k(\lambda r) + B\, N_k(\lambda r),$$

where $J_k(\lambda r)$ и $N_k(\lambda r)$—the Bessel and Neumann functions respectively, A, B—arbitrary constants. The number λ is determined from the boundary condition (7) as the root of the system of two equations

$$\lambda r_1 \left[A\, J_k'(\lambda r_1) + B\, N_k'(\lambda r_1) \right] + \frac{kf}{\sigma} \left[A\, J_k(\lambda r_1) + B\, N_k(\lambda r_1) \right] = 0,$$
$$\lambda r_2 \left[A\, J_k'(\lambda r_2) + B\, N_k'(\lambda r_2) \right] + \frac{kf}{\sigma} \left[A\, J_k(\lambda r_2) + B\, N_k(\lambda r_2) \right] = 0. \tag{10}$$

System (10) is a system of two linear homogeneous equations for A and B. Nontrivial solutions for A and B exist only when the determinant of system (10) vanishes. After some transformations, we obtain a transcendental equation for the eigenvalues λ:

$$\lambda^2\, r_1\, r_2\, U_2(\lambda) + k\lambda \left(\frac{f}{\sigma} - 1 \right) [r_1\, U_1(\lambda) + r_2\, V_2(\lambda)] + k^2 \left(\frac{f}{\sigma} - 1 \right)^2 V_1(\lambda) = 0, \tag{11}$$

where

$$U_1(\lambda) = J_{k-1}(\lambda r_1)\, N_k(\lambda r_2) - J_k(\lambda r_2)\, N_{k-1}(\lambda r_1),$$

$$U_2(\lambda) = J_{k-1}(\lambda r_1)\, N_{k-1}(\lambda r_2) - J_{k-1}(\lambda r_2)\, N_{k-1}(\lambda r_1),$$

$$V_1(\lambda) = J_k(\lambda r_1)\, N_k(\lambda r_2) - J_k(\lambda r_2)\, N_k(\lambda r_1),$$

$$V_2(\lambda) = J_k(\lambda r_1)\, N_{k-1}(\lambda r_2) - J_{k-1}(\lambda r_2)\, N_k(\lambda r_1).$$

It follows from Eq. (9) $\sigma^2 = f^2 + gH\lambda^2$ or $\sigma^2 = \pm\sqrt{f^2 + gH\lambda^2}$.

We get the Poincaré waves. The presence of solutions σ of the same magnitude but different signs means that Poincaré waves can propagate in both positive and negative directions ϕ. Note that the frequency is always greater than f. Let's find the characteristics of motion. For each mode we have:

$$\eta = P(\lambda r)\, \cos(k\phi + \sigma t),$$

$$v_r = \frac{g}{f^2 - \sigma^2} \left[\sigma\lambda\, Q(\lambda r) + \frac{k(f - \sigma)}{r} P(\lambda r) \right] \sin(k\phi + \sigma t),$$

$$v_\phi = \frac{g}{f^2 - \sigma^2}\left[f\lambda\,Q(\lambda r) + \frac{k(\sigma - f)}{r}P(\lambda r)\right]\cos(k\phi + \sigma t),$$

where

$$P(\lambda r) = A\,J_k(\lambda r) + B\,N_k(\lambda r), \qquad Q(\lambda r) = A\,J_{k-1}(\lambda r) + B\,N_{k-1}(\lambda r).$$

Differential equations for determining the speed v_r, v_ϕ can be found in (Aleshkov 1990; Sretensky 1972).

2. Suppose that the depth of the basin gradually decreases according to the law

$$H(r) = D\left(1 - S\frac{(r - r_1)^2}{(r - r_{21})^2}\right),$$

where the slope S is much less than unity, D—constant. Equation (6) takes the form

$$\left(1 - S\frac{(r - r_1)^2}{(r - r_2)^2}\right)\left[R'' + \frac{1}{r}R' - \frac{k^2}{r^2}R\right] - 2S\frac{(r - r_1)}{(r - r_2)^2}\left(R' + \frac{kf}{\sigma r}R\right) + \frac{\sigma^2 - f^2}{gD}R = 0.$$

$$(12)$$

Assuming further

$$R(r) = \exp\left(\frac{S(r - r_1)^2}{2a^2}\right)M(r),$$

where $a = r_2 - r_1$, and given that the value $\frac{r-r_1}{a}$ does not exceed unity, for small S, that is, small changes in depth, problem (12), (7) can be written as

$$\left[M'' + \frac{1}{r}M' - \frac{k^2}{r^2}M\right] + \left[\frac{2S}{a^2} - \frac{2Skf}{\sigma a^2} + \frac{\sigma^2 - f^2}{gD}\right]M = 0, \qquad (13)$$

$$r_1 M'(r_1) + \frac{kf}{\sigma}M(r_1) = 0, \qquad r_2 M'(r_2) + \left(\frac{kf}{\sigma} + \frac{Sr_2}{r_2 - r_1}\right)M(r_2) = 0. \quad (14)$$

Note that we have neglected the slope S only in those terms where it is directly compared with quantities of the order of unity. We write Eq. (13) as

$$r^2 M'' + r\,M' + \left[\kappa^2 r^2 - k^2\right]M = 0.$$

Its solution, as in the case of a horizontal bottom, has the form

$$M = A\,J_k(\kappa r) + B\,N_k(\kappa r),$$

where

$$\kappa^2 = \frac{\sigma^2 - f^2}{gD} - \frac{2Skf}{\sigma a^2} + \frac{2S}{a^2}. \tag{15}$$

In the last expression, the right side can be less than zero. In this case

$$M = C\, I_k(\kappa r) + D\, K_k(\kappa r),$$

where $K_k(\kappa r)$—Macdonald function, $I_k(\kappa r)$—Infeld function, A, B, C, D—arbitrary constants. The use of boundary conditions (13) generates an equation for the eigenvalues:

$$\kappa^2 r_1 r_2 U_2(\kappa) + \kappa k \left(\frac{f}{\sigma} - 1\right) [r_1 U_1(\kappa) + r_2 V_2(\kappa)] + \kappa r_1 \frac{Sr_2}{a} U_1(\kappa) + $$
$$+ k^2 \left(\frac{f}{\sigma} - 1\right)^2 V_1(\kappa) + \frac{kSr_2}{a}\left(\frac{f}{\sigma} - 1\right) V_1(\kappa) = 0. \tag{16}$$

It is important to note that Eq. (16) differs from Eq. (11) (for the case of a flat bottom) by adding two terms with the bottom slope parameter S. There are two distinct classes of solutions to Eq. (15). The first class includes frequencies exceeding f. In this case, up to an accuracy of $O(S)$, we obtain the Poincaré modes, i.e.,

$$\sigma^2 = f^2 + gD\kappa^2 + O(S),$$

so the high-frequency Poincaré waves are essentially unaffected by the slight tilt of the bottom. An important new decision, ie. the third root of the cubic Eq. (15) has a frequency $\sigma = O(S)$, for which the term σ^2 is negligible, while the second term σ^2 has the order $O(1)$. This leads to the dispersion relation for the topographic Rossby wave, i.e.

$$\sigma = -\frac{2Sf}{a^2}\frac{k}{\kappa^2 - \frac{2S}{a^2} + \frac{f^2}{gD}}.$$

The frequency of the Rossby wave reaches its maximum value when

$$k = \sqrt{\kappa^2 - \frac{2S}{a^2} + \frac{f^2}{gD}},$$

where

$$\sigma = \sigma_{max} = -\frac{2S}{a^2}\frac{f}{\sqrt{\kappa^2 - \frac{2S}{a^2} + \frac{f^2}{gD}}},$$

So that for small frequency of the Rossby wave is always less than f.

Thus, the Rossby wave, for the existence of which it is necessary that and f and S are nonzero, is a oscillation, i.e. its low-frequency wave period is greater than the period of rotation of the coordinate system. The elevation and velocity for the Rossby wave are:

$$\eta = W(r)\, P(\kappa r)\, \cos(k\phi + \sigma t),$$

$$v_r = \frac{g\sigma W(r)}{f^2 - \sigma^2}\left[\kappa\, Q(\kappa r) + \left(\frac{S(r - r_1)}{a^2} + \frac{k(f - \sigma)}{r\sigma}\right)P(\kappa r)\right]\sin(k\phi + \sigma t),$$

$$v_\phi = \frac{g f W(r)}{f^2 - \sigma^2}\left[\kappa\, Q(\kappa r) + \left(\frac{S(r - r_1)}{a^2} + \frac{k(\sigma - f)}{r f}\right)P(\kappa r)\right]\cos(k\phi + \sigma t),$$

where $W(r) = \exp\!\big(S(r - r_1)^2/2a^2\big)$.

Arbitrary constants A, B are determined from conditions (14).

3. Let the basin depth be an arbitrary function depending only on r. Then problem (6), (7) will take the form:

$$\frac{1}{r}\frac{\partial}{\partial r}\left(r\tilde{R}'\right) + \left(\alpha(r) - \frac{\kappa^2}{r^2}\right)\tilde{R} = 0, \qquad r\tilde{R}' + \gamma\tilde{R} = 0, \quad r = r_1,\, r_2.$$

where

$\alpha(r) = \frac{\sigma^2 - f^2}{gH} + \frac{(2kf - \sigma)H'}{2\sigma r H} + \frac{H'^2}{2H^2} - \frac{H''}{2H}, \quad \gamma(r) = \frac{kf}{\sigma} - \frac{rH'}{2H}, \quad r = r_1,\, r_2, \quad R = \tilde{R}/\sqrt{H}$. If the $\alpha(r)$ is equal constant, we will have the problems considered in paragraphs 1 and 2. Let we have

$$\alpha(r) = \begin{cases} \kappa_1^2, & r_1 \le r \le r_3, \\ \kappa_2^2, & r_3 \le r \le r_2. \end{cases}$$

Then the general solutions for functions R in each layer will be as follows: $R_1 = A J_k(\kappa_1 r) + B N_k(\kappa_1 r)$, $r_1 \le r \le r_3$, $\qquad R_2 = C J_k(\kappa_2 r) + D N_k(\kappa_2 r)$, $r_3 \le r \le r_2$., $r_3 \le r \le r_2$.

At $r = r_3$ the solution must be continuously differentiable. Two pairs of arbitrary constants must satisfy four homogeneous conditions, from which follows the dispersion relation

$$\begin{vmatrix} S_1(J) & S_1(N) & 0 & 0 \\ J_k(\kappa_1 r_1) & N_k(\kappa_1 r_3) & -J_k(\kappa_2 r_3) & -N_k(\kappa_2 r_3) \\ T_1(J) & T_1(N) & -T_2(J) & -T_2(N) \\ 0 & 0 & S_2(J) & S_2(N) \end{vmatrix} = 0,$$

where

$$S_j(J) = \kappa_j r_j J_{k-1}(\kappa_j r_j) + \left(\gamma_j - k\right) J_k(\kappa_j r_j),$$

$$T_j(J) = \kappa_j J_{k-1}(\kappa_j r_3) - \frac{k}{r_3} J_k(\kappa_j r_3), \quad j = 1, 2,$$

γ_1, γ_2—values $\gamma(r)$ at r_1 and r_2 respectively.

The boundary value problem for $W(r) = \exp\left(S(r - r_1)^2/2a^2\right)$ can be reduced to an integral equation, particular cases of which can be solved analytically and numerically (Kholodova 2009; Peregudin and Kholodova 2011; Peregudin et al. 2021; Peregudin et al. 2019; Kholodova 2008; Shamin 2019)

$$\tilde{R}(r) = \int\limits_{r_1}^{r_2} G(r, \zeta) \, \alpha(\zeta) \, \tilde{R}(\zeta) \, d\zeta,$$

where

$$G(r, \zeta) = \begin{cases} \left(r^k + r_1^{2k} \frac{k+\gamma_1}{k-\gamma_1} r^{-k}\right) \left(\zeta^k + r_2^{2k} \frac{k+\gamma_2}{k-\gamma_2} \zeta^{-k}\right), & r \leq \zeta, \\ \left(r^k + r_2^{2k} \frac{k+\gamma_2}{k-\gamma_2} r^{-k}\right) \left(\zeta^k + r_1^{2k} \frac{k+\gamma_1}{k-\gamma_1} \zeta^{-k}\right), & r \geq \zeta. \end{cases}$$

References

Aleshkov YuZ (1990) Theory of interaction of waves with barriers. L.: Publishing house of Leningrad State University, 327 p

Kholodova SE (2008) Dynamics of a rotating layer of an ideal electrically conductive incompressible fluid. Comput Math Mathem Phys 48(5):882–898

Kholodova SE (2009) Wave motions in a stratified electrically conductive rotating fluid. J Comput Math Math Phys 49(5):916–922

Peregudin SI, Kholodova SE (2011) On the features of the propagation of non-stationary waves in a rotating spherical layer of an ideal incompressible stratified electrically conductive fluid in the equatorial latitudinal belt. Appl Mech Techn Phys 52(2):(306):44–51

Peregudin S, Peregudina E, Kholodova S (2021) Mathematical modeling of the dynamics of a rotating layer of an electrically conducting fluid with magnetic field diffusion effects. Springer Geology, pp 277–87

Peregudin S, Peregudina E, Kholodova S (2019) The influence of dissipative effects on ynamic processes in a rotating electrically conductive liquid medium. J Phys Conf Series 1359(1):012118

Shamin RV (2019) Machine learning in economic problems. M: Green Print, 140 p

Sretensky LN (1972) Transition of long waves from one depth to another in a rotating basin. Continuum mechanics and related problems of analysis. M.: Science, pp 473–494

The Problem of Propagation of Plane Long Waves in a Nonhomogeneous Liquid over a Deformable Bottom

S. I. Peregudin and S. E. Kholodova

Abstract The problem of the impact of waves on the bottom, which can be deformed and changed, is considered. This task is an urgent problem of military-technical construction at the design stage in order to protect the supports of hydraulic structures from the possible force impact of the water element. Along with the theoretical study, an important applied aspect is the result of numerical simulation with possible further implementation through modern integrated software development environments.

Keywords Waves of finite amplitude · Force impact · Offshore hydrotechnical construction · Nonlinear wave dynamics · Support hydraulic structures

The bottom of natural water areas is not always solid and undeformable. If it is a mixture consisting of sand, clay, silt or gravel, the bottom is a complex three-dimensional surface on which wave-like sediments are formed under the influence of a liquid flow. They can be observed on the shallows of rivers after the water drains, in deserts, dunes and dunes are formed under the influence of air currents.

Studies of the phenomenon of periodicity of deformations of the interface of easily deformable media originate in Helmholtz studies (Shulyak 1971). The hypothesis formulated initially only for media with infinitesimal friction was later generalized for the case when the amount of friction in one of the contiguous media is finite. Periodic deformations occurring on the surface of the sections of media with infinitesimal friction—waves on the liquid-atmosphere interface and internal waves in the liquid at the top of the density jump are similar to periodic deformations on the interface of two media, in one of which the friction value is small, and in the other finite— observed in nature sand waves arising at the interface of the bulk medium with the atmosphere or hydrosphere.

S. I. Peregudin (✉)
St. Petersburg State University, Saint Petersburg, Russia
e-mail: s.peregudin@spbu.ru

S. E. Kholodova
ITMO University, Saint Petersburg, Russia

© The Author(s), under exclusive license to Springer Nature Singapore Pte Ltd. 2023   179
T. Chaplina (ed.), *Processes in GeoMedia—Volume VII*, Springer Geology,
https://doi.org/10.1007/978-981-99-6575-5_17

Despite the apparent external difference in the forms of both types of waves, the nature of the causes of their occurrence has common patterns determined by the similarity of the conditions of formation and the identity of the forces acting on the surface of the deformable medium. The lesser degree of knowledge of sand waves, in contrast to waves on water, is due to the lack of a complete mathematical apparatus for theoretical developments and an insignificant number of experiments in this field.

The first experiments of this phenomenon were carried out by Dikon (Velikanov 1949), he obtained an approximate relationship between the flow velocities and the movement of the crest of sand waves, introduced the so-called critical fluid flow velocities, which undoubtedly depend on the size of the particles of the bottom surface, at which sand waves arise and disappear. The first theoretical results belong to F. Exner (1920) (Velikanov 1949; Aleshkov 2001; Peregudin 2003), who, based on his approximate theory, as well as from his experiments in a tray with a water flow and in a wind tunnel, basically correctly described the mechanical side of the phenomenon.

Moreover, for the flat case, he derived an equation linking the flow rate $Q(x, t)$ of this substance with the shape of the surface of the separation of the liquid and bottom layers $\eta(x, t)$.

This flow rate is characterized by the rheology of the soil. To overcome this difficulty F. Exner accepts the hypothesis of a linear dependence of the flow rate on a given velocity u_b, i.e. $Q = \kappa u_b$. From the assumption of the constancy of the flow rate in the water layer, the equality of the horizontal components of the bottom and water velocities follows. Velikanov (1949) generalizes the hypothesis of F. Exner, assuming an arbitrary dependence of the flow rate Q on a given speed, i.e. $Q = Q(u_b(x, t))$.

The scientist's research was continued by F. I. Frankl, who considered the problem of the flat motion of sand waves with more complete consideration of the hydrodynamics of the water layer (Frankl 1953). In the article by Aleshkov (2001), a general case is considered—the non-potential motion of a layer of inhomogeneous fluid over a loose medium. In Aleshkov (1990), Ilyichev et al. (2011), the issues of propagation of long waves over a solid bottom are studied. In this article, a two-dimensional problem of the propagation of long waves in a two-layer liquid over a deformable bottom is investigated.

Let's consider a three—layer medium in the vertical plane—two layers of a homogeneous incompressible ideal liquid, a layer of soil. The liquid is bounded from above by a free surface $z = H_2 + \zeta(x, t)$, bottom—the surface of the section bottom layer-soil $z = -H_0 + \eta(x, t)$. The horizontal axis coincides with the undisturbed interface of liquid layers $z = \eta_1(x, t)$, the axis z is directed vertically upwards.

As a result of the movement of the liquid, the bottom deforms and changes, its surface deviates from the horizontal level $z = -H_0$. If we consider the fluid motion to be potential, the boundary value problem describing this dynamic process in dimensionless quantities will have the form

$$\mu \phi_{jxx} + \phi_{jzz} = 0, \qquad \mu = \left(\frac{H_*}{L}\right)^2,$$

$$\mu(\zeta_t + \zeta_x \phi_{2x}) = \phi_{2z}, \; \mu\left(\phi_{2t} + \frac{1}{2}\phi_{2x}^2 + \zeta\right) + \frac{1}{2}\phi_{2z}^2 = \mu F_2(t), z = H_2 + \zeta,$$

$$\mu(\eta_{1t} + \eta_{1x}\phi_{jx}) = \phi_{jz}, \; g\eta_1(\rho_1 - \rho_2) + \rho_1\left[\phi_{1t} + \frac{1}{2}\left(\phi_{1x}^2 + \frac{1}{\mu}\phi_{1z}^2\right)\right]$$

$$- \rho_2\left[\phi_{2t} + \frac{1}{2}\left(\phi_{2x}^2 + \frac{1}{\mu}\phi_{2z}^2\right)\right] = \rho_1 F_1(t) - \rho_2 F_2(t), z = \eta_1(x, t),$$

$$\mu(\eta_t + \eta_x\phi_{1x}) = \phi_{1z}, \quad \eta_t + Q_x = 0, \quad z = -H_0 + \eta(x, t).$$

Here $\phi_j(x, z, t)$, $F_j(t)$, H_*, L —the velocity potential of the fluid in each layer, an arbitrary function of time, vertical and horizontal scales, respectively.

The formulated boundary value problem as a result of the assumption of a linear dependence of the flow rate of a given substance on the bottom velocity

$$Q(x, t) = \kappa u_{1b}(x, t), \qquad u_{1b}(x, t) = \phi_{1x}|_{z=-H_0+\eta(x,t)}$$

and transformations corresponding to the transformations used in the theory of long waves can be reduced to a system of five partial differential equations with respect to unknown functions $\alpha(x, t)$, $\beta(x, t)$, $\eta(x, t)$, $\eta_1(x, t)$, $\zeta(x, t)$— horizontal components of the velocity potential in each layer and deformable surfaces.

The resulting system of equations will contain a small parameter μ characterizing the dispersion of waves. If a small amplitude parameter ε is introduced into consideration, the system presented above will include special cases: linear models without variance and with variance, a nonlinear model without variance (Ilyichev 2003; Peregudin et al. 2019, 2021; Kholodova 2008).

References

Aleshkov YuZ (1990) Theory of interaction of waves with obstacles. L, 372 p

Aleshkov YuZ (2001) Waves on the surface of loose media caused by fluid flow. Vestn. St. Petersburg. university Ser. 1: Mathematics, mechanics, astronomy Issue. 4 (No. 25), pp 35–43

Frankl FI (1953) On the motion of sand waves. DAN SSSR 89(1):9–32

Ilyichev AT (2003) Solitary waves in hydromechanics models. M, 256 p 8; Peregudin SI, Kholodova SE (2011) On the features of the propagation of non-stationary waves in a rotating spherical layer of an ideal incompressible stratified electrically conductive fluid in the equatorial latitudinal belt. ApplMech Techn Phys 52(2):(306):44–51

Kholodova SE (2008) Dynamics of a rotating layer of an ideal electrically conductive incompressible fluid. Comput Math Mathem Phys 48(5):882–898

Peregudin SI (2003) Spatial wave motions on the surface of loose media. Proceedings of Srednevolzhsk. mat. about-va. V 5, No 1, pp 130–138

Peregudin S, Peregudina E, Kholodova S (2019) The Influence of dissipative effects on dynamic processes in a rotating electrically conductive liquid medium. J Phys Conf Series 1359(1):012118

Peregudin S, Peregudina E, Kholodova S (2021) Mathematical modeling of the dynamics of a rotating layer of an electrically conducting fluid with magnetic field diffusion effects. Springer Geology, pp 277–87

Shulyak BA (1971) Physics of waves on the surface of a loose medium and liquid, 400 p

Velikanov MA (1949) Dynamics of channel flows, 474 p

Technogenic Gold-Bearing Mineralization of Eastern Orenburg Region

I. V. Kudelina, T. V. Leontiena, and M. V. Fatyunina

Abstract Technogenic deposits have a high resource potential. In the mining areas of gold and other precious metals, dump complexes have been formed that contain a conditioned amount of metal and can be used for re-mining. According to the results of surface geochemical survey in the southeastern part of the Aydyrlinsky area, a technogenic gold deposit was revealed, represented by the ephels of the former Aydyrlinsky gold extraction factory. The purpose of this study is to characterize the features of the technogenic deposit and determine the prospects for its development. According to the granulometric composition, silty and sandy types of ephels are distinguished. The gold content in siltstone ephels is almost 1.5 times higher than in the sandy fraction (the average content is 1.28 g/t and 0.89 g/t, respectively). The obtained indicators of gold extraction indicate a fairly good enrichment of ephels. The Aydyrli technogenic gold deposit is very promising for industrial development in the coming years.

Keywords Technogenic deposit of gold · Eiffel · Halos of gold · Silver · Arsenic · Tungsten and bismuth

1 Introduction

Technogenic deposits have a sufficiently high resource potential (Alekseev 2019). In the areas of gold and other noble mining, dump complexes were formed—ephel, pebble dumps, peat, etc. Gold content of dump complexes according to modern economic indicators is comparable to the metal content in natural placers (Litvintsev 2017). Technogenic deposits can be used for re-mining. To do this, it is necessary to conduct research aimed at studying the properties of dump complexes of technogenic deposits, identifying the causes of gold losses during primary mining, as well as choosing a technology for re-extraction (Mirzekhanov 2018; Litvintsev 2013).

I. V. Kudelina (✉) · T. V. Leontiena · M. V. Fatyunina
Orenburg State University, Orenburg, Russia
e-mail: kudelina.inna@mail.ru

© The Author(s), under exclusive license to Springer Nature Singapore Pte Ltd. 2023
T. Chaplina (ed.), *Processes in GeoMedia—Volume VII*, Springer Geology,
https://doi.org/10.1007/978-981-99-6575-5_18

The Aydyrlinsky technogenic gold deposit is located in the Kvarkensky district of the Orenburg region (Kazakov 2009). In the south-eastern part of the gold field there are ruins of the former village of Aidyrlinsky (Aidyrlya mine) and a gold extraction factory (GEF).

Numerous old mining and exploration and mining-exploitation workings (pits, ditches, mines) in the form of unpaved trenches, sinkholes and deep (up to 10–15 m) craters have been preserved on the square. In the logs and valleys, numerous dumps, small quarries and burrows of former prospecting workings of small spoon placers of gold have been preserved.

The beginning of operation of the Aidyrlinsky ore deposit dates back to 1897, when prospectors worked out the richest upper parts of the veins ("verkhoviki") to a depth of 15–25 m, rarely 46 m– In the pre–revolutionary period from 1912 to 1917, the largest veins were worked by mines: up to a depth of 130 vertically, there was an Oblique vein, up to a depth of 148 m on the western flank, there was a Miass vein, up to 64 m there was a Safonovskaya vein.

The restoration of the mines after the revolution began in 1933 and was worked out by prospecting teams until 1942, and on the 2 largest veins—Evasive and Miass mining was carried out by the Chkalovzoloto combine. All the extracted ore was carried to the Aydyrli runner factory, where it was crushed and washed. In 1942, due to lack of fuel, materials and human resources, the mine was closed. In the post-war period, mining operations at the field were practically not carried out.

Within the Aidyrli site, to the west of the Aidyrli technogenic deposit, the partially developed Aidyrli ore gold deposit has been developed, which is represented by a series of gold quartz veins located in the granodiorite massif in the southern endo-contact. The largest veins (Oblionnaya, Miasskaya, Safonovskaya) contained up to 75% of the total gold reserves explored at the deposit during its mining (Kazakov 2009; Gray 2018).

Thus, at the moment, the Aydyrlinskoye technogenic deposit is an under-studied and possibly promising object of the gold mining industry.

2 Results and Discussions

The Aidyrlinsky technogenic gold deposit is located in the southeastern part of the Aidyrlinsky area on the development area of the porphyrite (Aidyrlinsky) strata (D_{2zv2}).

The thickness is represented by greenish-gray lavas of almond-stone uralite basalts, andesite-basalts and andesites with interlayers of tuffs of the main composition. The porphyrite thickness is 500–800 m (Kazakov 2009).

The Aydyrli technogenic gold deposit was artificially formed during a long period of mining of the ore gold deposit of the same name. Therefore, the mineral composition of the ores of the technogenic deposit is directly related to the composition of the processed ores of the indigenous deposit. Ore bodies are 70% quartz. In addition to vein quartz, beresitized granodiorites are present in the ores: pyrite, galena and sphalerite are the most common ore minerals, arsenopyrite, chalcopyrite and pale ores are less common. The content of sulfides in ores does not exceed 2%. Gold is mostly free, often associated with sulfides (Alekseev 2019).

According to the results of the 1:10000 scale surface lithogeochemical survey, high-contrast halos of gold, silver, arsenic, tungsten and bismuth were detected, forming a large anomalous zone. During the route survey of the anomaly on the ground and its subsequent certification by pits, previously unknown gold-bearing formations were revealed, represented by the ephels of ore processing of the former Aidyrli gold extraction factory in the periods 1897–1917 and 1936–1942. Together with the previously known tailings dump, they form a technogenic gold deposit.

During the processing of gold-bearing ore, coarse-grained gravity-enriched material was stored in small dumps near the factory on elevated terrain areas, and a finer fraction of crushing (so-called "flushing") was concentrated in lower terrain areas, in sedimentation ponds, where the richest ores of the technogenic deposit were deposited.

According to the granulometric composition of the Eiffel, the deposits are represented by two types:

Silty eiffel. Eifels of this type, as a rule, are clearly layered, yellowish-orange, light gray and grayish-yellow with layers of fine quartz sands. The layering is smooth, wavy, sometimes with turbulence such as "turbid flows" and is emphasized mainly by the color scheme of the deposits and, to a lesser extent, by the granulometric composition. The thickness of the layers is from the first mm to the first discovered, rarely up to 10–20 cm. This type of ephel was previously unknown and was discovered by GeoUral-Resurs LLC in 2008 when checking the geochemical anomaly of gold identified by the results of geochemical survey of the site by secondary scattering halos (Kazakov 2009).

The siltstone fraction ephels are confined to the most lowered parts of the terrain of the site—hollows and gullies, blocked by artificial dams. In the spring-summer period, the places of distribution of this type of eiffel are flooded, and in some areas (near dams) were completely under water. For this reason, they have apparently not been previously discovered. It was only because of the very hot and arid summers of the last few years that the moistened areas and artificial reservoirs dried up, and therefore the ephelas were exposed and became available for study.

The sandy type of ephels is represented by finely ground (up to 0.63 mm) yellowish-light gray sands. The sands are composed mainly of quartz grains with a small amount of pulverized clay particles. These deposits form an artificially created low-power (up to 1.0 m) cloak-like deposit on the gentle slope of the relief depressions in the eastern part of the work site. In addition, small dumps in the southern part of the site and a relatively large dump in the southwestern part of the site are stacked with sandy ephels.

For the first time, the evaluation of the sandy ephels of the former Aidyrli GEF was carried out in 1991 by the Party of mass searches of the Zelenogorsk expedition of the GGP "Zelenogorskgeologiya" (Kazakov 2009). The volumes and reserves of ore were determined by an eye survey. In the eastern part of the dump, 6 samples were taken, in which the average gold content of 1.18 g/t was determined by assay analysis. The disadvantage of the work performed was the lack of a specific binding of sampling points.

In 1992, the assessment of the ash dumps was carried out by the Eastern GRE PGO "Orenburggeologiya" (Kazakov 2009). Testing of the efel blade was carried out on 3 profiles with a distance between them of 100 m, between points in the profile—50 m. Samples were taken from the pits 0.5 m deep, the weight of the samples was about 20 kg. The samples were reduced by quartering to 2 carvings of 300 g each—the main and control, which were subjected to assay analysis. The remaining part of the sample was washed. The results of washing gave negative results.

The gold content according to the analysis ranged from 0.2 to 1.4 g/t and averaged 0.3 g/t. The area of the efel dump was determined approximately and amounted to about 58,000 m^3. With a volume mass of 2.0 t/m^3 of sand, gold reserves amounted to 34.8 kg. According to the results of the evaluation, the author concluded that since the gold in the ephels is finely dispersed and it will have to be extracted by chemical means, gold mining is impractical.

The disadvantage of the work was that the samples were taken from the upper part of the dump. Taking into account the fact that gold can migrate ("sink") to the lower horizons under the influence of natural (water) and gravitational factors during a long stay in such loose media as ephels, the resulting low concentrations do not characterize its true content. A negative fact is the purely conditional distribution area and the power of the ephels, which was confirmed by subsequent work.

The following assessment of ephels in the dump, their material composition and the degree of gold content was carried out in 1994 by the Eastern GRE. To do this, small-scale core wells were drilled over the entire area of the efel dump over a network of 60 × 60 m, the true capacity of the efels was determined, samples were taken from the core of the wells. According to the drilling results, the capacity of the tunnels in the dump varied from 0.5 to 5.2 m and, on average, amounted to 2.35 m. According to the assay data, the gold content in the samples ranged from 0.12 to 1.69 g/t, the average was 0.50 g/t.

In 2007–2008, during the prospecting works of Geo Ural Resource LLC, previously unknown gold-bearing formations were identified, represented by the ephels of ore gold processing of the former Aidyrli gold extraction factory. Together with the previously known dump of processing tailings, they form a technogenic gold deposit. The deposit was explored in 2007–2011 by pits and screw wells on a network of 80–100 × 40–50 m with thickenings of 50–60 × 20–40 m. Based on the materials of exploration and previous work, stocks of ephels with an average content of 1.36 g/t and stocks of ephels with an average content of 0.68 g/t were determined.

The gold content in siltstone ephels is almost 1.5 times higher than in the sandy fraction (the average content is 1.28 g/t and 0.89 g/t, respectively). The obtained indicators of gold extraction indicate a fairly good enrichment of ephels.

When developing technogenic placer deposits, it is necessary to take into account the causes of metal losses during primary development (Alekseev 2019). Most often, these are insignificant sizes and shape features of the gold pieces, the formation of hydrophobic films on them. During gravitational enrichment, the gold in oxidized forms is poorly captured and washed off into the tailings. For a long time, the presence of such gold in dumps leads to physical and chemical transformation, which is expressed in a decrease in hydraulic size and unsuitability for gravity enrichment.

Innovative technologies have been developed and tested to extract placer gold. These are leaching technologies (Alekseev 2019), jigging machines, centrifugal concentrators, screw separators, which allow to increase the extraction of precious metals with a fineness class of less than 0.2 mm. In Russia and abroad, the bulk of precious metals are extracted on flushing devices equipped with sluices (Gray 2018). Sluice installations are easily upgraded for specific mining and geological conditions and are capable of extracting fine and fine gold.

When concentrates are enriched in sludge concentration plants (SCP), it is possible to use their treatment with a halogen-containing reagent. This adds hydrophilic properties to thin-plate, porous gold, as well as gold with collected films and coatings, which does not contain metal-washed metal in the tailings (Alekseev 2018). The use of surfactants (surfactants) in conditions of gravity enrichment (gravity table, tray) reduces the effect of surface stress of water, and gold particles with a high coefficient of flattening settle into the concentrate (Alekseev 2021; Naumov 2010).

3 Conclusion

The study of the characteristics of the Aydyrli deposit of technogenic gold makes it possible to determine the prospects for its production. This requires research in the following areas:

- complete and detailed testing and study of technogenic formations with determination of gold contents to assess the possibility of their extraction;
- identification of the causes of metal losses during the initial development of the deposit;
- the use of innovative technologies for the re-extraction of gold, taking into account certain morphotypes and sizes of gold pieces.

References

Alekseev VS (2019) Increasing the extraction of fine gold on a sluice-type flushing device. Alekseev VS, Gray RS, Sobolev AA (eds). Enrichment of ores. № 5, pp 13–18. https://doi.org/10.17580/or.2019.05.03

Alekseev VS (2021) Peculiarities of technogenic gold from dump complexes of the Solovyovskiy gold-bearing cluster. Alekseev VS, Banshchikova TS, Alekseeva EV (eds). Mining Inf Analyt Bull 7:134–145

Alekseev VS (2018) Application of non-traditional technologies for the extraction of gold from technogenic formations of placers in the Amur region. Alekseev VS, Banshchikova TS (eds). Mining J 10:52–57. https://doi.org/10.17580/gzh.2018.10.10

Alekseev YaV (2019) Quantitative aspects of the development of SME gold in the Russian Federation. Alekseev YaV, Konkina OM, Pivovarova TA (eds) Collection of theses of reports IX Int. scientific and practical. conf. "Scientific and methodological foundations of forecasting, prospecting, evaluation of deposits of diamonds, precious and non-ferrous metals." Moscow: TsNIGRI, S.64

Gray RS (2018) Reducing the loss of precious metals on flushing devices by optimizing the operation of fine filling locks. Gray RS, Taganov VV, Gevalo KV (eds). Mining J 10:49–52. https://doi.org/10.17580/gzh.2018.10.09

Kazakov SV (2009) Report on the object "Prospecting for ore gold within the Aidyrlinsko-Sineshikhansky ore district (Orenburg region) in 2006–2009. Kazakov SV et al. (eds) Orenburg branch of the FBU "TFGI for the Volga Federal District". Orenburg. 2009 Inv. No 9909, 567 p

Litvintsev VS (2013) The main directions of the strategy for the development of technogenic ore and placer deposits of noble metals. Litvintsev VS (eds). Mining Magaz 10:38–41

Litvintsev VS (2017) Resource potential of technogenic gold-placer deposits and strategy of their large-scale development. Litvintsev VS, Alekseev VS, Stolen IA (eds). Mine surveying and subsoil use. № 5, pp 21–29

Mirzekhanov GS (2018) Status and problems of development of technogenic placer deposits of noble metals in the Far East region. Mirzekhanov GS, Litvintsev VS (eds). Mining magazine. № 10. S 25–30. https://doi.org/10.17580/gzh.2018.10.04

Naumov VA (2010) The concept of managing the formation of deposits on the example of technogenic placers of gold. Naumov VA (eds). Nat Techn Sci 2:262–265

Morpholitosystem of Piltun Lagoon (Sakhalin Region)

Victor V. Afanas'ev, A. V. Uba, A. I. Levickij, and A. B. Faustova

Abstract The results of the analysis of geospatial and geological-geomorphological information on the structure and dynamics of the morpholithic system of the Piltun lagoon are presented. Based on certain parameters of abrasion-accumulative processes of the barrier form, internal banks and the system of the lagoon strait for the period 1952–2014. a conceptual morpholithodynamic model has been constructed, which will form the basis for the analysis of the processes of organogenic lagoonal sedimentogenesis. At this stage, the questions of the structure and dynamics of intertidal mudflat and salt marches are considered.

Keywords Coast dynamics · Inlet migration · Spatial structure of the erosion · Salt marsh · Intertidal mudflat · Carbon sequestration

1 Introduction

Piltun Bay is the largest of all the lagoons on the northeastern coast of Sakhalin Island. The length of Piltun Bay is 57 km, and including Astokh Bay it is 71 km. The maximum width of the bay is 12 km; the mirror area for 2020 was 438.9 km^2, which is approximately 3.6 times larger than the area of the Chaivo lagoon with which they were associated in the first phase of the Middle Holocene coastal-marine accumulation. The average depth of the bay is from 1 to 2 m. The thickness of the exposed organogenic silts in the areas of runoff concentration reaches 6 m. The hydrological regime of the bay is determined by the influence of sea waters coming from the Sea of Okhotsk and river runoff. The average monthly inflow of river waters during the low-water period is 40–50 million m^3, however, during spring floods and autumn floods during cyclones, the figure is significantly higher. The most intensive

V. V. Afanas'ev (✉) · A. V. Uba · A. I. Levickij
Institute of Marine Geology and Geophysics, FEB RAS, Yuzhno-Sakhalinsk, Russia
e-mail: vvasand@mail.ru

A. B. Faustova
Sakhalin State University, Yuzhno-Sakhalinsk, Russia

water exchange with the sea through a channel about 12 km long and up to 19 m deep in the mouth part is observed during the low-water period during spring tides and storm surges reaching a height of 0.5–1.0 m.

The intricately constructed coastal barrier separating Piltun Bay from the Sea of Okhotsk consists of areas with high coastal ledges composed of Neogene-Pleistocene sediments (mainly unconsolidated, sandy-pelitic fractions) that alternate with sandy-pebble Holocene barrens proper. There are no areas with high levels in the zone of Late Holocene migrations of the straits. They were destroyed here in the first phases of coastal-marine accumulation. An analysis of the features of the morpholithodynamics of complexly built and "classical" sandy bays separating lagoons from the open sea showed that for complexly built barrier forms, a higher variability of deformations of the sea edge of the Holocene areas is characteristic than for completely Holocene bays. Such barrier forms are also characterized by a greater variety of morphometric parameters of the upper part of the coastal profile (Afanas'ev 2020).

The height of the Holocene level above the river line varies from 3.5 to 8–9 m. The level of Upper Miocene sediments of the Okobykai formation (N13ok)) is 11–25 m.

The thickness of lagoonal aleurite-pelitic and fine-grained sandy deposits with a high content of organic matter in the section of the coastal ledge of the inner shores of the lagoon is 2–4 m. The exposed thickness of Holocene lagoonal silts is 18 m.

2   Methods and Results

The work is based on the analysis of arrays of aerial photographs of 1952 and satellite images of 2013–2020, which was performed in the Quantum GIS geoinformation system. The calculations were made on the WGS84 EPSG:7030 ellipse. The resulting attributes were exported to spreadsheets for further processing. The average error in the location of objects on this map is about 5 m, the minimum is 0.2 m, the maximum is 18 m, the standard deviation is about 3 m. changes in contours and areas for the period 1952–2020 The published data and the authors' own results were used in calculating the carbon stocks in the sediments of marshes, silty desiccations, and bottom sediments.

Based on the processing in the GIS environment of aerial photographic information (flights of 1952, 1974) and space photographic information (2014 and 2015), followed by an analysis of the plots of the spectral density of the erosion intensity, the values of the average annual rates of coastal erosion over a period of 60 years were determined and the spatial structure of the erosion of a complexly built coastal barrier form of the Piltun bay (Figs. 1 and 3).

Areas with high coastal ledges composed of Pleistocene sediments alternate here with Holocene bay bars proper. Average long-term erosion rates never exceed 2 m/year.

The maximum erosion rates are noted in areas that represent high (14–28 m) Pleistocene terrace levels. The same areas are characterized by the maximum volumes

Fig. 1 Abrasive-accumulative processes on the sea side of the Piltun bay barrier form (averaging over 500-m intervals)

of sediments that annually enter the coastal zone during the erosion of the coastal ledge with a length of one meter (5.6–7.7 m³).

The average long-term volume of erosion of the barrier form is 211.75 ·10³ m³/year, and the accumulation is 117.73 ·10³ m³/year.

The main patterns of the spatial structure of the destruction of the coasts of north-eastern Sakhalin determine the phenomena of rhythm in the manifestation of erosion processes. Erosion rhythms with wavelengths of 1.3, 2.0, 4.0, 5.6, 8.5, 11.6, 25.6, and 64 km have been established (Afanasiev 2018). The high-order rhythmicity of erosion processes (in our case, ≈30–60 km) is associated with stable zones of divergence of sediment flows, characterized by a reduced volume of beach sediments. The erosion rhythm of 1.3 km corresponds to the dimension of large megafestoon beach structures.

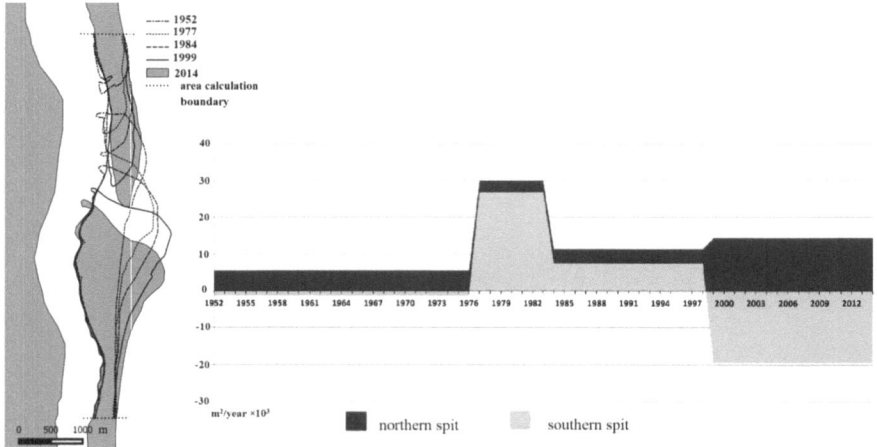

Fig. 2 Changing the contour of the coastline in the area of the strait connecting the Sea of Okhotsk and Piltun Lagoon in plan and area in calculated boundaries by years for the period 1952–2014

It has been established that the greatest morpholithodynamic changes in the coastal zone of northeastern Sakhalin occur precisely in the strait systems (Afanasiev 2019). The maximum volume of sediments in the active layer of the coastal profile is also observed in the areas of the straits. The dynamics of the strait of the Piltun lagoon, along with the shift to the south, is characterized by a change in the shape of the distal ends of the streamers (Fig. 2).

Therefore, despite the displacement of the strait of the Piltun Lagoon in a southerly direction in 1977–1999 by about 870 m, there was an increase in the area of both the growing northern spit and the receding southern one.

The situation has changed in the last twenty years, when the area of the receding southern spit began to decrease by about 2 hectares annually, however, on average, over the period 1952–2014, sediments in the mouth of the lagoon accumulated annually in the amount of $121.48 \cdot 10^3$ m^3/year.

An analysis of sedimentation processes in the mouth zone of Piltun Bay gives a figure of 196,000 m^3/year of sediment moving south in the strait system, of which $74.52 \cdot 10^3$ m^3/yearexits the morpholithosystem in a southerly direction.

Figure 3 and Table 1 present the results of measuring changes in the area of the Piltun Lagoon marshes. The area of the marches of the lagoon for 1952–2020 increased by an average of 29,471 m^2/year.

This value roughly corresponds to the growth rate of the marshes of the Chaivo lagoon, with a significantly smaller area in which 28,765 m^2 of marshes grow annually. And in general, the rate of rise of marches in the Piltun lagoon, when converted to the area of the water surface, is lower than in all the studied lagoons of the island. Sakhalin (Afanas'ev and Faustova 2022). Significantly lower in terms of the area of the lagoon and the volume of solid runoff of rivers flowing into the Piltun lagoon (Bobrik and Brovko 1986; Gosudarstvennyy vodnyy kadastr 1987, 1979). Given the

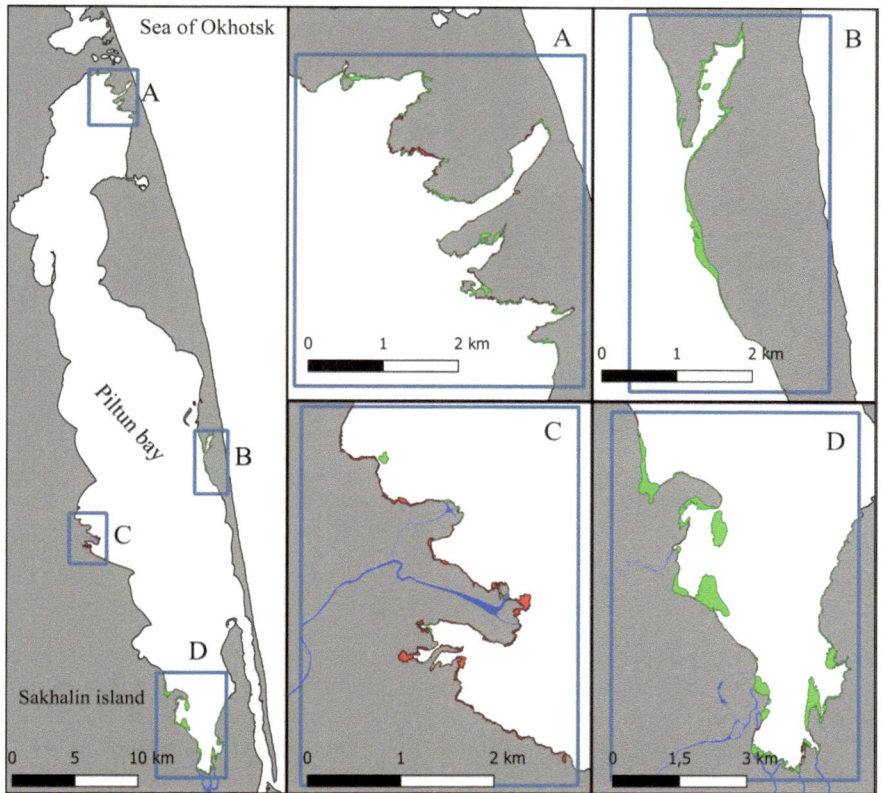

Fig. 3 Dynamics of the inner shores of the Piltun lagoon

Table 1 Results of measuring changes in the area of marches for the polygons shown in Fig. 3

Piltun Lagoon sections km²	Formed	Lost	Total	Average km²/year
A (1978–2020)	0,143,987,421	0,045,382,217	0,099	0,002,415
B (1977–2020)	0,241,486,781	0,001,459,164	0,240	0,005,714
C (1977–2020)	0,019,280,299	0,124,059,315	−0,105	−0, 002,500
D (1952–2020)	1,63,443,879	0,025,696,558	1,609	0,024,015
				0,029,471

high bioproductivity of the lagoon, such low marsh formation rates require serious study (Fig. 4). It is likely that with such a small solid runoff, the dying organic matter does not have time to be buried and is carried out by floods and floods.

Fig. 4 Formation of coastal vegetation cover from the dying zostera on the muddy drainage of the Piltun Lagoon Strait

3 Conclusion

Increasing the carbon sink capacity of estuarine-lagoonal geosystems, which can be used to mitigate the effects of climate change, is impossible without understanding the conditions of organogenic sedimentation. Therefore, along with the assessment of the parameters of organogenic sedimentation of marshes and silty lands, it is very important to understand the mechanisms of organogenic sedimentogenesis under various morpholithodynamic and climatic conditions (Rosentreter et al. 2021; Curado et al. 2014). For example, if rainfall cannot form a surface with the flood-free duration required to support budding vegetation, littoral mudflat development does not progress to a march stage even with adequate nutrient inputs, etc.

With regard to concerns about the adaptation of estuarine-lagoonal formations to climate change, it should be noted that the rate of observed sea level rise is still not a critical threat to tidal estuarine and coastal wetlands (Kirwan et al. 2016; Leontiev and Afanasiev 2016).

References

Afanas'ev VV (2020) Morpholithodynamic processes and development of the coast of the contact zone of the subarctic and temperate seas of the North Pacific. Yuzhno-Sakhalinsk: IMGiG FEBRAS, p 234. ISBN 978–5–6040621–8–0. https://doi.org/10.30730/978-5-6040621-8-0.2020-1

Afanasiev VV Erosion of the sea coasts of northeastern Sakhalin; Afanasiev VV, Uba AV (2018) Geomorphology 4:25–35. https://doi.org/10.7868/S043542811804003X

Afanasiev VV Morpholithodynamics of lagoonal straits of northeastern Sakhalin; Afanasiev VV (2019) Geomorphology 2:79–94

Afanas'ev VV, Faustova AB (2022) The first results of the study of sequestration properties of coastal marine biomorpholithosystems (Sakhalin Region) Physica land mathematical modeling of earth and environment processes—2022. Springer Proceedings in Earth and Environmental Sciences. https://doi.org/10.1007/978-3-031-25962-3_46

Bobrik KP, Brovko PF (1986) River runoff and sedimentation in the lagoons of Northern Sakhalin in Materialy po gidrologii rek zony BAM i Dal'nego Vostoka (Materials on the hydrology of the rivers of the BAM zone and the Far East) T. XX, no 3. Gidrometeoizdat (Publ.), L, pp 439–444 (in Russ)

Curado G, Rubio-Casal AE, Figueroa E, Castillo JM (2014) Plant zonation in restored, nonrestored, and preserved Spartina maritima salt marshes. J Coast Res T. 30(3). C:629–634.https://doi.org/10.2112/JCOASTRES-D-12-00089.1

Gosudarstvennyy vodnyy kadastr (1979) Osnovnyye gidrologicheskiye kharakteristiki (za 1971–1975 gg. i ves' period nablyudeniy), t. 18. vyp. 4. Sakhalin i Kurily (State water cadastre. The main hydrological characteristics (for 1971–1975 and the whole period of observations), vol 18, p 4. Sakhalin and the Kurils). Gidrometeoizdat (Publ.), L, p 156 (in Russ)

Gosudarstvennyy vodnyy kadastr (1987) Mnogoletniye dannyye o rezhime i resursakh poverkhnostnykh vod sushi. T. 1 RSFSR. Vyp. 22. Basseyny rek Sakhalinskoy oblasti (State water cadastre. Long-term data on the regime and resources of surface waters of the land. T. 1 of the RSFSR. Issue. 22. Basins of the rivers of the Sakhalin region). Gidrometeoizdat (Publ.), L, p 227. (in Russ)

Kirwan ML, Temmerman S, Skeehan EE, Guntenspergen GR, Fagherazzi S (2016) Overestimation of marsh vulnerability to sea level rise. Nat Clim Chang 6(3):253–260

Leontiev IO, Afanasiev VV (2016) Dynamics of the lagoon coast of northeastern Sakhalin on the example of the Nyisky Bay system and the Plastun spit. Oceanology 56(4):618–626

Rosentreter JA, Al-Haj AN, Fulweiler RW, Williamson P (2021) Methane and nitrous oxide emissions complicate coastal blue carbon assessments. Global Biogeochem Cycles 35:e2020GB006858. https://doi.org/10.1029/2020GB006858

Propagation of Tsunami Waves in the Forked Bays

A. Yu. Belokon, D. I. Lazorenko, and V. V. Fomin

Abstract Within the framework of numerical simulation, a study was made of the penetration of tsunami waves into a forked bay, which has the characteristic dimensions and configuration of Dvoynaya Bay, located in the system of Sevastopol bays. Two variants of forked bays were considered: symmetrical—consisting of two identical bays; asymmetric—consisting of bays of various lengths. The SWASH nonlinear hydrodynamic model was used to model the tsunami propagation. To determine the boundary conditions on the liquid boundary of the computational domain, using the Black Sea tsunami model, the level fluctuations near Sevastopol in the region of depths of 90 m were calculated during the propagation of tsunami waves from three potential tsunami sources caused by underwater earthquakes of magnitude 7. It was found that in the case of tsunami penetration into an asymmetric bay, level fluctuations at the tops of bays of different lengths occur in antiphase. The highest sea level rise in the bays as a result of tsunami penetration into them from the nearest focus was 2.5–2.8 m; for the focus located in the Yalta seismically active zone—0.6–0.8 m; the focus near coast of Turkey—0.4–0.6 m.

Keywords Tsunami in bays · Tsunami in the Black Sea · Dvoynaya Bay · Mathematical modeling · SWASH model

1 Introduction

The tsunami in the Black Sea region is known mainly from literary sources (Dotsenko 1995; Nikonov et al. 2018; Papadopoulos et al. 2011) and, in some cases, from mareographic data (Grigorash 1959, 1972). Due to the small amount of information about this destructive phenomenon in this region, it becomes possible to study this problem using numerical methods. The study of the tsunami in the Black Sea region using mathematical modeling was carried out in Dotsenko (2012), Dotsenko and

A. Yu. Belokon (✉) · D. I. Lazorenko · V. V. Fomin
Marine Hydrophysical Institute of the Russian Academy of Sciences, 299011, Kapitanskaya Street, 2, Sevastopol, Russia
e-mail: aleksa.44.33@gmail.com

© The Author(s), under exclusive license to Springer Nature Singapore Pte Ltd. 2023   197
T. Chaplina (ed.), *Processes in GeoMedia—Volume VII*, Springer Geology,
https://doi.org/10.1007/978-981-99-6575-5_20

Ingerov (2013), Yalciner et al. (2004), Pelinovsky and Zaitsev (2011), Zaitsev and Pelinovsky (2011). For a more detailed study of the nature of tsunami propagation and determination of the most dangerous areas, it becomes necessary to consider individual sections of water areas. Tsunamis pose the greatest danger directly in the coastal zone, especially when waves penetrate into narrow bays, gulfs and straits. The works (Lobkovsky et al. 2018; Baranova and Mazova 2020; Belokon and Fomin 2021) simulated the propagation of tsunami waves in the Kerch Strait. Tsunami penetration into Balaklava Bay was studied in Fomin et al. (2022).

For the Crimean coast, one of the areas with complex geometry is the system of Sevastopol bays. Several cases of tsunamis near Sevastopol are known from the literature. According to Nikonov (1997), during a strong earthquake with magnitude $M \geq 7$ that occurred in 103 in the area of the Sevastopol Bay, the water receded towards the sea at a distance of up to 3–4 km, while the height of the arriving waves was at least 2 m. In (Nikonov 1995) it is mentioned an earthquake on April 29, 1650 in the northwestern part of the Black Sea, which caused waves about 3 m high near Sevastopol. A destructive underwater earthquake of magnitude $M \geq 6.5$ occurred on September 12, 1927, 30 km southeast of the city of Yalta. Level rises during calm were observed in different places and after strong shocks. The subsequent weaker shocks with sources near Sevastopol and Balaklava were accompanied by the withdrawal of water from the shore and the run-up of single waves on the shore (Bezushko et al. 2015). On December 26, 1939, a strong earthquake $M = 8$ occurred in the city of Fatsa (the coast of Turkey). According to eyewitnesses, the sea receded by 50 m, and then flooded the coast by 20 m. Tsunami waves crossed the Black Sea and were recorded by level gauges in Sevastopol, where the wave height was 50 cm (Nikonov 1997).

One of the most complex in the system of Sevastopol bays is Dvoynaya bay. It is located 15 km west of the center of Sevastopol, between Kamyshovaya Bay and Cape Khersones, and consists of Kazachya and Solenaya bays, different in length. The study of the patterns of wave propagation in such bays is of scientific and practical interest.

In this article, we studied a model forked bay, which has the characteristic dimensions, average depth, and configuration of the Dvoynaya Bay, located in the system of Sevastopol bays. The results of numerical simulation of the propagation of tsunami waves into the bay are presented. Using a nonlinear model of long waves, the problem of the evolution of tsunami waves in the Black Sea from three seismic foci was solved. Dependences on the time of sea level fluctuations at the entrance to the Dvoynaya Bay are obtained. These dependences were used as boundary conditions on the liquid boundary of the computational domain for the model bay. The Simulating WAves till SHore (SWASH) model (SWASH User Manual 2012) was used to numerically simulate the propagation of tsunami waves in the bay.

2   Mathematical Model

To study a tsunami in a forked bay, the nonlinear hydrodynamic model Simulating WAves till SHore (SWASH) (Fomin et al. 2022; Bazykina et al. 2018) was used. The computational domain was a basin with three liquid boundaries, having the configuration and topography of the bottom of the Dvoynaya Bay. The average length of the bay was about 12 m. Two variants of the computational domain were considered: the first one with a symmetrical bay, which consists of two forked bays of the same length; the second—with an asymmetric bay, which includes bays of various lengths. These computational regions are shown in Fig. 1, where the numbers 1–6 also denote the points (virtual mariographs) at which the sea level fluctuations caused by the tsunami were analyzed.

On the left boundary of the computational domain ($x = 0$), sea level fluctuations were set, obtained using the tsunami model for the entire Black Sea (Bazykina et al. 2018), which has a spatial resolution of 500 m and a time step of 1 s. Three cases of tsunami occurrence in the Black Sea as a result of an underwater earthquake with a magnitude of 7 were modeled. The parameters of tsunami generation foci

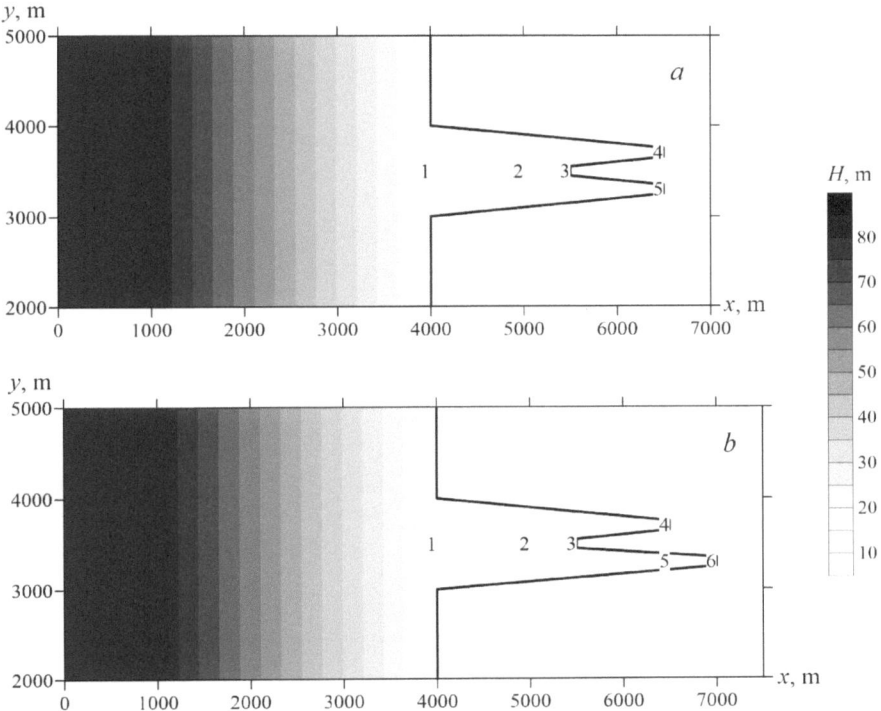

Fig. 1 Bottom relief of the computational area with symmetric (**a**) and asymmetric (**b**) bays and the position of virtual mariographs 1–6

were determined using empirical formulas from Ulomov et al. (1993). The initial displacements of the free surface of the sea, caused by earthquakes of magnitude 7, have a height of 1 m, the major and minor axes of the elliptical region are 50 and 29 km, respectively. The longitudinal axes of the ellipses are oriented along the 1500 m isobath, since all known Black Sea earthquakes that led to tsunamis occurred on the continental slope at depths not exceeding 1500 m. The position of the considered model tsunami sources is shown in Fig. 2. All of them are located in seismically active zones. Focus 1 is the closest to Sevastopol, where Dvoynaya Bay is located, focus 2 is similar to the one that caused the Yalta earthquake on September 12, 1927. Focus 3 is a remote focus, which is located in a seismically active zone near the Turkish coast. As calculations have shown, in the process of lowering the initial sea level elevation, an annular wave is formed, which propagates over time over the entire Black Sea area, and when it reaches the shelf, the tsunami front becomes almost plane. Thus, setting the boundary conditions on the left boundary of the computational domain (Fig. 1) in the form of the obtained mareograms is quite justified.

The radiation condition was used on the liquid boundaries of the computational domain. On solid sections of the boundary, the impermeability condition was set. The bottom roughness was taken into account using the Manning parameter $n = 0.019$ s/m 1/3. The rotation of the Earth was not taken into account. All calculations

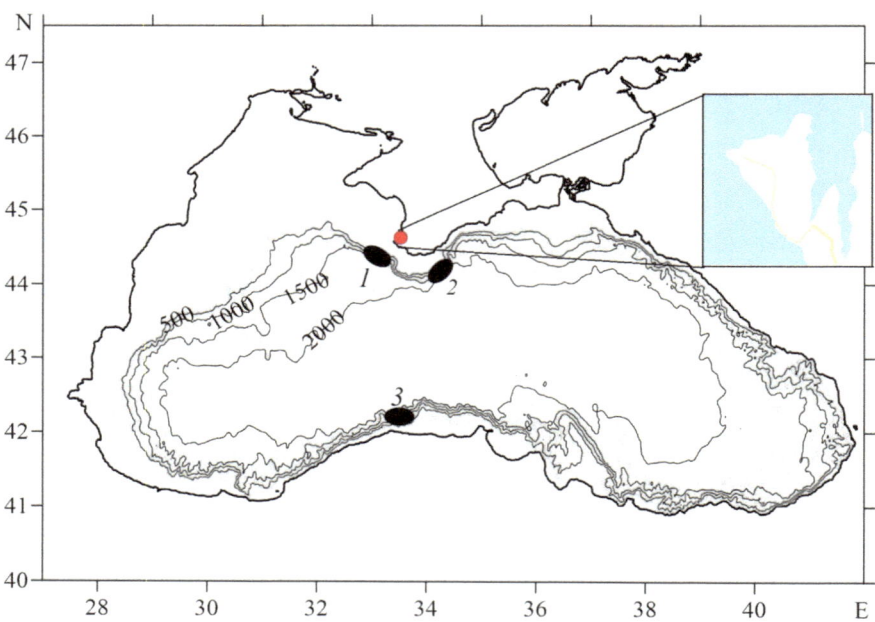

Fig. 2 The position of three hypothetical tsunami foci in the Black Sea: 1—the nearest focus in relation to Dvoynaya Bay; 2—focus similar to the one that caused the Yalta earthquake on September 12, 1927; 3—remote focus. The red circle marks the position of Dvoynaya Bay

were performed for a period of 6 h with a step of 10 m in space and a step in time of 0.5 s. Sea level fluctuations in the bays were recorded by virtual mariographs 1–6 (Fig. 1).

3 Results of Numerical Experiments

Sea level fluctuations in the coastal zone of Dvoynaya Bay for three tsunami generation foci are shown in Fig. 3 (left boundary of the computational domain, depth 90 m). A comparison of the graphs shows that for the nearest tsunami focus (1 in Fig. 3), the maximum sea level rise on the approach to the bay is 0.44 m, the maximum decrease is –0.24 m; for the focus located in the Yalta seismically active zone and the distant focus of the tsunami (2 and 3 in Fig. 3), the sea level deviations were about ±0.07–±0.09 m. Such small amplitudes of level fluctuations for foci 2 and 3 are explained by the fact that the seashore of the bay is protected by Cape Khersones from waves coming from the south and southeast. Thus, the maximum tsunami energy from these foci falls on the southern coast of Crimea. The seashores of the Dvoynaya Bay are reached only by the waves that go around Cape Khersones. In the case of the near focus 1, the energy maximum falls on the head wave, followed by oscillations of smaller amplitude.

The travel time of tsunami waves from generation source 1 to the entrance to the bay is about 13 min, from source 2–27 min, from source 3–41 min.

On Fig. 4 shows the evolution of a tsunami in a symmetrical bay from focus 1. As can be seen from the figure, the wave enters the computational region, where its amplitude increases and its length decreases due to a decrease in depth (Fig. 4a), when approaching the coast, the wave is partially reflected, and partially penetrates into the water area of the bay (Fig. 4b). Further, wave disturbances propagate to the tops of two identical bays (Fig. 4c), where they are amplified upon reflection from a vertical wall (Fig. 4d). The rise in the level in the bays is replaced by a decrease in the sea level (Fig. 4e). These oscillations continue for quite a long time, their amplitude decays with time (Fig. 4f–j).

Let us consider the propagation of tsunami waves in an asymmetric bay. The evolution of a tsunami in such a bay is shown in Fig. 5. The initial process of tsunami wave entry into an asymmetric bay (Fig. 5a–c) is similar to that in a symmetrical bay (Fig. 4a–c). When waves penetrate into the tops of bays of different lengths, the level rises, and in a relatively long bay, an increase in the amplitude of sea level fluctuations is observed (Fig. 5d). An increase in a relatively short bay is replaced by a decrease in sea level (Fig. 5e), while the wave energy moves to the top of a long bay. Then, the opposite picture is observed: in a relatively short bay, the sea level rises, and in a relatively long one, it decreases (Fig. 5f). Thus, sea level fluctuations in both bays are in antiphase during the entire duration of the tsunami action (Fig. 5g–j).

Figure 6 shows the calculated sea level fluctuations at points 1–6 during the penetration of tsunami waves into symmetric and asymmetric bays from the nearest focus (focus 1 in Fig. 2). Point 1 is located at the entrance to

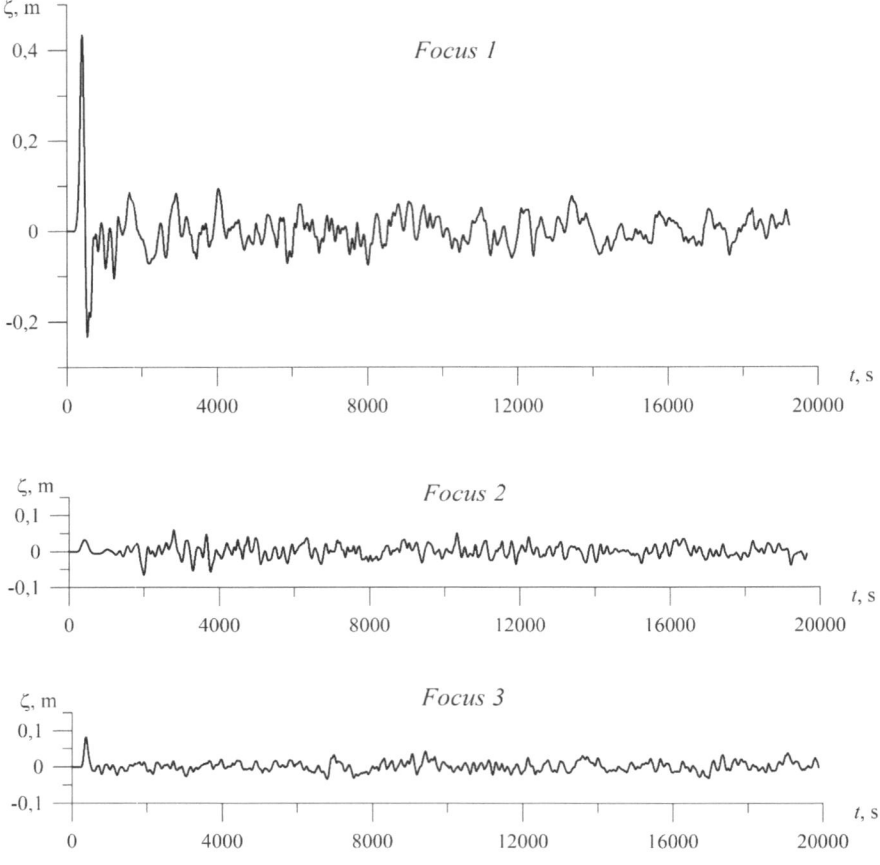

Fig. 3 Sea level fluctuations near Dvoynaya Bay caused by tsunami foci 1–3

each of the considered bays. It can be seen from Fig. 6 that at the entrance to both bays, the wave amplitude increases two times compared to the amplitude at the left boundary of the computational domain and reaches 0.8 m. Then, in the middle part of the bays (points 2, 3), the amplitude of level fluctuations increases to 1–1.5 m due to the narrowing of the bays. Points 4 and 5 are located at the tops of the forked of a symmetrical bay, so the level fluctuations in them are the same. Here, the wave amplitude can increase up to 2.5 m. Point 4 for an asymmetric bay is located at the top of a relatively short bay, and point 5 is in the middle part of a relatively long bay. It should be noted that for point 4 in the asymmetric bay, the level fluctuations somewhat decrease compared to those in the symmetrical bay, while a phase shift is observed when comparing the mareograms for two bays. The amplitude of the head wave at point 5 of the asymmetric bay coincides with that in the case of a symmetrical bay, then there is some amplification of the oscillation amplitude at this point for the

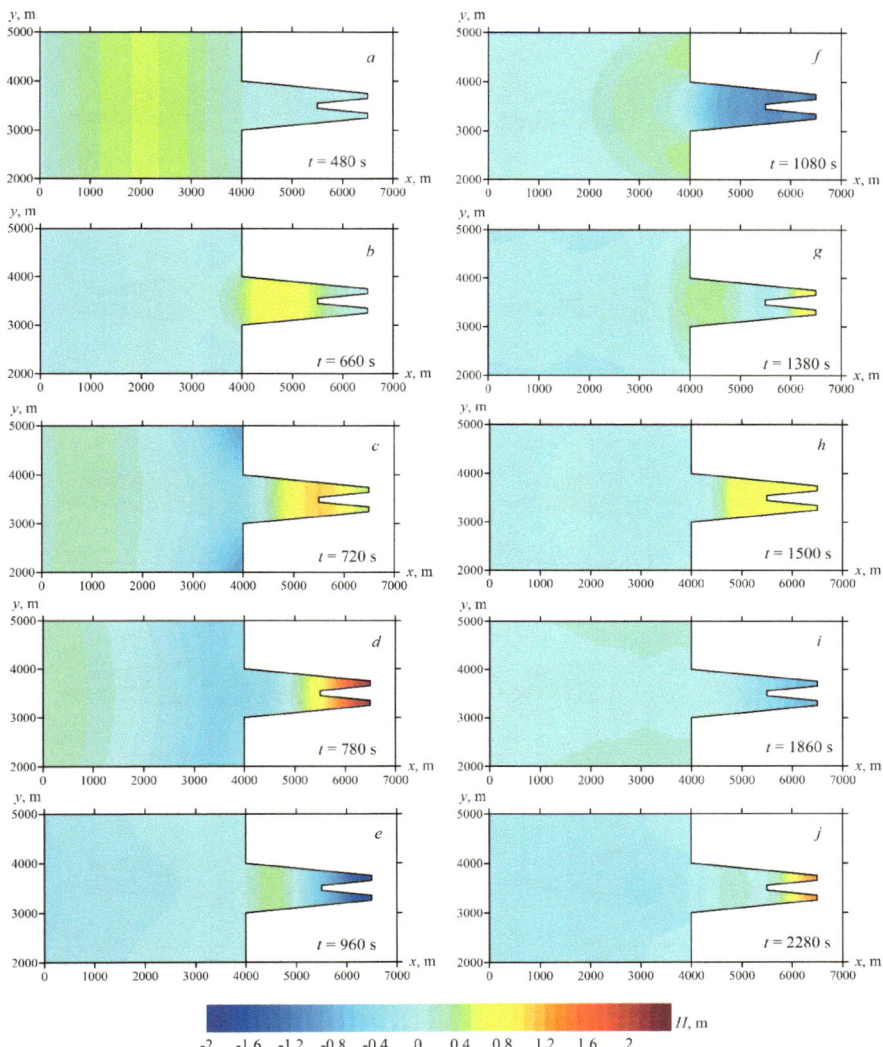

Fig. 4 Tsunami evolution in a symmetrical bay for focus 1

asymmetric bay. Point 6 refers only to the asymmetric bay and is located at the top of the long bay. Here, the level rise can reach 2.8 m, and the amplitude over 1 m can be observed for 2 h.

Mareograms for tsunami waves from the Yalta focus (focus 2 in Fig. 2) are shown in Fig. 7. In this case, the largest amplitudes of level fluctuations in bays amounted to 0.5–0.6 m, and at the top of a relatively long bay (point 6) they can reach 0.8 m. At the same time, the head wave is not maximum.

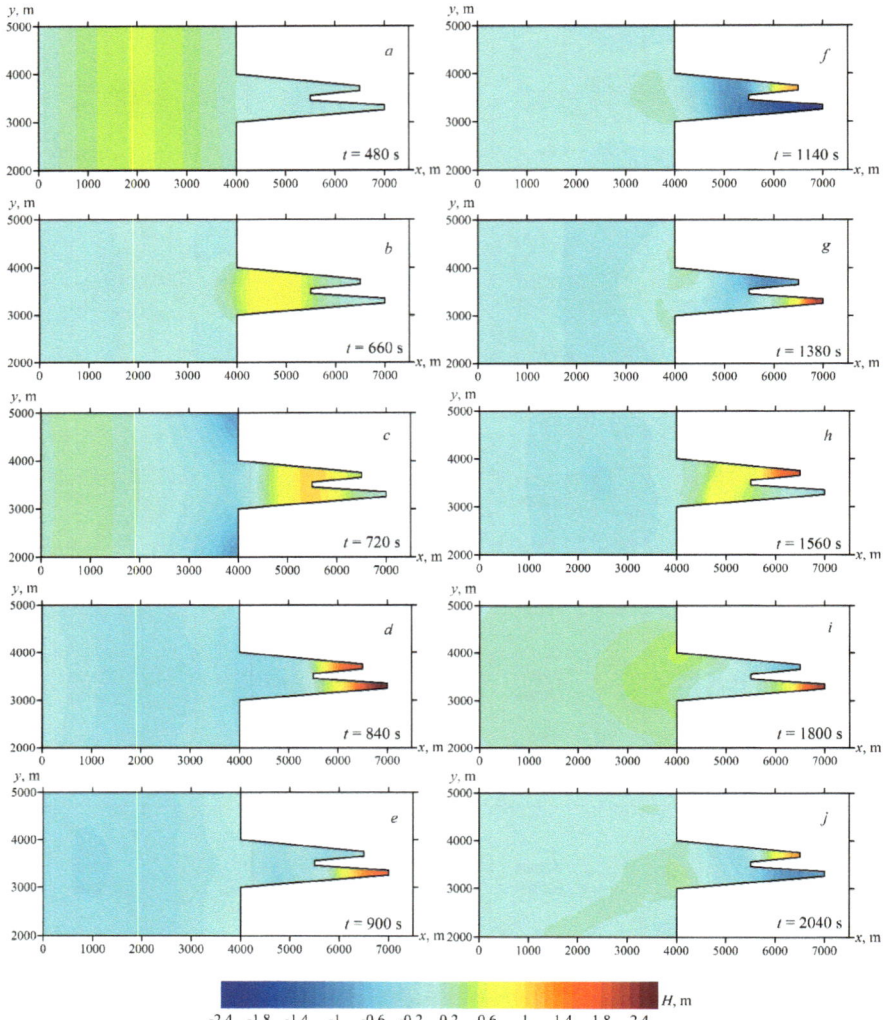

Fig. 5 Tsunami evolution in an asymmetric bay for focus 1

In the case of a remote focus located off the coast of Turkey (focus 3 in Fig. 2), the sea level rise in the bays can reach 0.4–0.6 m.

For all the studied foci, the largest amplitudes of sea level fluctuations in bays occur during the first three hours of tsunami action, while the wave amplitudes can increase up to 6–7 times compared to the amplitudes at the entrance to the computational domain (Fig. 8).

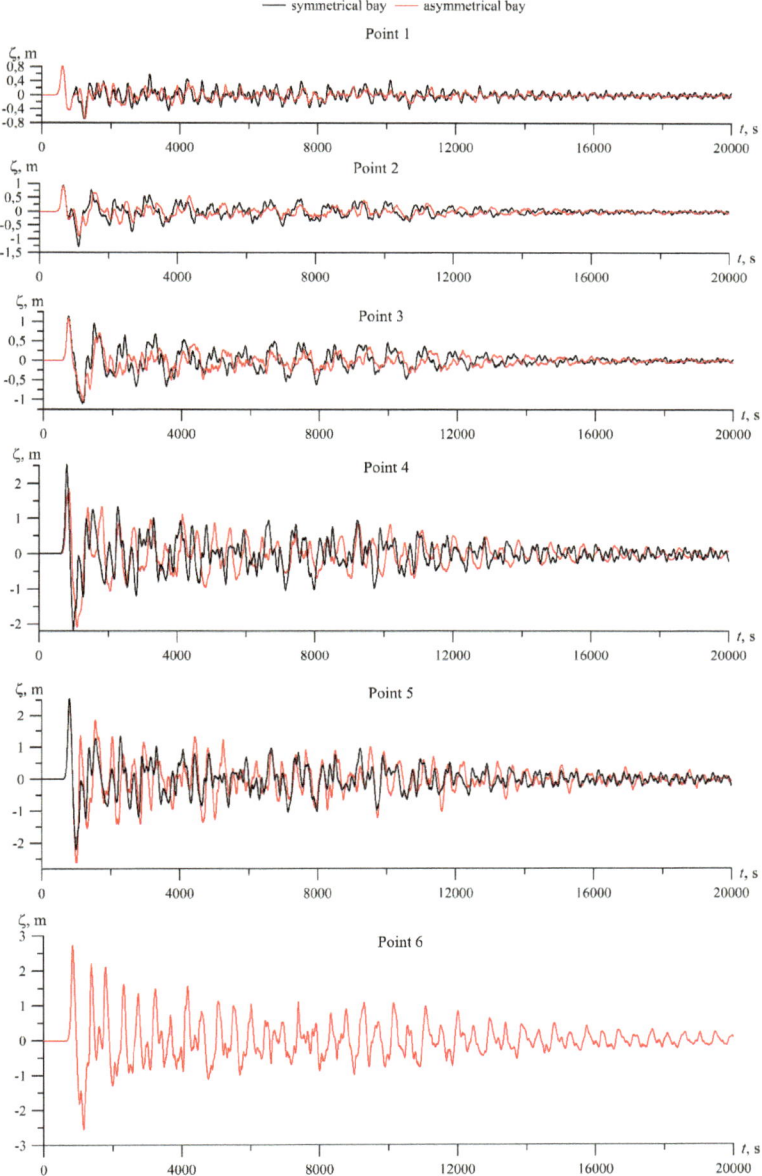

Fig. 6 Sea level fluctuations at points 1–6 in bays caused by focus 1

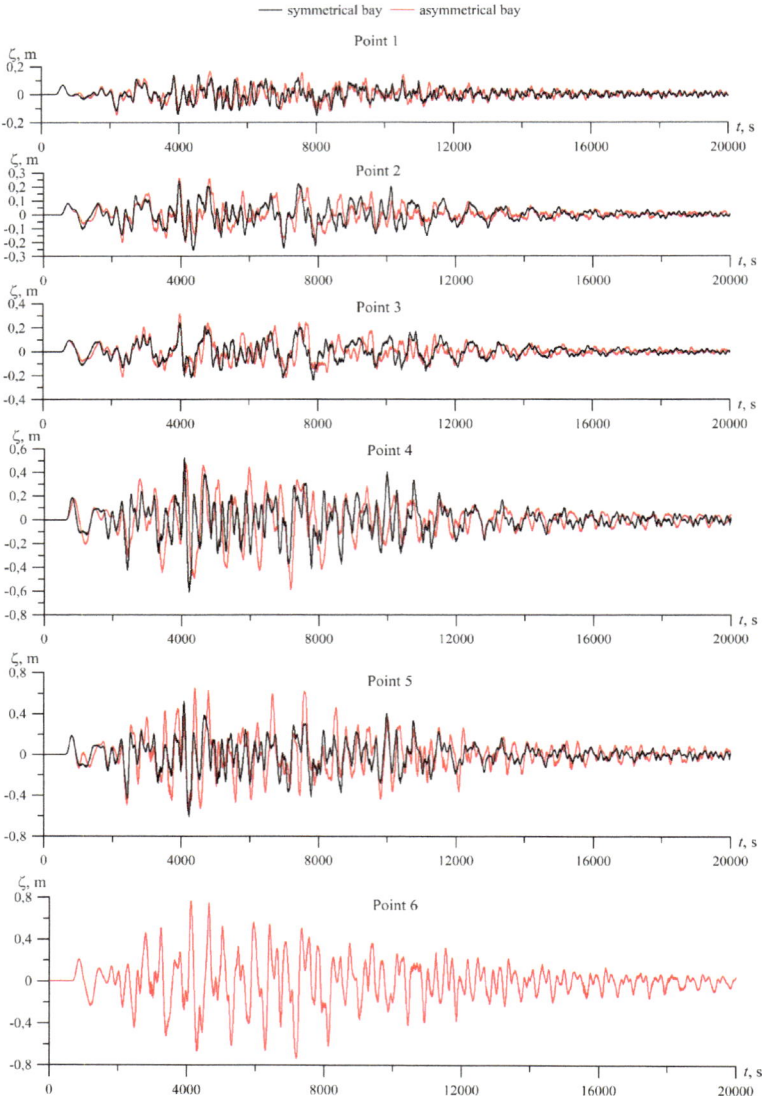

Fig. 7 Sea level fluctuations at points 1–6 in bays caused by focus 2

4 Conclusions

The results of numerical modeling of the penetration of tsunami waves into a forked bay, which has the characteristic dimensions and configuration of Dvoynaya Bay, located in the system of Sevastopol bays, are presented. Two variants of bays were

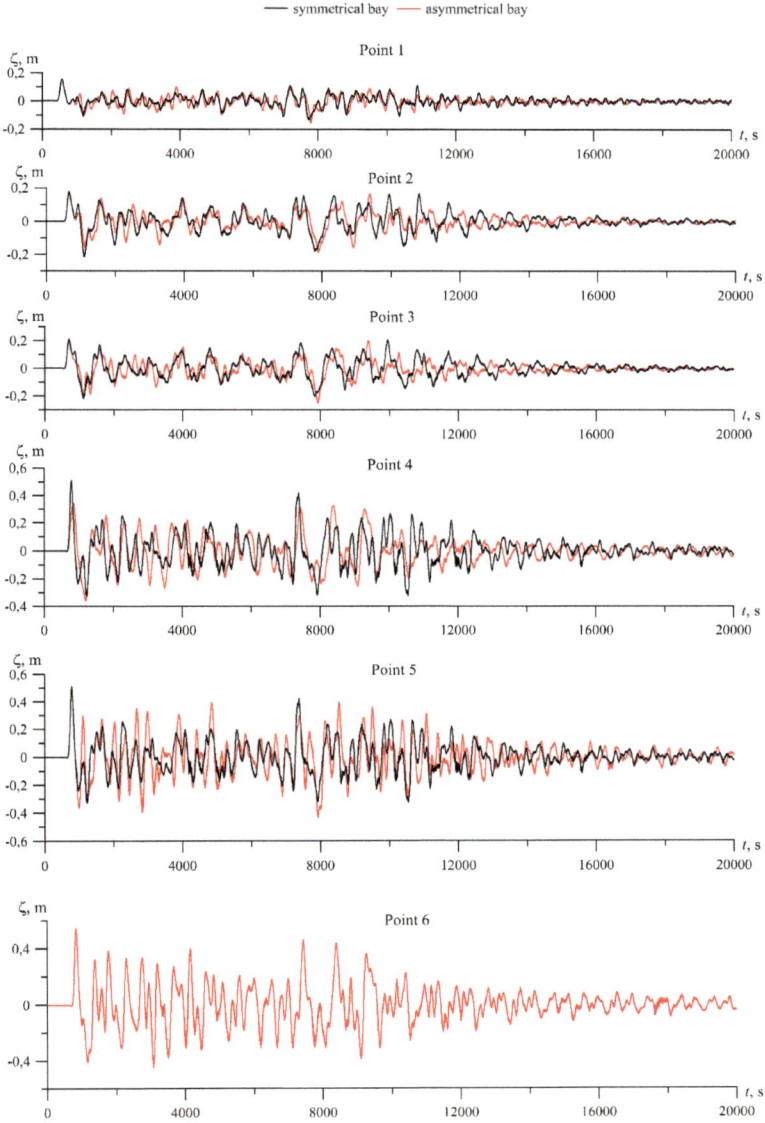

Fig. 8 Sea level fluctuations at points 1–6 in bays caused by focus 3

considered: symmetrical—consisting of two identical bays; asymmetric—consisting of bays of various lengths. At the first stage, using the Black Sea tsunami model, the evolution of tsunami waves from three potential foci caused by underwater earthquakes of magnitude 7 was studied. Time dependences of sea level fluctuations in the coastal zone of Sevastopol were calculated. At the second stage, the obtained mareograms were used as boundary conditions on the liquid boundary of the computational domain, for which, using the SWASH model, numerical simulation of the propagation of tsunami waves into a forked bay was performed.

It is shown that the amplitudes of tsunami waves at the tops of both bays can increase up to 6–7 times compared to the amplitudes of the waves at the entrance to the computational domain. For a symmetrical bay, sea level fluctuations in two branched bays are, as expected, the same. In the case of wave propagation in an asymmetric bay, level fluctuations at the tops of bays of different lengths occur in antiphase.

The highest sea level rise in the bays as a result of tsunami penetration into them from the nearest focus was 2.5–2.8 m; for the focus located in the Yalta seismically active zone—0.6–0.8 m; focus near coast of Turkey—0.4–0.6 m.

Acknowledgements The work was carried out within the framework of the scientific theme of MHI No. FNNN-2021-0005. Mathematical modeling was performed on the MHI computing cluster (www.hpc-mhi.org).

References

Baranova EA, Mazova RKh (2020) Tsunami hazard for the Crimean Coast of the Black Sea and the Kerch Strait at the catastrophic tsunamigenic earthquakes, the locations of which are close to that of the historical yalta earthquake on September 12, 1927. Phys Oceanogr [e-journal] 27(2):110–125. https://doi.org/10.22449/1573-160X-2020-2-110-125

Bazykina AY, Mikhailichenko SY, Fomin VV (2018) Numerical simulation of tsunami in the Black Sea caused by the earthquake on September 12, 1927. Phys Oceanogr 25(4):295–304. https://doi.org/10.22449/1573-160X-2018-4-295-304

Belokon AY, Fomin VV (2021) Modeling of tsunami wave propagation in the Kerch Strait. Fundamentalnaya i prikladnaya gidrofizika 14(1):67–78. https://doi.org/10.7868/S20736673 2101007X

Bezushko DI, Mironenko IN, Murashko AV (2015) Tsunami of the Black Sea coast of Ukraine. Vestnik Odesskogo Natsionalnogo morskogo universiteta 1(43): 82–90. (in Russian). https://meb.com.ua/onmu/201543.pdf

Dotsenko SF (1995) Tsunami in the Black Sea. Izvestiya RAN. Fizika Atmosfery i Okeana 30:513–519 (in Russian)

Dotsenko SF, Ingerov AV (2013) Characteristics of tsunami waves of seismic origin in the Black Sea basin based on the results of numerical modeling. Morskoy Gidrofizicheskiy Zhurnal 3:25–34 (in Russian)

Dotsenko SF (2012) Numerical modeling of tsunamis in the Black, Azov and Caspian Seas as a necessary element of regional tsunami early warning systems. Ekologicheskaya Bezopasnost' Pribrezhnykh i Shel'fovykh Zon i Kompleksnoe Ispol'zovanie Resursov Shel'fa. Sevastopol, ECOSI-Gidrofizika, vol 2, no 26, pp 287–300 (in Russian)

Fomin VV, Belokon AY, Kharitonova LV, Alekseev DV (2022) Numerical simulation of tsunami wave propagation to the balaklava bay. Phys Oceanogr 29(4):379–394. https://doi.org/10.22449/1573-160X-2022-4-379-394

Grigorash ZK (1959) Black Sea tsunami of 1927 according to mareographic records. Trudy MGI AN SSSR 17:59–67 (in Russian)

Grigorash ZK (1972) Review of remote mareograms of some tsunamis in the Black Sea. Trudy SakhKNII DVO AN SSSR. iss.29:271–278. (in Russian)

Lobkovsky LI, Mazova RKh, Baranova EA, Tugaryov AM (2018) Numerical simulation of propagation of the Black Sea and the Azov Sea tsunami through the Kerch strait. Phys Oceanogr [e-journal] 25(2):102–113. https://doi.org/10.22449/1573-160X-2018-2-102-113

Nikonov AA (1995) Unknown earthquake in crimea. Priroda 8:88–93 (in Russian)

Nikonov AA (1997) Tsunami frequency on the shores of the Black and Azov Seas. Izvestiya RAN. Fizika Zemli 33:72–87 (in Russian)

Nikonov AA, Gusiakov VK, Fleifel LD (2018) Assessment of the tsunami hazard on the Russian Coast based on a new catalogue of tsunamis in the Black Sea and the Sea of Azov. Russ Geol Geophys 59(2):193–205. https://doi.org/10.1016/j.rgg.2018.01.016

Papadopoulos G, Diakogianni G, Fokaefs A, Ranguelov B (2011) Tsunami hazard in the Black Sea and the Azov Sea: a new tsunami catalogue. Nat Hazards Earth Syst Sci 11945–11963

Pelinovsky EN, Zaitsev AI (2011) The estimation and mapping of tsunami dangerous on the Ukrainian Black Sea coast. Trudy Nizhegorodskogo Gosudarstvennogo Tekhnicheskogo Universiteta Im. R.E. Alekseeva 3(90):44–50 (in Russian)

SWASH User Manual (2012) SWASH version 7.01/The SWASH team. Delft University of Technology, Delft, p 144

Ulomov VI, Polyakova TP, Shumilina LS, Chernysheva GV, Medvedeva NS, Savarenskaya OE, Stepanova MB (1993) Experience in earthquake foci mapping. Seismichnost i Seismicheskoe Rayonirovanie Severnoy Evrazii, no 1. Institut Fiziki Zemli RAN, Moscow, pp 99–108 (in Russian)

Yalciner A, Pelinovsky E, Talipova T, Kurkin A, Kozelkov A, Zaitsev A (2004) Tsunamis in the Black Sea: comparison of the historical, instrumental and numerical data. J Geophys Res Oceans 109(C12):C12023. https://doi.org/10.1029/2003JC002113

Zaitsev AI, Pelinovsky EN (2011) Forecasting of tsunami wave heights at the Russian coast of the Black Sea. Oceanology 51:907–915. https://doi.org/10.1134/S0001437011050225

Assessment of the State of the Hydrosphere in the Zone of Development of Oil Fields of the Western Orenburg Region

I. V. Kudelina, T. V. Leontiena, and M. V. Fatyunina

Abstract The Orenburg region occupies one of the first places in the Russian Federation in terms of natural resources. More than 180 deposits of a wide variety of ores, oil, gas, and mineral raw materials have been explored on its territory. Strengthening in recent years of mining and processing of minerals, lead to an increase in the anthropogenic load on the underground hydrosphere and its main component–groundwater. Technogenic impact causes a change in the quality of groundwater and the depletion of their reserves. The purpose of this study is to assess the state of groundwater in the area of intensive development of the Kurmanaevskoye and Shulaevskoye oil fields, based on the study of the results of regime observations on observational networks of these fields. It was revealed that there are excesses of such indicators of the chemical composition of groundwater as: iron, manganese, oil products and chlorides. It is recommended to expand the groundwater monitoring network, modernize production with its transfer to low-waste technologies, conduct territorial planning when placing productive forces and production facilities.

Keywords Hydrosphere · Groundwater · Mineralization · Oil fields

1 Introduction

Oil production is accompanied by a complex and negative impact on the ecosystem as a whole and on its individual components, since there are no such technologies for prospecting, exploration and production of hydrocarbon raw materials at the present stage of development of science and technology that would not have a negative impact on the geosphere (Kudelina 2019a; Leontieva 2018).

The negative technogenic impact on the environment depends on the following factors: features of the geological structure, development method, qualitative composition of oil and gas, equipment of oilfield equipment, its condition, intensity of

I. V. Kudelina (✉) · T. V. Leontiena · M. V. Fatyunina
Orenburg State University, Orenburg, Russia
e-mail: kudelina.inna@mail.ru

© The Author(s), under exclusive license to Springer Nature Singapore Pte Ltd. 2023   211
T. Chaplina (ed.), *Processes in GeoMedia—Volume VII*, Springer Geology,
https://doi.org/10.1007/978-981-99-6575-5_21

production of formation fluids, associated changes in formation pressure, volumes of pollutant emissions, degree of stability components of the natural environment, etc. (Kudelina 2019b; Kudelina 2021).

The volumes of production and processing of oil and gas have been increasing in recent years, which leads to an increase in the technogenic load on the hydrosphere and its main component–groundwater. These impacts cause changes in the quality of groundwater, including pollution and depletion of their reserves (Khusainova and Leontieva 2020; Ecological state of the hydrosphere of the Buzuluk pine forest 2022).

Given the critical role of groundwater, primarily for domestic drinking water supply of the population, their close relationship with other components of the natural environment, it is necessary to study the state of groundwater, their qualitative and quantitative characteristics on the basis of a single State concept, which provides for mandatory pairing with the monitoring system of other components of the natural environment (Kudelina 2021; Klimenko 2019).

The purpose of this study is to assess the state of groundwater in the area of intensive development of the Kurmanaevsky and Shulaevsky oil fields in the Western Orenburg region, based on the study of the results of regime observations on the observation networks of these fields.

2   Results and Discussions

The most environmentally significant impact during mining is the pollution of environmental components with oil, oil products and highly mineralized waters. According to expert estimates, up to 0.5% of all produced crude oil is lost in the oil fields.

The exploitation of deposits leads to the formation of the following main types of waste: reservoir wastewater; associated petroleum gas; oily sludge from tank cleaning; domestic waste water; surface sewage; construction waste (Khusainova and Leontieva 2020; Klimenko 2019; Sanitary rules and regulations 2001).

Measures to protect the geological environment are to preserve the natural regime of ground and groundwater, reduce pollution of rocks in the aeration zone. Groundwater is the most mobile component of the subsoil and is the first to respond to technogenic changes. The creation and maintenance of local monitoring of groundwater and surface water at the Kurmanaevskoye and Shulaevskoye oil fields is the receipt of prompt and timely information about negative trends and changes.

3   Kurmanaevskoye Oil Field

In the area of the Kurmanaevskoye oil field, the aeration zone is subject to pollution, represented by an anhydrous permeable Quaternary alluvial-deluvial horizon (adQ), the first Upper Pleistocene-Holocene alluvial horizon from the surface (aQIII-H) and the main aquiferous Lower Triassic complex (T1), which is widespread and mainly economically–drinking value. Observations are organized for these aquifers and complexes (Analysis of the development of the Kurmanaevsky oil field in the Orenburg region 2005; Report on the results of work on the creation of a regime network of observation wells on the territory of the Shulaevskoye oil field 2003; Gatskov 1999).

Anhydrous permeable Quaternary alluvial-deluvial horizon (adQ)

Five observation wells have been equipped for this horizon. According to the chemical composition, groundwater is mixed with a mineralization of 1.7 mg/dm^3. Of the microcomponents, a significant excess of the norm was observed for iron – 1.95 mg/dm^3 (at MPC-0.3 mg/dm^3) and bromine –1.20 mg/dm^3 (at MPC-0.2 mg/ dm^3), oil products were found in insignificant quantity –0.37 mg/dm^3 (at MPC-0.1 mg/dm^3).

In all observation wells, an excess of dry residue from 1.2 MPC to 3 MPC (from 1213 mg/dm^3 to 3014 mg/dm^3) was found at MPC-1000 mg/dm^3. The chemical composition of water is bicarbonate-calcium, and in a separate well, where an increased mineralization (from 1299 to 1622 g/dm^3) is noted, underground water is hydrocarbonate-chloride magnesium-calcium. With an increase in mineralization, the content of calcium and magnesium increases, respectively, the hardness of the underground water of the considered aquifer increases. This well is located in a zone of increased man-made load, downstream from the central pumping station and the commodity park, and, apparently, the recorded pollution is associated with this.

In some wells, the excess of manganese content can be traced from 0.22 to 1.46 mg/ dm3 (at MPC 0.1).

In all wells, the content of turbidity and color is recorded from 3.4 to 57.0 mg/l (at MPC 1.5), which may either be the result of poor pumping before sampling, which cannot be done due to the small water column in shallow wells, or is the result of the oxidation of ferrous iron, the content of which exceeds the MPC. The latter may be due to the natural background.

Exceeding the MPC for salinity and water hardness by 1.5–2.3 times may be the result of anthropogenic impact. The increase in hardness is observed mainly due to an increase in the calcium content in underground water, respectively, the removable hardness increases. In accordance with the general requirements for the composition and properties of water in water bodies, the pH value should not go beyond the range of 6.5–8.5. The pH value in the water of the evaluated wells varies from 7.2 to 8.14. The waters are neutral.

Aquifer Upper Neopleistocene-Holocene alluvial horizon (aQIII -H) is distributed mainly in the valley of the Buzuluk river within the low, high floodplain and in the valley of the Tarpanovka river (right tributary of the Buzuluk river), as well as in the mouths of numerous ravines and dens, in a ravine-beam network.

Monitoring wells of the regime network are also equipped for this aquifer. Water-bearing rocks are fine-grained, fractional sands, with the inclusion of gravel-pebble and sand-gravel deposits. The thickness of the water-bearing rocks is 10.0–15.0 m. The waters are free-flowing.

According to the chemical composition, underground waters are variegated, along with hydrocarbonate waters, there were hydrocarbonate-sulfate, hydrocarbonate-chloride and mixed waters. Of the cations, sodium and calcium ions predominate. Mineralization of underground waters is 0.3–0.8 g/dm^3 and in isolated cases exceeds 1.0 g/dm^3. In the observation well, located downstream from the booster pumping station (BPS), the water salinity was 1.2 g/dm^3. In terms of water hardness, they are mostly moderately hard, and in the well near the CPS, the water is very hard– 14.8 mmol/dm^3 (at MPC–7 mmol/dm^3). Of the microcomponents, the excess of the norm is recorded for iron from 1.92 to 6.50 mg/dm^3 (at MPC–0.3 mg/dm^3), manganese from 0.43 to 1.11 mg/dm^3) and bromine from 0.21 to 1.0 mg/dm^3 (with MPC–0.2 mg/dm^3), the total hardness varies from 2.36 to 14.8 mmol/dm^3, most often it was 3.0–4.0 mmol/dm^3.

Exceeding the MPC for salinity and hardness of underground water by 1.5–2.3 times in some wells may be a consequence of anthropogenic impact. These wells are located in the zone of increased technogenic load, downstream from the booster pumping station (BPS) and, probably, the pollution of the observed aquifer is associated with this. So the content of oil products in almost all samples in these wells is above the permissible limits of MPC, from 0.15 to 0.36 mg/dm^3. The content of iron and manganese in underground water varies from 1.5 to 12.8 MPC, which may be the result of poor pumping before sampling or is a natural background.

The excess in turbidity and color is recorded for all wells of the monitoring network of observation wells of the Kurmanaevskoye oil field, which may be the result of poor pumping before sampling.

The aquiferous Lower Triassic complex (T1) is ubiquitous in the area of the described deposit. Depending on the depth of immersion, it is the second or third aquifer from the surface.

The water-bearing complex was tested by monitoring wells of the regime network. The depth of the wells is 85.0 m.

In groundwater, there is an increased content of oil products in individual wells from 0.16 to 0.5 mg/dm^3 at MPC 0.1 mg/dm^3, an increased content of iron from 0.59 to 1.37 mg/dm^3 at MPC-0.3 mg/dm^3, increased content of dry residue 2839 mg/dm^3 at MPC 1000 mg/dm^3. With increased mineralization, the chemical composition of underground water is hydrocarbonate-chloride magnesium-calcium. With an increase in mineralization, the content of calcium and magnesium increases and, accordingly, the hardness of the water increases, which is 33.5 mg-eq /dm^3 and the water is characterized as very hard. Exceeding the MPC for salinity and water hardness by 1.5–2.3 times may be the result of anthropogenic impact, and it is in those observation wells where the development of neighboring oil fields affects. These observation wells were drilled by the project as background wells to control the qualitative composition of groundwater from the possible impact of the development of the Bobrovsky and Shulayevsky oil fields.

The described water-bearing complex is the most promising for the organization of both household and drinking and technical water supply of the field. In erosion incisions, that is, where the complex occurs close to the day surface, it can be used for drinking water supply in its upper and middle parts. In the lower parts of the water-bearing complex, where the salinity increases above 1.0 g/dm^3, it is recommended for use for technical water supply.

In the process of further activities, it is necessary to control the limit of the use of drinking water and technical quality and forecast the development of possible pollution and depletion of groundwater.

An analysis of the quality of drinking water in rural settlements of the Kurmanaevsky district revealed an excess of the standards for such indicators as total hardness, iron, oil products, ammonia, and copper.

So, in s. Romashkino exceeded the standards for total hardness (1.56–8 MPC,), in the village of Kurmanaevka for ammonia nitrogen (2.1 MPC) and oil products (5.1 MPC). Exceeding the MPC standards for domestic and drinking water supply in terms of total hardness is noted in the villages of Semenovka (1.1–1.3 MPC), Petrovka- (1.1–1.2 MPC), Danilovka - (1.4 MPC) and Volzhsky (1.0–1,3 MPC).

Exceeding the standards for iron was noted in the villages of Petrovka (3.4–6.1 MPC) and Kandaurovka (1.4–3.3 MPC), for copper–in the village. Volzhsky (3.5 MPC), total mineralization–in the village. Danilovka (1.3 MPC). All excess components can be influenced not only by the Kurmanaevskoye field, but also by other nearby oil fields.

To assess the current state of surface waters, the results of departmental monitoring in the area of the Kurmanaevskoye field on the Buzuluk River (2 sampling points: before the field and after the field) and Tarpanovka (1 point within the field) were used. Water samples were taken monthly and examined for a number of indicators standardized by SanPiN 2.1.5.980–00 "Hygienic requirements for the protection of surface waters". The quality of water bodies was assessed by general chemical analysis and oil products.

According to the results of the analyzes, the chemical composition of the water in the river. Buzuluk hydrocarbonate magnesium-calcium, calcium-magnesium. Fresh water–dry residue is 586–917 mg/dm^3. In terms of total hardness, which varies within the range of 5.1–8.24 mg-eq/dm^3, water is moderately hard and hard. The active reaction of the medium is slightly alkaline (pH 7.7–8.43). The content of oil products is mainly contained within the MPC for fishery rivers (0.5 mg/dm^3) or not found in water samples, in isolated cases the excess of oil products is 1.6–2.8 MPC.

According to the results of analyzes, the water in the Tarpanovka River is fresh with water salinity from 596–831 to 1004–1286 mg/dm^3. According to the chemical composition, the water is sulfate-hydrocarbonate chloride-hydrocarbonate, according to cations, mainly calcium-magnesium. The pH value in the water is within the normal range. The waters can be classified as neutral and slightly alkaline (pH 6.94–8.36). The total hardness of water varies within 4.1–12.0 mg-eq/dm^3–water from moderately hard to very hard. The content of oil products is within the MPC or not detected. As a result of consideration and analysis of the content of chemicals in the surface waters of rivers, according to monitoring studies, it can be concluded that the composition of water in the rivers of the Kurmanaevskoye field corresponds to the original natural composition.

Water pollution in reservoirs in the Kurmanaevsky district is mainly due to runoff from agricultural fields and discharges from livestock farms. These pollutions are seasonal and occur mainly during the warm season.

The results of laboratory monitoring of water in reservoirs on the territory of the Kurmanaevskoye oil field indicate the absence of samples exceeding sanitary standards, both in terms of bacteriological and chemical indicators, it does not exceed the MPC.

Shulaevskoye oil field. The object of observation in the analysis of the qualitative composition of groundwater is the Lower Triassic aquifer complex, the first from the surface (Report on the results of work on the creation of a regime network of observation wells on the territory of the Shulaevskoye oil field 2003). Within the field there is an observation network of 10 wells for the aquiferous Lower Triassic complex. Well depth–from 50.0 to 202.9 m.

The upper interval of the Lower Triassic aquifer is exposed and observed by wells with a depth of 50.0–60.0 m. The chemical composition of groundwater is varied– from sulfate-hydrocarbonate and mixed with a mineralization of 0.6 g/dm 0.3 to 1.6 g/dm^3. By cations, the water is mixed and calcium-magnesium. Exceedances of MPC were noted for dry residue and total stiffness in a well located near an oil well, for oil products and oxidizability in a well located in the BPS area.

The quality of groundwater in the upper part of the complex does not meet the requirements of SanPiN 2.1.4.1074–01 "Drinking water …." in all wells, perhaps this is due to the anthropogenic factor in the formation of the chemical composition (Sanitary rules and regulations 2001; Koronkevich 2022). Dry residue, chlorides, hardness in individual observation wells. Also, excesses were noted for color, turbidity and iron in observation wells.

The color of groundwater refers to normalized physical properties, the definition of which is required by regulatory documents. Color is a secondary organoleptic indicator of groundwater quality and cannot be formed without certain geochemical processes. The formation of increased color of groundwater is possible when they are contaminated with micro-amounts of specific organic compounds (Maskooni et al. 2020; Ushakova et al. 2021). An increased color of groundwater is also possible with an increased concentration of ferrous iron.

The turbidity index also refers to the normalized physical properties of groundwater. The turbidity of groundwater is formed in the same way as the color in cases where they contain suspended, colloidal particles. The amount of turbidity depends on the amount of suspended particles, a large content of which is in the water of an insufficiently pumped well. Therefore, before sampling water from observation wells, it is necessary to carry out pumping according to the existing method (Yu et al. 2021; Zakir et al. 2020).

At the Shulaevskoye field, there were no excesses for oil products in the upper aquifer at a depth of 50.0–60.0 m. Only in one observation well, at the date of drilling, the content of oil products was recorded–0.21 mg/dm^3 (at 0.1 MPC).

In the groundwater of the upper aquifer in an observation well with a depth of 50.0 m, the highest content of manganese was found; at MPC 0.1, the concentration of manganese was 1.29 mg/dm^3. Perhaps this is due to surface contamination and spillage of oily waters.

The presence of ammonium ions in amounts above the MPC was recorded in one sample in the interval of the observed aquifer at the date of drilling. Basically, in the groundwater of the interval under consideration, the ammonium content is below the MPC.

The average interval of the Lower Triassic aquifer was opened at a depth of 115.0 m. mixed calcium-magnesium with a mineralization of 0.6–0.9 g/dm^3. Exceeding the MPC of normalized components in terms of dry residue, chloride content, hardness in samples taken in the observed interval was not recorded.

The content of oil products above the MPC and ammonium in this water-bearing interval was observed in individual wells.

In the observed interval of the aquifer, as in all observation intervals, an increased content of color, turbidity, iron, and manganese was recorded.

An increased content of ammonium ions was recorded in observation wells, but in subsequent years, a decrease in the content of this component is observed in the interval of the observed aquifer.

The lower interval of the Lower Triassic aquifer was opened at a depth of 200.9 and 202.9 m, respectively. Underground waters are bicarbonate-chloride with a mineralization of 0.6–0.7 g/dm^3. No excesses of the MPC of normalized components were noted. The nature of the change in the content of iron, color, turbidity, manganese, hardness, oil products, ammonium in the observation well is almost identical to the content of these components of the second observation interval, and only by the content of chlorides and color are opposite.

In addition, observations of the quality of groundwater in the Lower Triassic aquifer complex are carried out at a spring located in the upper part of the Guryev ravine (near the uninhabited village of Shulaevka). According to the results of observations, the chemical composition of the water in the spring is hydrocarbonate calcium and chloride-hydrocarbonate calcium. The water is fresh with a mineralization value of 0.5–0.6 g/dm^3, very hard, the reaction of the aquatic environment is neutral (pH 6.47–7.21). Exceedances of MPC are noted in terms of color by 2.1 times, once the content of oil products exceeded the norm by 2 times. The data of the subsequent selection of excesses for oil products did not show.

4   Conclusions

Analyzing all the data of chemical analyzes, it can be noted that the water in most observation wells is fresh. The chemical composition of water in the wells is hydrocarbonate, hydrocarbonate-sulfate, chloride-hydrocarbonate, sodium predominates in the cationic composition. The water is soft to moderately hard. The reaction of the environment is most often neutral and slightly alkaline, less often alkaline. Exceedings of MPCs for oil products, ammonium nitrogen, and manganese were noted periodically.

Practically in all water samples from observation wells there is an increased content of iron, color and turbidity. Exceeding the MPC for color and turbidity is most likely due to insufficient well pumping before water sampling.

In observation wells located in the area of oil wells, an increase in dry residue is noted. According to the chemical composition, the water is bicarbonate-chloride and mixed with a predominance of chlorides, brackish, very hard.

In conclusion, we can say that objective information has been obtained on the impact of the economic activities of the oil production at the Kurmanaevsky and Shulayevsky oil fields on the environment. As a result of regime observations, the situation with the state of groundwater at these fields was revealed, priority indicators of pollution were reflected. A stable source of pollution in the groundwater of the main aquifer used for domestic and drinking water supply is temporary and is associated with oil production facilities.

Based on the analysis carried out, for further monitoring work, it is recommended to carry out repair work on all observation wells to continue local monitoring. It is necessary to expand the groundwater monitoring network, modernize oil production with its transfer to low-waste technologies, conduct territorial planning when placing productive forces and production facilities.

References

Analysis of the development of the Kurmanaevsky oil field in the Orenburg region (2005) CJSC Izhevsk Oil Research Center, Izhevsk, p 360

Ecological state of the hydrosphere of the Buzuluk pine forest; Kudelina IV, Leontieva TV, Fatyunina MV, Khusainova MF (2022) Problems of regional ecology, no 6, pp 55–59

Gatskov VG (1999) Assessment of the state of the natural environment on the territory of the enterprises of JSC "Orenburgneft"; Gatskov VG, Lukinykh EN (1999) Orenburg, p 267

Khusainova LF, Leontieva TV (2020) Prospecting and exploration for oil and gas; University complex as a regional center of education, science and culture: materials of All-Russia. Scientific method conference (with international participation), 23–25 Jan 2020, Orenburg, pp 940–943

Klimenko AV (2019) Associated petroleum gas and methods of its processing; Klimenko AV, Leonteva TV (2019) New directions of work on oil and gas, innovative technologies for the development of their deposits, prospects for the production of unconventional hydrocarbon raw materials: materials of Vseros. Scientific-practical conference; resp. Kolomoets AV (ed). Press Agency, Electron. Dan. - Orenburg, pp 40–42

Koronkevich NI (2022) Assessment of the impact of urbanization on the annual runoff and water quality in the world and on the continents; Koronkevich NI, Baranova EA, Zaitseva IS et al (2022) Izvestia of the Russian academy of sciences. Geogr Ser 86(3):470–480

Kudelina IV (2019) Influence of oil field development on the underground hydrosphere of the Orenburg region. Regional problems of geology, geography, technospheric and environmental safety: collection of articles. Proceedings of the scientific-practical conference, 18–20 Nov 2019, Orenburg; Ministry, pp 19–23

Kudelina IV (2019) To the methodology of hydrogeological research in the Southern Urals; Kudelina IV (2019) New directions of work on oil and gas, innovative technologies for the development of their fields, prospects for the extraction of non-traditional carbon raw materials: materials of Vseros. Scientific-practical conference; otv. Kolomoets AV (ed). Electron. Dan. Press Agency, Orenburg, pp 57–59

Kudelina IV (2021) Impact of the development of the Vakhitovskoye oil field on the underground hydrosphere. Int Res J 7(109):150–153. Part 1

Leontieva TV (2018) Issues of methods of hydrogeological research and substantiation of the possibility of replenishing groundwater reserves, no 2. Izvestia of universities in Kyrgyzstan, pp 14–17

Maskooni EK, Naseri-Rad M, Berndtsson R, Nakagawa K (2020) Use of heavy metal content and modified water quality index to assess groundwater quality in a semiarid area. Water 12:1115

Report on the results of work on the creation of a regime network of observation wells on the territory of the Shulaevskoye oil field (2003) Monitoring Regulations for 2004–2007 Rep. Use Levanina SS, Chechetkin TV. CJSC "Vostochnaya GRE". Orsk, p 236

Sanitary rules and regulations (2001) Drinking water. Hygienic requirements for water quality of centralized drinking water supply systems. Quality control" SanPiN 2.1.4.1074–01. Goskomsanepidnadzor of Russia, M, p 103

Ushakova E, Menshikova E, Karavaeva T, Puzik A (2021) Trace element distribution in the snow cover of different functional zones in Berezniki-Solikamskindustrial hub. Russia. J Ecol Eng T. 22(10):28–39

Yu L, Zhang F, Zang K, He L, Wan F, Liu H, Zhang X, Shi Z (2021) Potential ecological risk assessment of heavy metals in cultivated land basedon soil geochemical zoning: Yishui County, North China Case Study. Water 13:3322

Zakir HM, Sharmin S, Akter A, Rahman MS (2020) Assessment of health risk of heavy metals and water quality indices for irrigation and drinking suitability of waters: a case study of Jamalpur Sadar area, Bangladesh. Environ Adv 2:10000

Biogenic Morpholitogenesis on Sea Coasts Sakhalin Island

Victor V. Afanas'ev

Abstract Biogeomorphological morphogenetic systems—a special type of geomorphological systems, the development of which is determined by biogeocenosis—a natural system of functionally interconnected living organisms and components of their abiotic environment, characterized by a certain energy state, type and rate of metabolism and information. Hydro-biogeomorphological systems are characterized by "accumulative" biogeochemical type of lithogenesis with "energy-intensive" physical and chemical properties. These are freshwater and marine geoecosystems (estuary-lagoon, swamp, lake, etc.) where organogenic energy-intensive deposits are formed: sapropels, peat, amber. The paper considers the role of geomorphological conditions for the development of these systems of formation, accumulation, transformation of energy-intensive deposits. When solving the problems of managing coastal wetlands through their accelerated programmable formation, heterogeneous hydro-biogeomorphological systems merge; absorption and transformation of existing and the emergence in their place of natural-anthropogenic systems with inherited or nature-like properties and structure.

Keywords Biogeomorphological system · Coastal-marine sedimentation · Salt marsh · Intertidal mudflat · Carbon sequestration

1 Introduction

Geomorphological approaches have already shown their effectiveness in analyzing the balance of producers and sinks of greenhouse gases, especially in relation to the dynamics of marches and silt lands (Rosentreter et al. 2021; Hu et al. 2015;

V. V. Afanas'ev (✉)
Institute of Marine Geology and Geophysics, FEB RAS, Yuzhno-Sakhalinsk, Russia
e-mail: vvasand@mail.ru

Sakhalin State University, Yuzhno-Sakhalinsk, Russia

Curado et al. 2014). The programmed formation of coastal blue carbon ecosystems, which can be reliably used to mitigate the effects of climate change, is impossible without understanding the conditions of sedimentation in the coastal zone and morpholithodynamically substantiated scenarios for their development.

What is the role of geomorphological conditions for the development of these systems of formation, accumulation, transformation of energy-intensive deposits, and what are the features of coastal morpholithogenesis? At the same time, issues related to the response of biomorpholithosystems to the projections of global climate changes, including temperature regime, sea level changes and salt water intrusion on the sequestration/outflow of greenhouse gases from the soils of marshes and sediments of silty drylands of coastal wetlands of the seas of Far East of Russia are of particular importance.

Accumulative biomorpholithosystems in their development go a long way from the basin of coastal-marine sedimentation to alluvial-marine plains with a polygenic and polychronic surface, and this surface can even be of the same geospatial level.

As for the largest basin of coastal-marine sedimentation in the Far East—the Amur Estuary and the northern part of the Tatar Strait and the Sakhalin Bay associated with it, the value of C org./year which the river takes out. Amur is 14 million tons (Karetnikova and Garetova 2015; Koltunov et al. 2009). The area of marshes and silt drains in northwestern Sakhalin exceeds 200 km^2 (Fig. 1).

The largest free-type accumulative formation on Sakhalin, the Tyk Spit, and the rhythmically constructed accumulative system of the Ikhdam, Noksi, and Nyide

Fig. 1 Intertidal mudflat and marches of the Nevelskoy Strait (https://ostrov2049.com/2017/12/18/2011)

spits of the Holocene age are confined to the marginal depression elements of the morphostructural plan of the coast of the Amur Estuary and the Nevelskoy Strait (Afanasiev 1989). The structure of sedimentary strata testifies to the maximum sedimentation rates for the region here (up to 8 km), throughout the Neogene-Quaternary period. And the absence of regional angular unconformities and dislocations in Quaternary deposits suggests that the formation of the structural plan, which began in the late Miocene, continued in the Pliocene, Quaternary period and has not ended at the present time (Galtsev-Bezyuk and Polunin 1975; Khudyakov 1977). Thus, these marine accumulative formations are considered by us as the result of the development, inherited from the Neogene period, of areas of intensive sedimentation in the middle late Holocene. The volume of sediments forming the accumulative forms is 152 million m^3 and 96 million m^3, respectively. The coastal-marine type of overcompensated sedimentation, expressed in the formation of the largest spits on Sakhalin, under conditions of predominantly alongshore sediment transport in the Middle Late Holocene, was realized by the mechanisms of the dynamic theory of Zenkovich (1962).

In addition, a comparative analysis of the results of multi-temporal measurements and geological and geophysical profiles of the underwater coastal slope in the area of the accumulative filling form of the reentrant angle north of the sheet piling dam in the area of Cape Pogibi (Nevelskoy Strait) showed that during the first five years of the dam's existence, the volume of sediment accumulation exceeded 7.5 million m^3 (Afanasyev 2020).

A significant contribution to the coastal-marine sedimentation, only with the opposite sign, is made by the erosion of coastal ledges of low Holocene marine terraces. Thus, the erosion of the coastal ledge of northwestern Sakhalin is $1.9 \cdot 10^5$ m^3/year, of which $\geq 50\%$ are peatlands and silty-pelite deposits of shallow water with a high content of Corg (Afanasiev 1991).

The decisive external factor in the evolution of biomorpholithosystems in the Middle-Late Holocene was sea level fluctuations, which determined the general direction of the evolution of coastal plains and barrier forms. One of the most dynamic components of coastal landscapes is the lithogenic base. For the coastal-marine accumulation of the Far Eastern seas in the Holocene, three rhythms associated with microfluctuations (2–4 m) of the sea level were established. Each rhythm begins with wave accumulation facies, which then give way to estuarine-lagoonal facies and ends with peat accumulation (Afanasiev 1992).

Depending on the nature of the manifestation of abrasion-accumulation processes in the Holocene, three main types of accumulative coasts can be distinguished. NTC, including the formations of all three rhythms of the Holocene accumulation, were noted on the island coast of Sakhalin Bay, Aniva and Patience Bays. In the case when the coastal zone during the Middle-Late Holocene did not respond to the accumulative phases of development or the accumulative formations were destroyed as a result of subsequent erosion, a coast of a different type is observed. It is characterized by intense erosion of the alluvial-marine Upper Pleistocene deposits, the formation of a ravine network and systems of drained lakes at the Upper Pleistocene levels. This type of coast is most common in northwestern Sakhalin, northern Okhotsk, and western Kamchatka.

An intermediate position between the two types described above is occupied by the coast, where the processes of erosion and accumulation are more or less balanced. Depending on the situation, morpholithocomplexes of one or two accumulation phases are usually represented here. A special case for coasts of this type is the lagoon coast. Quite often, ancient barrier forms created by wave accumulation are destroyed, while the corresponding estuarine-lagoonal levels are preserved.

2 Regime of Accumulation and Transformation of Nitrogen-Containing and Carbon-Containing Sediments

In the system of allochthonous flows of organic matter and autochthonous carbon turnover at the place of accumulation, the mode of accumulation and transformation is determined by the geomorphological position and biomorpholithodynamic features of the development of estuaries, deltas and lagoons of the coastal zone.

Obviously, the geomorphological position of the formed march, expressed mainly in the sediment budget, determines the ratio of the range of vegetation growth and tidal fluctuations. In the event that rainfall fails to form a surface with the flood-free duration required to support budding vegetation, littoral mudflat development does not progress to a march stage even with sufficient nutrient input.

The lithogenic base here acts as an active landscape-forming component, causing a restructuring of the landscape structure at the level of facies and natural boundaries. For pioneer landscapes, the age of the lithogenic base is, in fact, the time of the beginning of the separation of the NTC. Here, new habitats are formed during the growth of accumulative forms due to sedimentation and replacement of algal flora.

The determination of the lower limit of the potential pioneer zone, above which plants have a high probability of survival, is of the utmost importance (Wang and Temmerman 2013; Hu et al. 2015). Figures 2, 3 and 4 show the surfaces of estuarine-lagoonal mudflats, which began to be populated by marsh vegetation.

Understanding the mechanisms contributing to the transformation of a potential zone of pioneer halophytic plant communities into marshes is an important task for further research. As shown by morphometric studies of lagoon basins, approximately 70% of their area is intertidal mudflat, for which it is necessary to zoning the zones of potential marches in accordance with the values and features of changes in the level and budget of sediments (Tables 1 and 2).

Taking into account the peculiarities of the tidal regime, the data of survey work and drilling, the thickness of the march sediments pushed into the water area of the lagoons is 0.5–1.7 m here. Thus, annually in the march formations of lagoons, the formation of a sedimentary stratum with a total volume of 113,000–385,000 m^3/year occurs naturally.

Fig. 2 The surface of the mudflat of the southwestern part of the proposed landfill, exposed at low tide at 14:30 on October 6, 2022, the value of which is 25 cm less than the maximum possible

Fig. 3 The surface of mudflat with continuous marsh vegetation in the southwestern part of the proposed landfill, exposed at low tide at 14:20 on 10/06/2022, the value of which is 25 cm less than the maximum possible. The surface of the mudflat is 8 cm above the surface in Fig. 2

About 50% of this volume are slightly decomposed plant sediments with bulk density from 0.19 to 0.3 g/cm^3. According to our research, the content of organic carbon in them ranges from 30–40%.

Thus, when converted into pure carbon, these deposits in their natural state accumulate approximately 10,000–34,000 tons/year C_{org}. In the fine-grained facies of marsh deposits, averaging the fraction of organic carbon in bottom sediments to 2%, we obtain the accumulation of pure carbon in approximately the same volume. An assessment of the vertical growth of previously formed marching surfaces is planned.

Due to the larger area of the water area, the volumes of carbon annually recorded in bottom sediments are significantly higher than the volumes of C_{org} recorded in increasing marches. For example, in the closed Ainskoye lagoon and Nyiskaya lagoon, by about an order of magnitude, and in the Chaivo lagoon and in Salmon Bay, about one and a half times (Afanas'ev and Faustova 2022).

Fig. 4 The surface of mudflat with troughs in the central part of the proposed polygon on 05/11/2022 at a low tide level of 27–30 cm less than the maximum possible

Table 1 Surface areas of the water surface of the lagoons, measured for 2018–2020

Name	Area (km^2)	Length of coastline (km)	Name	Area (km^2)	Length of coastline (km)
Piltun Bay and Ostokh Bay	440.1	269.7	Kolendu Bay (Koldu)	21.6	37.7
Baikal Bay	430.4	147.5	Ekhabi Bay	15.4	16,0
Lake Nevskoe	184.6	179.5	Odoptu Bay	13.4	26,9
Hall. Nabilsky	178.8	168.9	Big and Small Chibisan lakes	13.3	22.2
Lake Tunaicha	176.6	93.8	Khanguza Bay	9.5	14.0
Pomr Bay	153.6	87.7	Lake Izmenchivoe	8.6	12.7
Chayvo Bay	112.9	204.5	Tront Bay (Tropto)	7.7	16.7
Nyisky Bay	105.0	186.8	Lake Lebyazhye	7.1	20.2
Lunsky Bay	54.2	77.1	Ketu Bay	6.8	19.8
Lake Big Vavayskoye Small Vavayskoye	44.9	42.1	Viakhtu Bay	6.7	21.2
Busse Lake and Vyselkovoe Lake	42.9	33.2	Lake Baklanye	5.2	17.6
Lake Ainskoe	32.2	47.4	Lake Bird	3.8	16.2
Kuegda Bay and Neurtu Bay	30.9	45.2	Over 80 small lagoons	62.3	361.3
Urkt Bay	24.8	27.2	**Total**	**2193.2**	**2213.3**

Table 2 Main parameters of lagoons with established march dynamics

Lagoon	Water surface area (km^2)		Change in water surface area (km^2)			Change in area (m^2/year)	Suspended sediment runoff (thousand tons/year)	Maximum tide
	1952	2018–2022	Rising marches	Blurred marches	Result			
Piltun	–	438,904	–	–	–	29.471	34.0	2.13
Chayvo	109.600	107.645	2.379	0.424	1.955	28.765	27.0	2.68
Nyisky	106.567	104.302	3.059	0.794	2.265	37.750	560.0	1.94
Nabil	164.700	162.380	2.388	0.068	2.320	39.320	19.0	1.92
Lunskaya	53.900	52.739	1.399	0.238	1.161	20.368	7.2	1.66
Lake Lagoon Nevsky	185.127	182.808	3.021	0.701	2.319	34.107	18.2	1.81
Lake Lagoon Ainskoe	32.075	31.778	0.856	0.559	0.297	4368	43.7	0.88
Salmon Bay	–	–	2.255	0.045	2.210	32.495	100.8	1.29
Total	651. 969	1 080.556	15. 357	2.829	10.317	226 644	809.9	–

3 Conclusion

One of the most dynamic components of coastal biomorpholithosystems is the lithogenic base. For the coastal-marine accumulation of the Far Eastern seas in the Holocene, three rhythms associated with microfluctuations (2–4 m) of the sea level were established. Each rhythm begins with wave accumulation facies, which are then replaced by estuarine-lagoonal facies and ends with peat accumulation.

The intensity of organogenic sedimentation under modern conditions, characterized by a rise in sea level by about 3 mm/year, has been established by us in general terms. Understanding the mechanisms contributing to the transformation of a potential zone of pioneer halophytic plant communities into marshes is an important task for further research. As shown by morphometric studies of lagoon basins, approximately 70% of their area is silty dry land, for which it is necessary to zoning the zones of potential marches in accordance with the values and features of changes in the level and budget of sediments.

References

Afanas'ev VV, Faustova AB (2022) The first results of the study of sequestration properties of coastal marine biomorpholithosystems (Sakhalin Region) physica land mathematical modeling of earth and environment processes. Springer proceedings in earth and environmental sciences. https://doi.org/10.1007/978-3-031-25962-3_46

Afanasiev VV (1989) On the question of the morphostructure of northwestern Sakhalin. Geological and geomorphological features of some regions of the Far East and the zone of transition to the Pacific Ocean. Vladivostok, DVPI, pp 100–106

Afanasiev VV (1992) Evolution of the coast of the Far Eastern seas in the Holocene. Moscow, pp 166–174

Afanasiev VV (1991) Coastal erosion of northwestern Sakhalin. Coastal zone of the Far Eastern seas. Publishing house of the GO USSR Moscow, pp 98–104

Afanasyev VV (2020) Morpholithodynamic processes and development of the coast of the contactzone of the subarctic and temperate seas of the North Pacific. Yuzhno-Sakhalinsk, IMGiG FEBRAS, p 234. ISBN 978-5-6040621-8-0, https://doi.org/10.30730/978-5-6040621-8-0.2020-1

Curado G, Rubio-Casal AE, Figueroa E, Castillo JM (2014) Plant zonation in restored, nonrestored, and preserved Spartina maritima salt marshes. J Coast Res T 30(3. C):629–634. https://doi.org/10.2112/JCOASTRES-D-12-00089.1

Galtsev-Bezyuk SD, Polunin GV (1975) On the newest border of the Quaternary system on Sakhalin. Natural resources of Sakhalin, their protection and use. Yuzhno-Sakhalinsk, SP pp 83–93

Hu Z, Van Belzen J, Van Der Wal D, Balke T, Wang ZB, Stive M, Bouma TJ (2015) Windows of opportunity for salt marsh vegetation establishment on bare tidal flats: the importance of temporal and spatial variability in hydrodynamic forcing. J Geophys Res 120(7):1450–1469. https://doi.org/10.1002/2014JG002870

Karetnikova EA, Garetova L (2015) Spatial and temporal distribution of bacterioplankton and bacteriobenthos in the Amur Estuary and adjacent marine areas. Oceanology T 55(5):776–776

Khudyakov GI (1977) Geomorphotectonics of the south of the Far East (Questions of theory). Moscow, Nauka, p 256

Koltunov AM et al (2009) Carbonate system of the Amur estuary and adjacent marine areas. Oceanology T 49(5):694–706

Rosentreter JA, Al-Haj AN, Fulweiler RW, Williamson P (2021) Methane and nitrous oxide emissions complicate coastal blue carbon assessments. Glob Biogeochem Cycles 35:e2020GB006858. https://doi.org/10.1029/2020GB006858

Wang C, Temmerman S (2013) Does biogeomorphic feedback lead to abrupt shifts between alternative landscape states?: An empirical study on intertidal flats and marshes. J Geophys Res Earth Surf 118(1):229–240

Zenkovich VP (1962) Fundamentals of the doctrine of the development of sea coasts. Publishing House of the Academy of Sciences of the USSR, Moscow, p 710

Organogenic Morpholitosystem of the Nabil Lagoon

Victor V. Afanas'ev, A. V. Uba, A. O. Gorbunov, A. I. Levickij, and A. B. Faustova

Abstract The results of the analysis of geospatial information for the period 1952–2020 are presented. The average multiyear growth rates of marches in the southern part of the Nabil lagoon were determined. The questions of the structure of intertidal mudflat and salt marshes are considered, the features of organogenic lagoonal sedimentogenesis are shown. The results of determining the C_{org} content in estuary-lagoon deposits are presented. Taking into account the fact that the study is essentially reconnaissance, the interpolation of the obtained data is of an estimated nature.

Keywords Sakhalin Island · Coast dynamics · Salt marsh · Intertidal mudflat · Organic carbon

1 Introduction

In terms of the area of the water surface (162.4 km^2), the Nabil lagoon is the fifth largest lagoon-type reservoir. The volume of solid runoff into the lagoon is approximately equal to that into the lagoon of Nevsky Island, the fourth largest lagoon in Sakhalin (Afanasiev 2020). The Nabil Lagoon is the only lagoon in northeastern Sakhalin where the Middle Pleistocene marine terrace is eroded as the strait moves. As noted by Mikishin and Gvozdeva in the history of the development of the lagoon, the weakening of the processes of sedimentation of organogenic deposits, up to a complete cessation, occurred in cold and dry segments of the Subboreal-Subatlantic, especially during the "Little Ice Age of peat accumulation" (Mikishin and Gvozdeva 2006). It should be noted that, in general, the results on the content of organic

V. V. Afanas'ev (✉) · A. V. Uba · A. O. Gorbunov · A. I. Levickij
Institute of Marine Geology and Geophysics, FEB RAS, Yuzhno-Sakhalinsk, Russia
e-mail: vvasand@mail.ru

A. V. Uba
e-mail: vvasand@mail.ru

A. O. Gorbunov · A. I. Levickij · A. B. Faustova
Sakhalin State University, Yuzhno-Sakhalinsk, Russia

© The Author(s), under exclusive license to Springer Nature Singapore Pte Ltd. 2023
T. Chaplina (ed.), *Processes in GeoMedia—Volume VII*, Springer Geology,
https://doi.org/10.1007/978-981-99-6575-5_23

Fig. 1 Sampling of bottom sediments at the mouth of the Nabil River, which flows into the lagoon of the same name, using a piston sampler (Eijkelkamp)

carbon in the sediments of the intertidal mudflat of the lagoon are slightly higher than the data obtained from the study of sections in the estuary zone of the river in southern Sakhalin (Afanas' ev et al. 2023a, 2023b). The work was carried out in August-October 2022 (Fig. 1).

2   Methods and Results

The work is based on the analysis of arrays of aerial photographs of 1952 and satellite images of 2020, which was performed in the Quantum GIS geoinformation system. The calculations were made on the WGS84 EPSG:7030 ellipse. The resulting attributes were exported to spreadsheets for further processing. At the next stage of the study, in the same QuantumGIS environment, using standard procedures for analyzing remote sensing materials, we reconstructed the changes in contours and areas for the period 1952–2020. Published data and the authors' own results were used in calculating carbon stocks in marshes and bottom sediments. The carbon content was analyzed in 2 sections of march deposits and 4 sections of bottom sediments. Sampling was carried out with a 150 cm geosliser and a Beaker sampler (Eijkelkamp).

Samples from the columns for the determination of total carbon, organic carbon, total nitrogen were taken at five- and two-centimeter intervals.

Sediment analyzes were carried out in the laboratory of the Forestry Institute. Sukacheva FRC KSC SB RAS. The content of carbon and nitrogen, as well as the ratio of stable carbon isotopes ($\delta13C$) was measured using an elemental analyzer (Vario Isotope Cube, Elementar Analysis Systems GmbH, Hanau, Germany) connected to an IRMS isotope mass spectrometer (IsoPrime100, UK, Elementar Analysis Systems GmbH, Hanau, Germany). Ratios of stable carbon isotopes are given using traditional δ-notation as ‰ deviation from the belemnite standard (VPDB).

Figure 2 presents the results of visualization of measurements of changes in the area of marshes in the delta parts of the rivers flowing into the southern part of the Nabil lagoon. The area of marches here increased over the period 1952–2020 by 2,319,842 m^2, or 34,115 m^2/year.

The low lagoon terrace studied by two sections is characterized by a two-layer structure. An example of the results of chemical analysis of deposits is presented in Table 1.

The upper layer is represented by floating peat bogs with a thickness of 25–40 cm. Taking into account the average value of the lowland peat skeleton density index— 0.144 g/cm^3, the content of organic carbon in the upper layer varies from 55.74 to 65 kg/cm^3. The average carbon content in the lower part of the section ranges from 35.19 to 39.61 kg/m^3. At the same time, bottom sediments are represented in the upper part by peat deposits 10–15 cm thick with a density of about 0.9 g/cm^3. Accordingly, the measured content of C_{org} in these sediments is 115.2–205.9 kg/m^3.

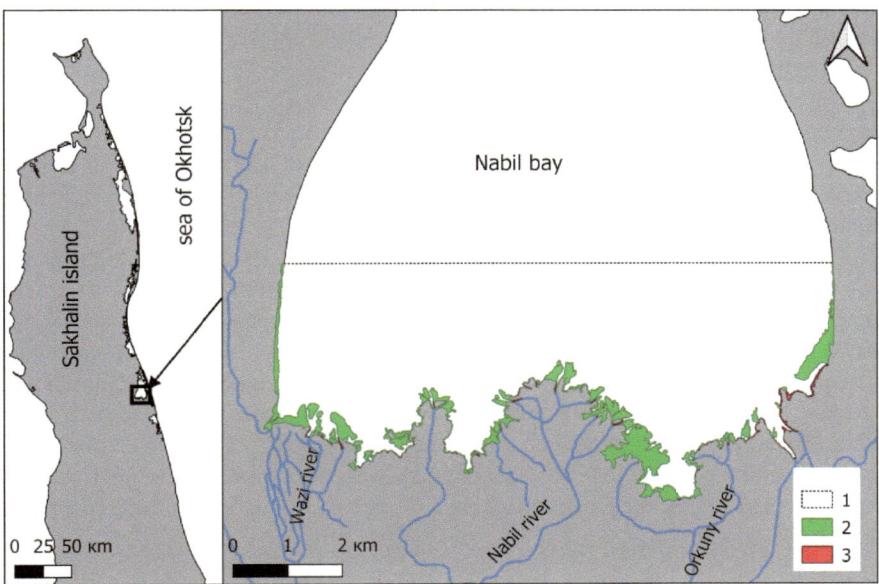

Fig. 2 Change in water surface area of the Nabil Lagoon; 1—area of calculation; 2—increase in the area of marches in the delta areas of the rivers flowing into the lagoon; 3—erosion of lagoon shores

Table 1 Results of chemical analysis of sediments from column No 6

Name	$\delta^{13}C$ (‰)	$\delta^{15}N$ (‰)	%N	%C	C/N	Weight
H6-2	−28.76	0.59	2.38	40.45	17.0	4.602
H6-3	−28.31	0.66	2.40	40.82	17.0	4.492
H6-4	−27.95	0.68	1.94	36.78	19.0	6.218
H6-5	−27.70	0.70	1.77	37.28	21.0	5.289
H6-6	−27.30	0.95	2.07	38.21	18.4	4.678
H6-7	−27.02	1.18	2.05	36.66	17.9	4.859
H6-8	−26.87	2.19	1.07	17.97	16.8	5.850
H6-9	−27.38	2.40	0.84	14.00	16.7	4.792
H6-10	−27.29	2.55	0.14	1.91	13.3	18.629
H6-11	−27.08	3.25	0.18	2.54	14.5	37.295
H6-12	−27.37	3.29	0.20	3.08	15.6	42.021
H6-13	−24.96	3.79	0.17	2.31	13.3	43.725
H6-14	−24.15	3.55	0.34	3.98	11.5	39.049
H6-15	−23.26	3.61	0.15	1.69	11.2	45.260
H6-16	−23.33	3.65	0.06	0.56	9.4	101.341
H6-17	−24.47	3.98	0.06	0.51	8.8	94.088

The content of organic carbon in the cores of bottom sediments in the southern part of the Nabil lagoon in the area of the mouth of the Nabil River varies from 24.99 kg/m^3 (H-4) to 27.88 kg/m^3 (H-3).

Moreover, in the upper 15 cm, the carbon content is 6.29 kg/m^3 and 8.16 kg/m^3, respectively. The average content of organic carbon in a 75 cm core 4.5 km east of the mouth of the Nabil River is 10.03 kg/m^3, but in the upper sediment layer, in contrast to the near mouth of the Nabil River, the carbon content is approximately 1.5 times higher than the average for the section.

Thus, the morpholithodynamic position has a significant effect on the distribution of organic matter in the upper dynamic layer of bottom sediments. In a more general form, this has already been noted earlier (Tokarchuk 1999; Efanov et al. 2013; Kafanov and Labay 2003). The section of bottom sediments of the lake was also analyzed. Protochnoe, located 16 km southeast of the mouth of the Nabil River. The lake is separated from the Sea of Okhotsk by an attached Holocene terrace 800 m wide. Waved dunes are marked on the surface of the terrace. The average content of organic carbon in the section was 26.35 kg/m^3. Moreover, in clearly defined "washed out" or eolian layers, the Corg content is 7.48 kg/m^3, respectively, in the rest of the section, 34.00 kg/m^3.

3 Conclusion

In the presented work, along with the geospatial characteristics of estuary-lagoonal biogeomorphological systems, the results of a geochemical analysis of the results of biogenic sedimentation processes are presented. Taking into account the fact that the study is essentially reconnaissance, the interpolation of the obtained data is of an estimated nature. Thus, when converted into pure carbon, 757 tons are accumulated annually in the peat deposits of the march. Dense deposits of the lower part of the march itself accumulate approximately 546 C_{org} t/year. The underlying silty sediments up to 1 m thick, when averaged over three cores, give 716 C_{org} t/year. Previously, we took as an average value of 1.5–1.7 m for the thickness of the march sediments pushed into the water area of the lagoon (Afanas' ev and Faustova 2023). Consequently, in the southern part of the Nabil lagoon, only due to the formation of new marshes, about 2000 tons of organic carbon annually accumulate. The obtained values generally correspond to those obtained by other methods (Afanasiev et al. 2022; Afanas' ev et al. 2023).

References

Afanas'ev VV (2020) Morpholithodynamic processes and development of the shores of the contact zone of the subarctic and temperate seas of the North Pacific. Yuzhno-Sakhalinsk, IMGiG FEB RAN, p 234

Afanas' ev VV, Faustova AB (2023) Sequestration of Organic Carbon in Salt Marsh Formations of Lagoons of Sakhalin. Physical and mathematical modeling of earth and environment processes—2022: proceedings of 8th international scientific conference-school for young scientists. Springer Nature Switzerland, Cham, pp 263–267

Afanas' ev VV, Uba AV, Levitsky AI, Faustova AB (2023a) Organogenic Sedimentation in the Nyivo Lagoon. Physical and mathematical modeling of earth and environment processes—2022: proceedings of 8th international scientific conference-school for young scientists. Springer Nature Switzerland, Cham, pp 503–508

Afanas'ev VV et al (2023b) Carbon sequestration in the coastal marine biomorpholitho systems of the Salmon Bay of Aniva Bay. Physical and mathematical modeling of earth and environment processes—2022: proceedings of 8th international scientific conference-school for young scientists. Springer Nature Switzerland, Cham, pp 495–501

Afanasiev VV, Uba AV, Levitsky AI, Faustova AB (2022) Peculiarities of organogenic sedimentation within the coastal zone of Aniva Bay. Process Geomedia 4(34):1828–1833

Efanov VN, Vypryazhkin EN, Latkovskaya EM (2013) The current state of bottom sediments of the Busse lagoon (Aniva Bay). Sci Bus Ways Dev 31–36

Kafanov AI, Labay VS, Pecheneva NV (2003) Biota and macrobenthic communities of the lagoons of northeastern Sakhalin. A. Yu - Sakh, SakhNIRO, p 176

Mikishin YA, Gvozdeva IG (2006) Stratigraphy of sediments and paleogeography of the coast of northeastern Sakhalin in the Holocene. Sci Rev (3):4–15

Tokarchuk TN (1999) Geochemistry of Sakhalin lagoons and rational use of their resources: diss. c and. Geographer. Sciences T.N. Tokarchuk. Vladivostok, FEGU, p 135

Concept of a Structural System of a Large-Panel Residential Building with a Free Layout Considering the Requirements for Earthquake Resistance During Construction in a Sharply Continental Climate

A. V. Zakharov ⓘ

Abstract In a sharply continental climate, it is necessary to minimize construction work, during or after the performance of which, there are technological interruptions caused by the nature of the materials used (wet processes). The conducted analysis shows the limited capabilities of the most widely used structural solutions of fully assembled panel buildings with transverse load-bearing walls in a sharply continental climate. Redevelopment or the freedom of planning decisions is not possible in such buildings. The article describes the proposal of a new constructive solution in panel residential buildings with longitudinal load-bearing walls of a pylon structural system, which allows for meeting the requirements for earthquake resistance, implementing the freedom of planning decisions, improving the efficiency of maintenance of engineering networks, and carrying out alterations in the building considering the functional purpose of the premises.

Keywords Sharply continental climate · Large-panel residential building · Free layout

1 Introduction

In a sharply continental climate, prefabricated large-panel housing construction is often used due to the possibility to conduct it at a high rate all year round (Markhayeva et al. 2023; Zaretskaya 2022; Grafkina and Sviridova 2022). This type of construction is associated with the use of reinforced concrete panels that can be quickly and easily connected to create walls and partitions (Bazhenov et al. 2021).

A. V. Zakharov (✉)
Department of Architecture, Research Moscow State University of Civil Engineering, Yaroslavskoe shosse, 26, Moscow 129337, Russia
e-mail: arkadiyzakharov@mail.ru

© The Author(s), under exclusive license to Springer Nature Singapore Pte Ltd. 2023
T. Chaplina (ed.), *Processes in GeoMedia—Volume VII*, Springer Geology,
https://doi.org/10.1007/978-981-99-6575-5_24

One of the main advantages of prefabricated large-panel construction is its high speed and ease of installation (Kashina et al. 2022). In a sharply continental climate, where severe frosts and temperature changes can complicate the construction process, the use of such technology can significantly reduce the time of construction of buildings and minimize the cost of resources and labor (Zakharov and Zabalueva 2022). Reinforced concrete, from which the panels are made, has high strength and resistance to external factors, which makes it especially suitable for use in severe frosts and temperature changes.

In addition, prefabricated large-panel construction allows (Glebova et al. 2022) creating buildings with high energy efficiency. Due to the high insulating capacity of reinforced concrete, such buildings can significantly reduce the cost of heating and air conditioning, which is especially relevant in a "extreme" continental climate (Sedova and Alekseev 2022; Baibussenov et al. 2021).

At present, the proposed design solutions have certain disadvantages that do not allow building a large array of panel residential buildings efficiently enough, meeting the needs of the time.

2 Research Background

According to the types of structures of intermediate floors, there are three types of systems of large-panel buildings (Drozdov and Sebekin 1967). The most common type is a closely-spaced building (about 3 m) of transverse wall panels bearing solid-section ceilings with a thickness of about 15 cm. The panels of the longitudinal walls on each floor are supported by the floor panels within their supporting edges in the platform joints, providing the **bonding** of vertical seams (joints) along the concrete body. This bonding, together with the connections of the metal-embedded parts of the panels, provides reliable stability for the entire frame of a large-panel building of this type under powerful natural and man-made force influences, such as an earthquake and an explosion of household gas. The transverse stability of the building with a margin is provided by a large number of often closely-spaced transverse load-bearing walls. Longitudinal stability, also with a margin, as stiffening diaphragms, is provided by panels of longitudinal, internal, and external walls. Such a rigid multicellular box structure allowed erecting 25-story large-panel buildings in the 1970s, and later in many cities of the world, from monolithic reinforced concrete—of over 200 m (Balova et al. 2021).

The thickness of the panels of the considered structural system for the technical level of construction of the twentieth century successfully combined enclosing and bearing functions. Thus, the thickness of the floor panels and internal walls made of heavy and durable reinforced concrete were approximately equal (14–16 cm), which provided sufficient sound insulation for inter-apartment fences, the load-bearing capacity of the floors with a span of about 3 m, and the wall panels with a thickness of 16 cm withstand the load of a 16-story building. The panels of the exterior walls, which perform a heat-protective function, were made of lightweight concrete. They transferred their weight floor-by-floor, through the supporting edges of the floor panels to the load-bearing internal panels and, therefore, loaded mainly only by their weight, they had equally low voltage on any floor. This allowed using the lightest concrete in them, which has the greatest thermal insulation. The close weight and overall parameters of the panels allowed effectively using transport and lifting equipment for the construction of the building (Kornilova et al. 2021). The described system with closely-spaced transverse load-bearing walls has become the basis of many serial projects used to date due to the technical advantages that allowed quickly erecting buildings at a relatively low price. The technology of construction has significantly improved, and the operational qualities and architectural appearance of buildings have also improved for more than 75 years of history (Volodin 2006; Zakharov and Leonteva 2016). Thus, their inherent shortcomings were revealed. One of them—is the overspending on reinforced concrete on internal walls, the thickness of which is assigned according to the conditions of the required inter-apartment sound insulation and the load-bearing capacity of the panel of the lowest floor of the building. These conditions correspond, for example, to a panel thickness of 16 cm on the first floor of a 16-story building.

The thickness of the interior (intra-apartment) walls with **sound insulation can be twice as thin** as the inter-apartment walls, the number of which is 2–3 times less in the building. Load-bearing panels are reduced as they are located on higher floors, which allows for reducing their thickness. However, they must be maintained under the operating conditions of the adopted structural system.

Over time, an even larger drawback was revealed with closely-spaced transverse load-bearing walls, a small-cell structure of the building frame, which excludes the possibility of free planning and redevelopment due to a change in the functional purpose of the building's areas and standard requirements.

The potential reliability of the building frame allows it to exist for much more than 100 years, even though at this stage of socio-economic world development, the demographic situation, and, consequently, the requirements for the number of apartments and their composition change greatly each generation. Over 100 years of service of the building frame, the obsolescence of the building occurs five times, requiring its redevelopment or the demolition and construction of a new building with a different apartment composition.

The freedom of planning is partially solved by the second (Nikolaev 2013a; Nikolaev et al. 2014) and third (Nikolaev 2013b, 2013c) constructive large-panel systems based on prefabricated floor slabs with a span of up to 6 m, with its width limited to 3 m, following the rules established by law for the transportation of bulky goods by road common use. The second system represents a large-panel building with widely pitched (6 m) transverse load-bearing walls, and the third—a large-panel building with longitudinal load-bearing walls. The floor panels here are prestressed hollow core slabs 22 cm thick, resting on the walls on two short sides. This ensures that they work according to the beam scheme and bonding of vertical seams on concrete wall panels only in the area of their support, which is significantly less than the area of bonding of floor panels and walls along the entire contour of each floor panel in the first structural system. This circumstance makes the building frames operating under the second and third structural systems less resistant to seismic and man-made impacts compared to the building frames operating under the first structural system, even with an increase in the number of embedded parts and fittings being welded during installation (Liu et al. 2022). The freedom of layout has increased: in one structural cell, it has become possible to arrange two rooms instead of one or one large room with an area of more than 30 m^2. With a decrease in the number of load-bearing wall panels, it became possible to install light interior partitions sufficient for sound insulation, which facilitated the whole building as a whole.

Buildings of the third structural system, with three longitudinal load-bearing walls, have even greater freedom of layout. However, to ensure transverse stability, they must have transverse stiffening diaphragms for the entire width of the building every 10 m of its length. There is an increasing need to compensate for the small bonding of the vertical seams of the panels by increasing the welded joints or by installing monolithic reinforced concrete dowels.

The purpose of the work is to develop a concept of a structural system of a large-panel residential building with a free layout, considering the requirements for earthquake resistance of buildings under construction in areas with a sharply continental climate.

3   Methods

To achieve this goal, a series of studies were conducted, in particular, by Zabalueva (2023) and Leontieva (Zakharov et al. 2014) and other employees of the Moscow State University of Civil Engineering, to identify opportunities for achieving free planning and redevelopment in a large (500 m^2) section of a residential large-panel building with longitudinal load-bearing walls. The experimental project was carried out based on the structural system, the plan of which is shown in Fig. 1.

The thermal insulation and facade layers of external wall panels are not shown in the article, due to the need to adhere to the stated focus of research and for ease of perception of the work of structures.

Calculated studies of the seismic resistance of the section in question with dimensions in terms of 42 × 18 m and a height of 17 floors were carried out based on (Kozyrev 2010a, b) to determine the correct type of foundation, structural strength, reinforcement of the main walls, and earthquake protection.

4   Results and Discussion

As can be seen on the plan (Fig. 1), the area of the section is limited only by external walls and walls enclosing the staircase and elevator node. The exterior walls are zigzagged—from flat panels (1) parallel to the facade and pylons perpendicular to it (2). Such an arrangement of pylons, with a width of 1 m, visually does not violate

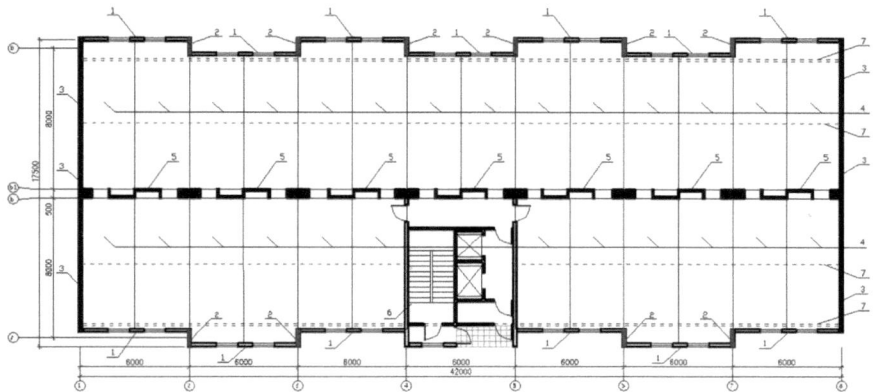

Fig. 1 Section plan: 1—a zigzag wall made of flat panels, 2—pylons, 3—end walls, 4—floor panels, 5—a middle wall made of three-dimensional panels, 6—staircase-elevator block, 7, 8—longitudinal reinforcement of the cover disk

the unity of the room space and provides transverse installation and local operational stability of the exterior walls and the building within 3–5 floors. The thickness of the pylons is determined by the number of floors of the building. The overall transverse stability of the building is provided by the end walls (3) and the staircase-elevator unit, all reinforced concrete elements of which are united into a core of rigidity.

Longitudinal rigidity is provided by longitudinal external walls and volumetric panels (5) of the middle wall, which accommodate the risers of all engineering and sanitary networks of the building in their cavities (Liu et al. 2022). All these rigidity elements are floor-by-floor combined by floor panels (4), gathered into a single disk using continuous reinforcement with rods (7) for the length of the section and filling the jelly-like profile of the joints with concrete on the stressed cement (Fig. 2).

The bearing function of the building structures is carried out by transferring the load of the floor panels through the panels of the inner wall directly to the foundation and floor-by-floor transmission through the outer wall panels, working as wall beams, to the pylons and, through them, to the foundation.

A necessary condition for the reliable functioning of a building made of prefabricated elements is the bonding of their vertical seams. Bonding in panel-reinforced concrete elements is carried out using monolithic reinforced concrete dowels, the

Fig. 2 The joint of panels of intermediate floors: 1—rods of longitudinal reinforcement of the disk of the intermediate floor, 2—heating pipes

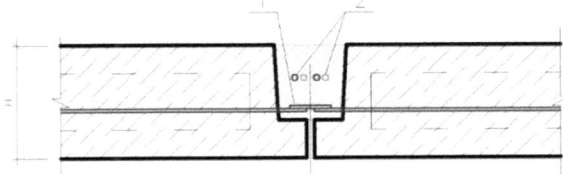

Fig. 3 The support of the
outer panels on the pylon:
1—the outer panel, 2—the
pylon, 3—the support
platform

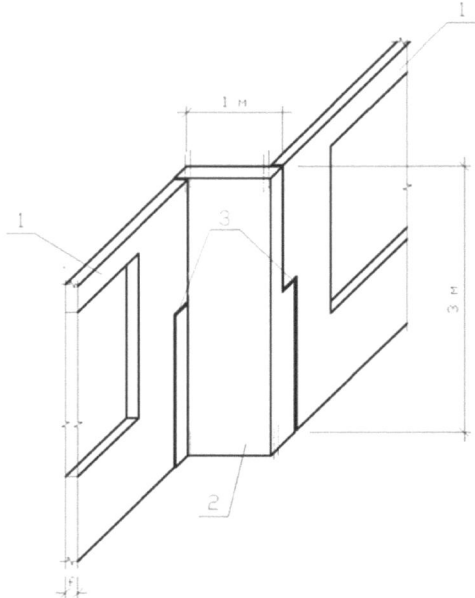

construction of which requires considerable time in construction conditions, especially at subzero temperatures in winter. In the project under consideration, it is proposed to carry out the bonding with ledges as shown in Fig. 3, the support of the bearing, inner, and outer layers of panels on the pylons.

The elements are connected in the following sequence.

The pylons are installed and fixed around the perimeter of the section at the level of the mounted floor, with the help of a conductor and templates. Then the panels, protruding supporting parts of their load-bearing layers, are lowered onto the support tables of the pylons (preferably equipped with plug-in docking nodes). The project should provide a level of support that is several cm higher than the center of gravity of the panel being mounted. Then the panel, under its weight, occupies the design position reducing the installation time by a multiple.

Floor panels with spans of 7 and 8 m with a width of 3 m with a construction height of 0.3 m have a box-shaped section, the free cavities of which are filled with a noise absorber to increase sound insulation. Along the entire perimeter, they have a solid edging with a height of 0.1 m. In the design position, adjacent panels of floors and walls along these perimeters form gutters with a depth of 0.2 m. Gutters are suitable for placing water pipes, sewerage, and heating pipes in them, coming from the connection points to the corresponding risers located in the cavities of the middle longitudinal wall. During the construction process, in the gutters and at the places of possible locations of sanitary appliances, connection nodes with pipes are placed and covered at the level of the surface prepared for the installation of floors. During the construction of apartments, the nodes that are only necessary for the layout adopted at this time are opened. Other necessary nodes are opened, and unnecessary ones are silenced during the subsequent redevelopment. Spending money on the construction of spare connection nodes pays off many times by reducing time and money during subsequent alterations.

All gutters are filled with concrete after the completion of the floor laying of the described sections of sanitary networks. To accelerate the strength gain of concrete in cold weather, it is possible to turn on the heating circuit mounted on the floor according to a temporary scheme.

After gaining the strength of concrete in the gutters, the intermediate floor is ready for the installation of walls and partitions following the project of the location of apartments on the floor. It is most expedient to install them according to the proven technology, using double plasterboard panels on a metal frame with an air gap containing a mineral wool sound absorber. In accordance with the requirements of sound insulation, fire resistance, and burglary protection, panel thicknesses are selected from the required number of drywall sheets. The use of plasterboard wall structures in the proposed structural system reduces the load from internal walls on intermediate floors by more than an order of magnitude compared to the use of single-layer concrete walls, which in general reduces the weight of the building by a third.

Calculated studies of seismic resistance have shown that the section can withstand a seismic load of 7 points with the thickness of the pylons and the bearing layer of the end wall panels 0.2 m.

5   Conclusion

The elements of the experimental project proposed in the article conceptually show the possibility of creating a constructive system of a multi-story large-panel building, providing the possibility of multivariate planning solutions for the number and composition of apartments within the section, as well as the possibility, at the request of the current time, of redeveloping them within the floor of the section, without the need for work in the premises of adjacent sections.

Therewith, the proposed concept is designed to be used for the construction of buildings in a sharply continental climate. This is especially important in regions with a long winter when the construction process can be difficult due to climatic and weather factors.

Calculated studies of seismic resistance have shown that a residential building can withstand a seismic load of 7 points, which means that it can withstand an earthquake with an intensity of up to 7 points on the Mercalli scale (or another similar scale of seismic hazard assessment), i.e. it can lead to significant damage to the building, but the building itself must remain standing.

References

Baibussenov K, Bekbaeva A, Azhbenov V, Sarbaev A, Yatsyuk S (2021) Investigation of factors influencing the reproduction of non-gregarious locust pests in northern Kazakhstan to substantiate the forecast of their number and planning of protective measures. OnLine J Biol Sci 21(1):144–153

Balova S, García De Velazco J, Polozhentseva I, Chernavsky M, Shubtsova L (2021) The formation of the concept of smart sustainable city with the purpose of environmental protection. J Environ Manag Tour 12(5):1269–1275

Bazhenov YM, Murtazaev S-AY, Bataev DK-S, Alaskhanov AH, Murtazaeva TSA, Saydumov MS (2021) High-strength concretes based on anthropogenic raw materials for earthquake resistant high-rise construction. Eng Solid Mech 9

Drozdov PF, Sebekin IM (1967) Design of large-panel buildings. Publishing house of literature on construction and architecture, Moscow

Glebova IS, Berman SS, Gribovskaya VA (2022) Housing construction: problems and prospects (As Exemplified by Russia). Revista Relações Internacionais do Mundo Atual 2(35)

Grafkina MV, Sviridova YE (2022) Application of risk-oriented approach for improvement of the environmental security of the urban area. Int J Saf Secur Eng 12(4):519–524

Kashina E, Yanovskaya G, Fedotkina E, Tesalovsky A, Vetrova E, Shaimerdenova A, Aitkazina M (2022) Impact of digital farming on sustainable development and planning in agriculture and increasing the competitiveness of the agricultural business. Int J Sustain Dev Plan 17(8):2413–2420

Kornilova AA, Khorovetskaya YM, Abdrashitova TA, Smagulova AB, Lapteva IV (2021) Modern model of a rural settlement: development of planning structure and reconstruction of villages. Civ Eng Arch 9(1):214–224

Kozyrev KV (2010) Limits of applicability of the theory of seismic stability. Seismicheskaya bezopasnost. http://seismics.kvkozyrev.org/index.php?option=com_content&view=section&layout=blog&id=5&Itemid=50&limitstart=10

Kozyrev KV (2010) Principles of the resonance approach in seismic science. Seismicheskaya bezopasnost. http://seismics.kvkozyrev.org/index.php?option=com_content&view=article&id=51:2&catid=40:articles&Itemid=18

Liu Y, Zub A, Zha S (2022) Impact of attracting intellectual capital on the innovative development of construction engineering enterprises. J Manag Technol 22(4):153–168

Markhayeva B, Ibrayev A, Beisenova M, Serikbayeva G, Arrieta-López M (2023) Green banking tools for the implementation of a state's environmental policy: comparative studythe active implementation of climate change projects, the latest resource- and nature-saving technologies, and environmental measures for the green growth of na. J Environ Manag Tour 14(1):160–167

Nikolaev SV (2013a) SPKD—housing construction system for future generations. Zhilishchnoe Stritelstvo 1:3–8

Nikolaev SV (2013b) Panel and frame buildings of a new generation. Zhilishchnoe Stroitelstvo 8:2–10

Nikolaev SV (2013c) Social housing at a new stage of improvement. Zhilishchnoe Stroitelstvo 3:2–8

Nikolaev SV, Shreiber AK, Khayutin YuG (2014) Innovativeness of the panel-frame housing construction system. Zhilishchnoe Stroitelstvo 5:3–8

Sedova OV, Alekseev AG (2022) Features of using mathematical models to calculate the effectiveness of a digital platform for ecological monitoring. J Theor Appl Inf Technol 100(3):856–869

Volodin V (2006) Panel building. Modern classic. SibDom. https://www.sibdom.ru/journal/188/

Zabalueva TR (2023) History and trends of modern museum architecture. Synesis 15(2):190–207

Zakharov AV, Leonteva MP (2016) Structural solutions for large-panel residential buildings of a new generation. Promyshlennoe i Grazhdanskoe Stroitelstvo 10:104–110

Zakharov AV, Zabalueva TR (2022) Identification of approaches to architectural and structural solutions in the design of sports buildings. Nexo Revista Científica 35(03):736–745

Zakharov AV, Zabalueva TR, Leonteva MP (2014) New approaches to the design of large-panel buildings with longitudinal load-bearing walls. Promyshlennoe i Grazhdanskoe Stroitelstvo 7:66–69

Zaretskaya M (2022) Assessment of geo-environmental consequences of oil and gas complex enterprises' extraction activities on the shelf. Math Model Eng Probl (9)6:1497–1502

Thermal Geophysical Investigation of the Snow Cover, Ice and Frozen Ground in the Lite of Climate Variation

D. M. Frolov, G. A. Rzhanitsyn, A. V. Koshurnikov, and V. E. Gagarin

Abstract The paper presents the results of the continuation of the thermal and geophysical researches of snow cover, ice glacier and frozen ground thickness and structure in the lite of climate variation. It presents results of investigations conducted in Moscow in winter period 2023. The previous research was conducted at the Garabashi glacier (Elbrus, Caucasus) in fieldworks periods of 2021 and 2022 and presents the results of using ground penetrating radar (GPR) to survey a ski slope prepared on a glacier. The ground temperature monitoring well is drilled to estimate the effect of snow cover on the ground freezing depth and temperature variation in the changing climate and to verify the applied for Russian permafrost monitoring ground temperature monitoring system. The newly obtained data also allows development of the revealed before results.

Keywords Temperature monitoring · Ice · Snow cover · Frozen ground · Spatial and temporal heterogeneities · Ground penetrating radar research

1 Introduction

According to the Copernicus Climate Change Service (https://climate.copernicu s.eu/) June 2023 was just over 0.5 °C warmer than the 1991–2020 average, beating the previous record of June 2019 and the Earth continues to break temperature records due to global warming. According to the National Center for Environmental Forecasting (USA), the average world temperature was 17 °C, which slightly exceeded the previous record of 16.9 °C recorded in August 2016 (Fig. 1).

This of cause may have some serious impact on humans and ecosystems. Now China faces waves of heat and breaking temperature records in Beijing. In India extreme heat led to death in some of the poorest regions of the country. Dangerous thermal dome covered Texas and northern Mexico and Great Britain experienced the

D. M. Frolov (✉) · G. A. Rzhanitsyn · A. V. Koshurnikov · V. E. Gagarin
Lomonosov Moscow State University, Moscow, Russia
e-mail: denisfrolovm@mail.ru

© The Author(s), under exclusive license to Springer Nature Singapore Pte Ltd. 2023
T. Chaplina (ed.), *Processes in GeoMedia—Volume VII*, Springer Geology,
https://doi.org/10.1007/978-981-99-6575-5_25

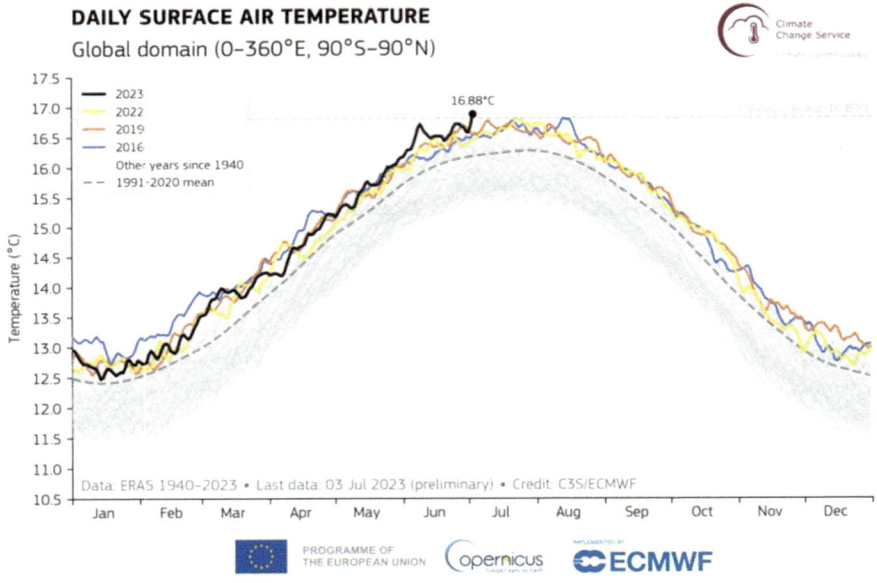

Fig. 1 Daily surface air word temperature

hottest June in the history of observation. In Germany the Rhine River water level sinks and river ships may carry only 40% of theirs cargo loads.

According to the NASA (https://www.nasa.gov/) the July of 2023 was also the hottest for the instrumental period of observations (Fig. 2).

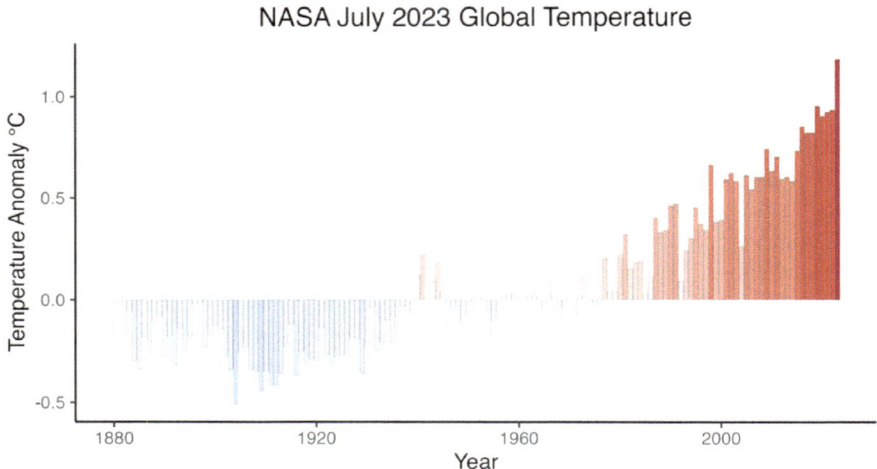

Fig. 2 July air temperature

According to the Japanese reanalysis JRA-55 (https://jra.kishou.go.jp/JRA-55/index_en.html), in August 2023 the average air temperature on the planet was + 16.91 °C. This is the highest value since at least 1836. In the Northern Hemisphere, the anomaly is +0.93 °C. Previously, August 2016 was the warmest. Then the average temperature was +16.62 °C. The average temperature over the past 8 months on the planet was the highest for all time of observations.

The World Meteorological Organization reported (https://public.wmo.int/en/media/news/) that for the first time in seven years El Niño conditions have emerged in the tropical Pacific Ocean, which contributes to a sharp increase in temperature. This creates a significant risk of exceeding global warming by 1.5 °C. And in a changing climate the snow, ice, glaciers and frozen ground resources of Russia are also highly influenced by the global warming. Since in summer 2020 when huge techno genic catastrophe in Norilsk with oil spill due to the thawing permafrost under the basement of the oil reservoir happened the new program of permafrost monitoring has been developed in Russia and on 28 June 2023 the deputies of the State Duma of Russia in the third final reading unanimously voted for the bill, thanks to which a system of state background monitoring of permafrost will appear in the country for the first time. This document was developed by the Ministry of Natural Resources of Russia on behalf of President Vladimir Putin.

The new permafrost monitoring system is being created on the basis of the observation network of Roshydromet, its data will be free and accessible to everyone. The Arctic and Antarctic Research Institute (AARI), a subordinate institution of Roshydromet, has been identified as the operator of the system. So the online data will be sent to the Permafrost Monitoring Center of the institute, and then submitted to the Unified State Data Fund on the State of the Environment, Its Pollution, to state authorities and other interested organizations. In total, the monitoring system concept provides for the equipment of 140 observation points for changes in permafrost all around cold regions of Russia.

So in order to test the applied ground temperature monitoring system it was installed at the Lomonosov MSU meteorological site and it operated there for the recent winter periods. Also the study included snow pit surveys at the meteorological site.

2 Materials and Methods

The paper presents the results of field studies conducted at the meteorological observatory site of Lomonosov Moscow State University for the recent winter period which included ground temperature monitoring in the well and also snow pits survey. For example in winter period 2022/23 the development of the snow mass was studied and its spatial variability over one winter season was revealed. Field research consisted of analyzing the stratigraphic layers of the snow mass and measuring their density. The winter of 2022–2023 turned out to be heterogeneous in terms of temperature regime.

The paper also presents results of investigations conducted at mountain slopes in winter period in Moscow and at Caucasus in 2023. The previous research was conducted at the Garabashi glacier (Elbrus, Caucasus) in fieldworks periods of 2021 and 2022 and presents the results of using ground penetrating radar (GPR) to survey a ski slope prepared on a glacier for summer competitions, where the ski slope was surveyed, and layers of snow, firn, ice and underlying rock were identified. Based on the analysis of the data obtained, a conclusion was made about the state of this route, and recommendations on the use of a GPR (Sokratov and Frolov 2023).

For example the natural opportunities of the Moscow region provide several objects with natural snow cover slopes in winter. The one of them is "Vorob'yovy gory" hills which snow cover slope was investigated by us in winter period 2022/23. The winter of 2022–2023 turned out to be heterogeneous in terms of temperature regime, with the average monthly temperature in December relatively close to the norm.

3 Results and Conclusion

The ground under the snow cover does not freeze during the winter period at the Lomonosov observatory site. The changes in temperature of the soil in the well at the Lomonosov observatory site are given in Fig. 3.

The observed thermal gradient in the well is about 3 °C/100 m. The displayed results of ground temperature monitoring in the well and also snow cover surveys

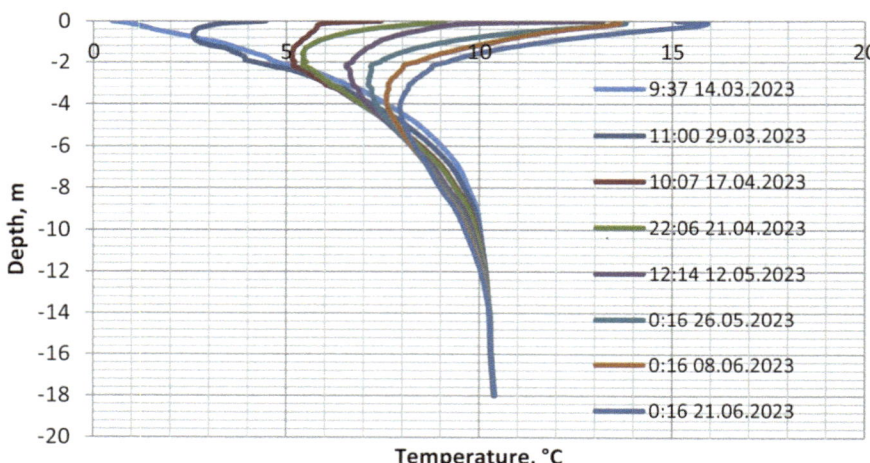

Fig. 3 Changes in the ground temperature in the well at the Lomonosov observatory site

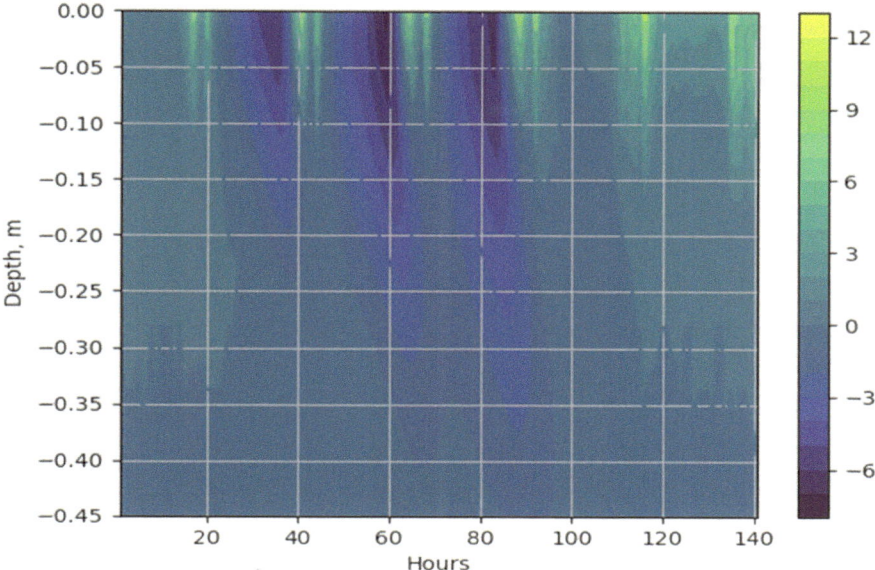

Fig. 4 The temperature variations in the slope snow pack at Vorob'yovy gory

allows to estimate the effect of snow cover on the ground freezing depth and temperature variation in the changing climate and to verify the applied for Russian permafrost monitoring ground temperature monitoring system.

The penetration of the day-night cold wave in the slope snow pack one can see at the temperature monitoring graph presented in the Fig. 4.

This may be compared to the snow pack temperature distribution at the Garabashi glacier during the summer period of 2021, where snow compaction work was carried out in that area and the quality of the snow made it possible to safely hold of sporting events and in the summer 2021, from June 25 to July 2 and a number of thermometric and geophysical works were carried out on that southeastern slopes of Elbrus (Sokratov and Frolov 2023).

Acknowledgements The work was carried out in accordance with the state budget theme "Evolution of the cryosphere under climate change and anthropogenic impact" (121051100164-0), "Danger and risk of natural processes and phenomena" (121051300175-4).

References

https://climate.copernicus.eu/
https://jra.kishou.go.jp/JRA-55/index_en.html
https://public.wmo.int/en/media/news/

https://www.nasa.gov/

Sokratov SA, Frolov DM (2023) Geophysical research in the preparation of ski slopes for compe-
    titions: case study of the Garabashi glacier (Elbrus, Caucasus). In: XV international scien-
    tific conference "INTERAGROMASH 2022". Lecture notes in networks and systems, vol 575.
    Springer, Cham, pp 2686–2693

Statement of Individual Problems of the Dynamics of Stratified Media with Allowance for Compressibility

V. V. Bulatov ⓘ

Abstract The paper considers issues related to the formulation of problems of the dynamics of stratified media, taking into account their compressibility. The physical formulations of problems in which wave oscillations can arise in such media are discussed. The properties of the main dispersion relations for various physical parameters are studied. Questions of the legitimacy of using the Boussinesq approximation in studying the dynamics of internal gravity waves in stratified compressible media are discussed. It is shown that taking into account the compressibility of the medium is important in the study of internal gravitational waves, since it makes it possible to remove the paradoxes of establishing the field of internal gravitational waves in various modes of wave generation. In a compressible stratified medium, the dynamics of internal gravity waves is qualitatively similar to the evolution of acoustic waves in a highly anisotropic medium.

Keywords Stratified compressible medium · Internal gravity waves · Boussinesq approximation · Buoyancy frequency · Dispersion curves

The excitation and propagation of internal gravity waves in real natural conditions are essentially nonlinear phenomena, in addition, compressibility effects play a significant role (Lightill 1978; Whitham 1974; Miropol'skii and Shishkina 2001; Vlasenko et al. 2005; Mei et al. 2017; Morozov 2018; Velarde et al. 2018; Ozsoy 2021; Bulatov and Vladimirov 2012, 2019, 2010; Garrett and Munk 1972; Haney and Young 2017; Broutman and Rottman 2004; Bulatov et al. 2019; Bulatov 2020; Lecoanet et al. 2015; Voelker et al. 2019). When studying internal gravity waves, taking into account the compressibility of a stratified medium, it is necessary to consider the Brunt-Väisälä frequency $N^2(z) = -\frac{g}{\rho_0} \frac{d\rho_0}{dz}$ in modified form $N_c^2(z) = -\frac{g}{\rho_0} \frac{d\rho_0}{dz} - g^2 / c^2 = N^2 - g^2 / c^2$, where c—sound velocity (Lightill 1978; Velarde et al. 2018; Meunier et al. 2018; Wang et al. 2017; Svirkunov and Kalashnik

V. V. Bulatov (✉)
Ishlinsky Institute for Problems in Mechanics RAS, Vernadskogo Ave. 101-1, 119526 Moscow, Russia
e-mail: internalwave@mail.ru

© The Author(s), under exclusive license to Springer Nature Singapore Pte Ltd. 2023
T. Chaplina (ed.), *Processes in GeoMedia—Volume VII*, Springer Geology,
https://doi.org/10.1007/978-981-99-6575-5_26

2014). Obviously, in conditions of weak stratification, this additional term must be taken into account (Lightill 1978; Whitham 1974; Ozsoy 2021; Broutman et al. 2021; Chai et al. 2022; Gnevyshev and Badulin 2020). However, the basic equation of internal gravity waves itself remains unchanged. Indeed, in the original system of hydrodynamic equations, the incompressibility equation is replaced by $\frac{1}{c^2}\frac{dp}{dt} = \frac{d\rho}{dt}$, which leads to an increase in the order of the corresponding dispersion equation, which has the form: $\omega^4 - \omega^2(N^2 + k^2 c^2) + k^2 c^2 N_c^2 \sin^2 \Theta = 0$, where $\sin \Theta = \omega/N$ (Lightill 1978; Whitham 1974; Miropol'skii and Shishkina 2001; Ozsoy 2021; Bulatov and Vladimirov 2012, 2019).

However, for wavelengths satisfying the condition $N^2 \lambda/g << 1$, which automatically leads to the condition of $\lambda << c^2/g = 200$ km, the dispersion equation splits into two: $\omega = kc$ and $\omega = N_c k_2/k$ (k_2—horizontal wave number) corresponding to sound and internal gravity waves (Lightill 1978; Whitham 1974; Velarde et al. 2018; Svirkunov and Kalashnik 2014; Gnevyshev and Badulin 2020).

In this paper, some issues of wave dynamics in stratified media will be considered, taking into account their compressibility, that is, the pulses of acoustic-gravity waves—waves in a compressible stratified medium will be investigated. Acoustic-gravity waves are generated during various atmospheric processes and play an important role in the dynamics of the atmosphere of the Earth and other planets (Ozsoy 2021; Bulatov and Vladimirov 2012, 2019; Svirkunov and Kalashnik 2014). Powerful pulsed sources of these waves can be, for example, industrial and nuclear explosions, volcanic eruptions. Taking into account compressibility is also important in the study of internal gravity waves in the ocean, as it allows us to remove some paradoxes of establishing the field of internal gravity waves under various modes of wave generation (Lightill 1978; Whitham 1974; Miropol'skii and Shishkina 2001; Lecoanet et al. 2015; Voelker et al. 2019). Indeed, in a compressible medium, internal gravity waves are to a certain extent "similar" to ordinary acoustic waves, although in a highly anisotropic medium (Lightill 1978; Whitham 1974; Ozsoy 2021). We will proceed from a linearized system of hydrodynamic equations of a stratified compressible medium (Lightill 1978; Vlasenko et al. 2005; Velarde et al. 2018; Ozsoy 2021)

$$\rho \frac{\partial U_1}{\partial t} + \frac{\partial p}{\partial x} = f_x$$

$$\rho \frac{\partial U_2}{\partial t} + \frac{\partial p}{\partial y} = f_y$$

$$\rho \frac{\partial W}{\partial t} + \frac{\partial p}{\partial z} + g\delta = f_z$$

$$\frac{\partial}{\partial t}\delta + \rho \frac{\partial W}{\partial z} + \rho\left(\frac{\partial U_1}{\partial x} + \frac{\partial U_2}{\partial y} + \frac{\partial W}{\partial z}\right) = \frac{\partial}{\partial t}\delta_0 \qquad (1)$$

$$c^{-2}\left(\frac{\partial}{\partial t}p - \rho g W\right) + \rho\left(\frac{\partial U_1}{\partial x} + \frac{\partial U_2}{\partial y} + \frac{\partial W}{\partial z}\right) = \rho\Omega$$

where $\rho(z)$—undisturbed density, $c(z)$—speed of sound, p—pressure perturbation, δ—density perturbation, (U_1, U_2, W)—velocity components, g—acceleration of gravity, f_x, f_y, f_z—components of external volumetric force, δ_0—source of density perturbations, Ω—source of volumetric velocity. The nature of the source is determined by the physical formulation of the problem (Vlasenko et al. 2005; Bulatov and Vladimirov 2010; Bulatov et al. 2019; Bulatov 2020; Svirkunov and Kalashnik 2014). For example, an explosion is a source of volumetric velocity and, due to the heating of the medium, a source of density disturbances. Characteristic sources of volumetric velocity in the atmosphere are the so-called "thermics"—superheated regions rising up and driving the layers of the atmosphere into motion (Lightill 1978; Velarde et al. 2018; Ozsoy 2021; Meunier et al. 2018).

Sources of external force can be associated with nonlinear interactions with other types of waves, as well as with the flow around bodies: areas of flow disruption during flow are sources of external volumetric forces (Velarde et al. 2018; Ozsoy 2021; Svirkunov and Kalashnik 2014; Broutman et al. 2021). System (1) can be reduced to one equation, for example, with respect to the vertical component of velocity W

$$L_1 W = \Pi \tag{2}$$

$$L_1 = \frac{\partial^2}{\partial t^2}\left(c^{-2}\frac{\partial^2}{\partial t^2} - \mu\frac{\partial}{\partial z} - \frac{\partial^2}{\partial z^2}\right) - \left(\frac{\partial^2}{\partial x^2} + \frac{\partial^2}{\partial y^2}\right)\left(\frac{\partial^2}{\partial t^2} + \mu g - g^2 c^{-2}\right),$$

$$\Pi. = \rho^{-1}\left[c^{-2}\frac{\partial^2}{\partial t^2} - \left(\frac{\partial^2}{\partial x^2} + \frac{\partial^2}{\partial y^2}\right)\left(\frac{\partial}{\partial t}f_z - g\left(\frac{\partial}{\partial t}\delta_0 - \rho\Omega\right)\right) - \right.$$
$$\left. - \left(\frac{\partial}{\partial z} + gc^{-2}\right)\left(\rho\frac{\partial^2}{\partial t^2}\Omega - \frac{\partial}{\partial t}\left(\frac{\partial}{\partial y}f_y + \frac{\partial}{\partial x}f_x\right)\right)\right]$$

where the value $\mu = -\rho\frac{\partial}{\partial z}\rho$ characterizes the rate of change of undisturbed density with height. Value $N = \left(\mu g - g^2/c^2\right)^{1/2}$—the Brunt-Väisälä frequency in a compressible stratified medium (Lightill 1978; Whitham 1974; Miropol'skii and Shishkina 2001; Bulatov and Vladimirov 2012, 2019). When studying internal gravity waves, the Boussinesq approximation is traditional: in the system of equations (2), the density gradient in those terms that are not multiplied by the value g (Miropol'skii and Shishkina 2001; Velarde et al. 2018; Ozsoy 2021) is neglected. In this case, the operator L_1 is reduced to the operator

$$L_2 = \frac{\partial^2}{\partial t^2} - \left(c^{-2}\frac{\partial^2}{\partial t^2} - \frac{\partial^2}{\partial z^2}\right) - \left(\frac{\partial^2}{\partial x^2} + \frac{\partial^2}{\partial y^2}\right)\left(\frac{\partial^2}{\partial t^2} + \mu g - g^2 c^{-2}\right) \tag{3}$$

If in (2) we formally direct the speed of sound to infinity, then from expression (3) we get the operator of internal gravity waves in an incompressible fluid.

$$L_3 = -\frac{\partial^2}{\partial t^2}\left(\mu\frac{\partial}{\partial z} + \frac{\partial^2}{\partial z^2}\right) - \left(\frac{\partial^2}{\partial x^2} + \frac{\partial^2}{\partial y^2}\right)\left(\frac{\partial^2}{\partial t^2} + \mu g\right).$$

If we use the Boussinesq approximation for an incompressible medium, then the operator takes the form

$$L_4 = -\left[\frac{\partial^2}{\partial t^2}\left(\frac{\partial^2}{\partial x^2} + \frac{\partial^2}{\partial y^2} + \frac{\partial^2}{\partial z^2}\right) - \mu g\left(\frac{\partial^2}{\partial x^2} + \frac{\partial^2}{\partial y^2}\right)\right].$$

If the medium is homogeneous ("in small"), but compressible, then in a homogeneous gravity field it necessarily has exponential stratification. This natural stratification has a parameter $\mu_0 = g/c^2$. At the $0 < \mu < \mu_0$ the medium turns out to be unstable, despite the decrease in density with height. Indeed, $N^2 g(\mu - \mu_0)$, from where it can be seen that solutions that grow exponentially over time appear. The wave operator for a "naturally stratified" medium has the form

$$L_5 = \frac{\partial^2}{\partial t^2}\left(\frac{\partial^2}{c^2\,\partial t^2} - \frac{\partial^2}{\partial x^2} - \frac{\partial^2}{\partial y^2} - \frac{\partial^2}{\partial z^2} - gc^{-2}\frac{\partial^2}{\partial z^2}\right)$$

Finally, in disregard of the gravity field ($g = 0$), the operator L_1 turns into a doubly differentiated wave operator (Whitham 1974)

$$L_6 = \frac{\partial^2}{\partial t^2}\left(\frac{\partial^2}{c^2\,\partial t^2} - \frac{\partial^2}{\partial x^2} - \frac{\partial^2}{\partial y^2} - \frac{\partial^2}{\partial z^2}\right)$$

Below we will discuss the validity of using all these approximations. In particular, it can be shown that using the Boussinesq approximation for a compressible medium (operator L_4) leads to physically absurd results. Consider a medium with exponential stratification $\rho = \rho_0 \exp(-\mu z)$ ($\mu = const$) and constant speed of sound, in this case the operator L_1 has constant coefficients, its fundamental solution G satisfies the equation: $L_1 G = \delta(x)\,\delta(y)\,\delta(z)\,\delta(t)$.

Then the solution of the problem with a specific source $(f_x, f_y, f_z, \delta_0, \Omega)$ is expressed as a convolution G with a function Π from (2). In particular, in the case of a point instantaneous source of volumetric velocity, we have: $\Pi = \delta''(t)\,\delta(x)\,\delta(y)\,\delta'(z)$, that is, the vertical component of velocity W is expressed in terms of a function G as follows: $W = \frac{\partial^3}{\partial t^2\partial z}G$. With the help of a suitable replacement of variables, it is convenient to get rid of the odd derivative $\frac{\partial}{\partial z}$ in (2). Indeed, we have

$$L_1 = L_7 \exp(-\mu z/2)$$

$$L_7 = \frac{\partial^2}{\partial t^2}\left(c^{-2}\frac{\partial^2}{\partial t^2} + \mu^2/4 - \frac{\partial^2}{\partial z^2}\right) - \left(\frac{\partial^2}{\partial x^2} + \frac{\partial^2}{\partial y^2}\right)\left(\frac{\partial^2}{\partial t^2} + N^2\right).$$

where $\exp(-\mu z/2)$—the multiplication operator by a function. Let's perform a one-way Fourier transform. The spectrum of the fundamental solution $\tilde{G}$ is written as

$$\tilde{G}(\omega, x, y, z) = \exp(\mu z/2)\tilde{G}'(\omega, x, y, z)$$

where $\tilde{G}'(\omega, x, y, z)$—fundamental solution of the Helmholtz operator in an anisotropic medium (Whitham 1974)

$$L_8\tilde{G}' = -\delta(x)\,\delta(y)\,\delta(z) \tag{4}$$

$$L_8 = \left(\omega^2 - N^2\right)\left(\frac{\partial^2}{\partial x^2} + \frac{\partial^2}{\partial y^2}\right) + \omega^2\frac{\partial^2}{\partial z^2} + \omega^4 c^{-2} - \omega^2\mu^2/4$$

In coordinates $x_1 = x$, $y_1 = y$, $z_1 = \omega\left(\omega^2 - N^2\right)^{-1/2}z$, Eq. (4) takes the form

$$\left[\Delta_1 + \omega^2\left(\omega^2 - N^2\right)^{-1/2}\left(\omega^2 c^{-2} - \mu^2/4\right)\right]\tilde{G}' = \\ = -\omega^{-1}\left(\omega^2 - N^2\right)^{-1/2}\delta(x_1)\,\delta(y_1)\,\delta(z_1) \tag{5}$$

$$\Delta_1 = \frac{\partial^2}{\partial x_1^2} + \frac{\partial^2}{\partial y_1^2} + \frac{\partial^2}{\partial z_1^2}$$

Using the well-known fundamental solution of the usual Helmholtz operator, from (5) we obtain

$$\tilde{G}' = -(4\pi\omega)^{-1}\left(\omega^2 - N^2\right)^{-1/2}R_1^{-1}\exp(i\kappa_1 R_1)$$

$$\kappa_1^2 = \omega^2\left(\omega^2 - N^2\right)^{-1}\left(\omega^2 c^{-2} - \mu^2/4\right) \qquad R_1 = R\left(1 - N^2\omega^{-2}\sin^2\theta\right)^{1/2}$$

$$R = \left(x^2 + y^2 + z^2\right)^{1/2}, \sin\theta = z/R$$

Thus, we have

$$\tilde{G}(\omega, x, y, z) = A_\theta(\omega)R^{-1}\exp\left[i\kappa_\theta(\omega)R + \mu z/2\right] \tag{6}$$

$$\kappa_\theta(\omega) = c^{-1}\sqrt{\frac{\left(\omega^2 - N_A^2\right)\left(\omega^2 - N_\theta^2\right)}{\omega^2 - N^2}}, \qquad A_\theta(\omega) = \frac{(4\pi)^{-1}}{\sqrt{\left(\omega^2 - N^2\right)\left(\omega^2 - N_\theta^2\right)}}$$

$$N_A^2 = \mu^2 c^2/4, \quad N_\theta^2 = N^2\sin^2\theta$$

Formula (6) describes the field of a harmonic source in a compressible stratified medium. Let's discuss the physical meaning of the quantities included in it. The function $\kappa_\theta(\omega)$ is the wave number of a plane acoustic-gravitational wave propagating at an angle θ to the horizon. Indeed, the plane wave field

$$W' = \exp\left[-i\omega t + i\kappa_\theta(\omega)\left(x\cos\theta + z\sin\theta\right)\right]$$

satisfies the equation $L_7 W' = 0$. The dependence of the wave number $\kappa_\theta(\omega)$ on the angle θ is due to the anisotropy of the medium. Acoustic-gravity waves have dispersion, their phase velocity depends on the frequency.

$$c_{ph} = \frac{\omega}{\kappa_\theta(\omega)} = \frac{\omega}{c}\sqrt{\frac{\omega^2 - N^2}{\left(\omega^2 - N^2\right)\left(\omega^2 - N_\theta^2\right)}}$$

The corresponding dispersion curves have two branches: upper and lower. The phase and group velocities of the upper curve, called the acoustic branch, tend to the speed of sound c at $\omega \to \infty$. As the waves described by this branch grow, they turn into ordinary sound waves. The lower critical frequency N_A of these waves is the lowest frequency of acoustic wave propagation. In the range called the upper locking range, the wave number $\kappa_\theta(\omega)$ is purely imaginary.

There are no propagating plane waves in this range, but only exponentially attenuating when moving away from the source. The lower branch of the dispersion curve describes the gravity branch of acoustic-gravity waves. The upper critical frequency of this branch is equal N, and the lower one depends on the angle θ. If the wave propagates strictly horizontally ($\theta = 0$), then there is no lower critical frequency.

The phase and group velocities of a horizontally propagating wave at $\omega \to 0$ tend to the value $c_0 = Nc/N_A < c$. If $\theta \neq 0$, then the group velocity has a maximum $c_* = c_{gr}(\omega_*(\theta))$ at some frequency other than zero, which we denote $\omega_*(\theta)$. The patterns of the wavefronts of the field excited by a harmonic source differ greatly in the cases $\omega > N_A$ and $\omega < N$ (when $N < \omega < N_A$ the field is in phase and decreases exponentially with distance from the source). At the $\omega > N_A$ fronts have the form of ellipsoids of rotation, and with increasing frequency their eccentricity decreases, and in the limit at $\omega \to \infty$ they turn into spheres, as for ordinary sound waves.

At $\omega \to N_A$ the ellipses become more and more flattened in the vertical direction. At $\omega < N$ the fronts are a family of single-cavity hyperboloids of rotation with a common asymptote at an angle $\theta_\omega = \arcsin(\omega/N)$ depending on the frequency. At $|\theta| > \theta_\omega$, the field is in phase and decreases exponentially with distance from the source. Inside the cone, $|\theta| < \theta_\omega$ the field is oscillating, the wave crests "move away" from the cone $|\theta| = \theta_\omega$. On a cone, at field of a harmonic point source has a root feature

$$A_\theta(\omega) = \frac{(4\pi)^{-1}}{NR\sqrt{\left(\omega^2 - N^2\right)\left(\sin^2\theta_\omega - \sin^2\theta\right)}}.$$

If the source has finite dimensions, then the singularity does not arise, the field is mainly concentrated in a narrow area near this cone, having transverse dimensions of the order of the source dimensions. Let's discuss the physical meaning of quantities N_θ, N, N_A. The frequency N_A is determined by the product of a small parameter μ by a large parameter c. Estimates show that for both the ocean and the atmosphere, the frequency N_A is close to the Brunt-Väisälä frequency N: for the atmosphere $N \approx 1.9 \cdot 10^{-2} \, c^{-1}$, $N_A \approx 2.1 \cdot 10^{-2} \, c^{-1}$, for the ocean $N \sim 10^{-2} \, c^{-1}$, $N_A \sim 1.5 \cdot 10^{-2} \, c^{-1}$ (Miropol'skii and Shishkina 2001; Mei et al. 2017; Velarde et al. 2018; Ozsoy 2021; Voelker et al. 2019).

From these estimates it can be seen that when analyzing the dispersion $\kappa_\theta(\omega)$ at frequencies close to N, it is necessary to take into account the compressibility of the medium. Note that $N_A \geq N$, $N_A^2 - N^2 = \left(\mu c / 2 - g / c \right)^2 = c^2 (\mu - 2\mu_0)^2 / 4$, where $\mu_0 = g / c^2$ is the natural stratification of a homogeneous medium in the gravity field. Thus, only when $\mu = 2\mu_0$ there is no upper locking range and a single dispersion curve is obtained. At $\mu = \mu_0$ the Brunt-Väisälä frequency vanishes and the lower branch of the dispersion curve disappears, unlike the previous case, this curve does not depend on θ (isotropy).

When considering low-frequency waves in the atmosphere, the rotation of the Earth is usually taken into account. It can be shown, in particular, that taking into account the rotation of the Earth leads only to a change in one parameter in (6), namely the values of N_θ: $N_\theta^2 = N^2 \sin^2 \theta + J_z^2 \cos^2 \theta$, where J_z—is the z-component of the angular velocity vector of the rotation of the Earth. It can be seen that the rotation of the Earth noticeably affects the field of internal gravitational waves only at very small θ : $\theta \leq J_z / N < 10^{-3}$ rads. The influence of the Earth's rotation leads to the fact that the dependence of the acoustic-gravitational waves field on R and t on the horizontal plane and near it is the same as for the field on a cone with an angle $\theta \approx J_z / N$ calculated without taking into account rotation. We will neglect the rotation of the Earth, it can always be taken into account by slightly changing the angular dependence N_θ.

Now let's discuss how the form of dispersion curves changes in the simplified models mentioned above. First, consider the incompressible medium approximation. In the limit $c \to \infty$ for $\kappa_\theta(\omega)$ we obtain the expression

$$K_\theta(\omega) = \frac{\mu}{2} \sqrt{\frac{\omega^2 - N_\theta^2}{N^2 - \omega^2}}$$

With such a law of dispersion, the acoustic branch of the dispersion curve disappears, and the gravity branch with critical frequencies N_θ and N does not change qualitatively (at $\omega << N$—quantitatively). It can be noted that the Boussinesq approximation in a compressible medium is physically absurd. In this case $N_A = 0$ and therefore $N_A < N$, that is not possible in stable environments. The dispersion curve has two branches, and the upper and lower boundaries of the locking range are distorted. Surfaces of equal phases at $\omega < N$ take the form of two-cavity hyperboloids of rotation. Note that in a compressible medium ($\mu = \mu_0$), the wave number κ_θ does not depend on the angle θ (isotropic case): $\kappa_\theta(\omega) = c^{-1}\sqrt{\omega^2 - N_A^2}$, $N_A = g/2c$, and therefore the anisotropy of the medium manifests itself only in the appearance of a multiplier $\exp(gz/2c^2)$ in expression (6).

When studying internal gravity waves in the ocean, the incompressible medium model and the Boussinesq approximation are usually used (Lightill 1978; Whitham 1974; Miropol'skii and Shishkina 2001; Mei et al. 2017; Morozov 2018; Velarde et al. 2018; Ozsoy 2021; Broutman et al. 2021; Chai et al. 2022). In this approximation, Eq. (5) "reduces" not to the Helmholtz equation, but to the Laplace equation, so that there is no wave propagation and the medium moves in phase, as when bodies vibrate in an incompressible homogeneous medium. However, when $\omega < N$ there is a phase shift by $\pi/2$ between the regions θ_ω and $|\theta| > \theta_\omega$. Another difference from the case of an unstratified fluid is the root feature of the amplitude at the boundary between the regions ($\theta = \theta_\omega$). So, to move to an incompressible medium in the Boussinesq approximation, it is enough to put in (6) $\kappa_\theta(\omega) = 0$.

Acknowledgements The work is carried out with financial support from the RSF project 23-21-00194.

References

Broutman D, Rottman J (2004) A simplified Fourier method for computing the internal wave field generated by an oscillating source in a horizontally moving, depth-dependent background. Phys Fluids 16:3682–3689

Broutman D, Brandt L, Rottman J, Taylor C (2021) A WKB derivation for internal waves generated by a horizontally moving body in a thermocline. Wave Motion 105:102759

Bulatov VV, Vladimirov YuV (2010) Estimate of the applicability limits of a linear theory of internal waves. Fluid Dyn 45:787–792

Bulatov VV, Vladimirov YuV (2012) Wave dynamics of stratified mediums. Nauka, Moscow

Bulatov VV, Vladimirov YuV (2019) A general approach to ocean wave dynamics research: modelling, asymptotics, measurements. OntoPrint Publishers, Moscow

Bulatov V, Vladimirov Yu (2020) Generation of internal gravity waves far from moving non-local source. Symmetry 12(11):1899

Bulatov VV, Vladimirov YuV, Vladimirov IYu (2019) Far fields of internal gravity waves from a source moving in the ocean with an arbitrary buoyancy frequency distribution. Russ J Earth Sci 19:ES5003

Chai J, Wang Z, Yang Z, Wang Z (2022) Investigation of internal wave wakes generated by a submerged body in a stratified flow. Ocean Eng 266:112840

Garrett C, Munk W (1972) Space-time scales of internal waves. Geophys Fluid Dyn 3:225–264

Gnevyshev V, Badulin S (2020) Wave patterns of gravity–capillary waves from moving localized sources. Fluids 5:219

Haney S, Young WR (2017) Radiation of internal waves from groups of surface gravity waves. J Fluid Mech 829:280–303

Lecoanet D, Le Bars M, Burns KJ, Vasil GM, Brown BP, Quataert E, Oishi JS (2015) Numerical simulations of internal wave generation by convection in water. Phys Rev E – Stat Nonlinear, Soft Matter Phys 9:1–10

Lightill J (1978) Waves in fluids. Cambridge University Press, Cambridge

Mei CC, Stiassnie M, Yue DK-P (2017) Theory and applications of ocean surface waves. Advanced series of ocean engineering, vol 42. World Scientific Publishing

Meunier P, Dizиs S, Redekopp L, Spedding G (2018) Internal waves generated by a stratified wake: experiment and theory. J Fluid Mech 846:752–788

Miropol'skii YuZ, Shishkina OV (2001) Dynamics of internal gravity waves in the ocean. Kluwer Academic Publishers, Boston

Morozov TG (2018) Oceanic internal tides. Observations, analysis and modeling. Springer, Berlin

Ozsoy E (2021) Geophysical fluid dynamics II. Stratified rotating fluid dynamics of the atmosphere-ocean. Springer textbook in earth sciences. Geography and environment. Springer Nature, Switzerland AG, Cham

Svirkunov PN, Kalashnik MV (2014) Phase patterns of dispersive waves from moving localized sources. Phys-Usp 57(1):80–91

Velarde MG, Tarakanov RYu, Marchenko AV (eds) (2018) The ocean in motion. Springer oceanography. Springer International Publishing AG, Berlin

Vlasenko V, Stashchuk N, Hutter K (2005) Baroclinic tides. Cambridge University Press, N.Y.

Voelker GS, Myers P, Walter M, Sutherland BR (2019) Generation of oceanic internal gravity waves by a cyclonic surface stress disturbance. Dyn Atm Oceans 86:16–133

Wang H, Chen K, You Y (2017) An investigation on internal waves generated by towed models under a strong halocline. Phys Fluids 29:065104

Whitham GB (1974) Linear and nonlinear waves. Willey-Interscience Publication, New York

Formation of Modern Technogenic Processes in Historical Urban Areas

A. G. Yanin, O. I. Yanina, Y. A. Yanina, and A. G. Chigarev

Abstract The aim of the research is to study the formation of modern technogenic processes in historical urban areas in complex engineering and geological conditions. These processes affect the safe operation of buildings—objects of cultural heritage located in these territories. Over time, it becomes necessary to monitor the foundation soils of buildings constructed in the 18–19 centuries. Monitoring of the foundation soils of the examined building showed a significant change in the physical properties of the soil base, associated with the soaking of sandy soil from engineering systems and the formation of man-made perched water. Calculations in the Midas GTS NX PC revealed the stabilization of the processes of the technogenic impact of a high-rise building on the foundation soils of the building's foundations of the nineteenth century. The formation of a modern technogenic process as a result of poor-quality conservation and damage to the engineering systems of the building has been established. It is proposed to use the results of the study in the development of projects for monitoring foundation soils during man-made processes in urban historical areas.

1 Introduction

New construction of high-rise buildings in the historical part of the city of Voronezh, practically concentrated in the central area. With dense urban development and complex engineering and geological conditions, a number of technogenic processes may appear. Modern technogenic conditions can cause the formation of negative engineering-geological processes that affect the safe operation of buildings—objects of cultural heritage, therefore, to preserve them, monitoring the foundation soils is mandatory (Zhuravlev and Marukyan 2020; Ilyichev et al. 2019; Karpova 2019; Telichenko and Roitman 2020).

A. G. Yanin · O. I. Yanina (✉) · Y. A. Yanina · A. G. Chigarev
Voronezh State Technical University, Street 20-Letia Oktyabrya, 84, 394006 Voronezh, Russian Federation
e-mail: yaninaoi@yandex.ru

© The Author(s), under exclusive license to Springer Nature Singapore Pte Ltd. 2023
T. Chaplina (ed.), *Processes in GeoMedia—Volume VII*, Springer Geology,
https://doi.org/10.1007/978-981-99-6575-5_27

2   Materials and Methods

The study of the formation of technogenic processes and changes in the properties of foundation soils in the historical urban area was carried out on the example of the building—the object of cultural heritage «House of the Cantonists» in Voronezh.

The building was built in 1816 and is a monument in the style of "late classicism" and one of the oldest buildings in the central part of the historical territory of the right bank of Voronezh. Initially, the building was educational, then in the 19–20 centuries it repeatedly changed its purpose and from 1950 to 2000 the building was used as a residential building. The building "House of Cantonists" is a two-story building with a basement and a mezzanine, of a complex U-shape and dimensions in terms of 26.28 × 18.68 m. The structural system of the building is frameless with external and internal load-bearing stone walls made of solid ceramic bricks. The spatial rigidity of the building is ensured by the joint work of longitudinal and transverse load-bearing stone walls and floor structures.

General view of the main facade of the building «House of Cantonists» is shown in Fig. 1.

Modern technogenic processes were identified visually during the reconnaissance survey in the process of inspecting the site of the building «House of Cantonists» and adjacent territories. To examine the condition and properties of the soils of the foundations, 3 pits were opened outside near the load-bearing walls of the building. Soil samples of undisturbed structure were taken from the pits for laboratory studies of physical properties. Based on the results of laboratory studies of foundation soils, local monitoring was carried out for the period 2017–2021. During a detailed survey

Fig. 1   Main facade of the building «House of Cantonists», Voronezh city

of the structures of the building, the main defects and damage to structures were identified and detailed; the actual loads and impacts on the building are established. Based on the survey materials, verification calculations of the bearing capacity of foundation soils were performed. The calculation of the limiting additional soil deformations was carried out by the finite element method in the Midas GTS NX software package (Belostotsky et al. 2017).

3 Results

Results of the study of technogenic conditions. Technogenic conditions of the site of the building «House of Cantonists» are determined by the conditions of urban development in the city of Voronezh that developed in the 19–21 centuries. The historical urban area is located on the high right bank of the Voronezh River, which has a significant slope of the surface.

At the beginning of the twenty-first century, a 22-storey residential building with a four-level underground parking was built on the adjacent territory to the building of the «House of the Cantonists». The construction of a high-rise building led to the formation of man-made processes that negatively affected the structures and the stress–strain state of the soils of the foundations of the «House of Cantonists» building, and as a result, the building was recognized as unsuitable for performing the functions of a residential building and settled. On a separate section of a non-residential building, damage to engineering networks led to many years of soaking of the foundation soils with technogenic waters. The blind area along the facades of the building is damaged and partially missing, which additionally creates conditions for soaking foundation structures and base soils.

Technogenic conditions of the location of the building-object of cultural heritage «House of Cantonists» in Voronezh are presented in Fig. 2.

Results of the study of soil conditions. According to the results of the opening of 3 pits at the base of the foundations of the building «House of Cantonists», the geological structure was studied and four engineering-geological elements (GTE) were identified: GTE-1, tQIV—bulk soil; GTE-2, aQIII—sand of medium size, dense, low degree of water saturation; GTE-3, aQIII—sand of medium size, dense, saturated with water; GTE-4, aQIII—sand of medium size, medium density, low degree of water saturation.

At the time of the study (2021), the presence of a "top water" was revealed in one of the sections of the building (pit No. 3) at a depth of 0.20 m from the floor surface of the basement floor with a thickness of 0.80 m. The water-bearing layer is medium-sized sand (GTE-3). The origin of the «top water» is technogenic and is associated with constant long-term leaks from damaged engineering systems in this area of a residential building after its resettlement. At the same time, no groundwater was found in pits No. 1 and No. 2 to a depth of 0.70.

Fig. 2 Technogenic conditions of the site of the building—«House of Cantonists», Voronezh city

Figure 3 shows a general view of «top water» in the basement of the building «House of Cantonists».

Based on the results of laboratory studies of the foundation soils of the building (2017 and 2021), local monitoring was carried out. Changes in the characteristics of the physical properties of the foundation soils for the period 2017–2021 shown in Fig. 4.

Diagrams of changes in the physical properties of soils in the base of sands of medium size (GTE-2, GTE-3, GTE-4) in 2017–2021 showed the following: for sands

Fig. 3 Technogenic «top water» on the ground floor of the building «House of Cantonists»

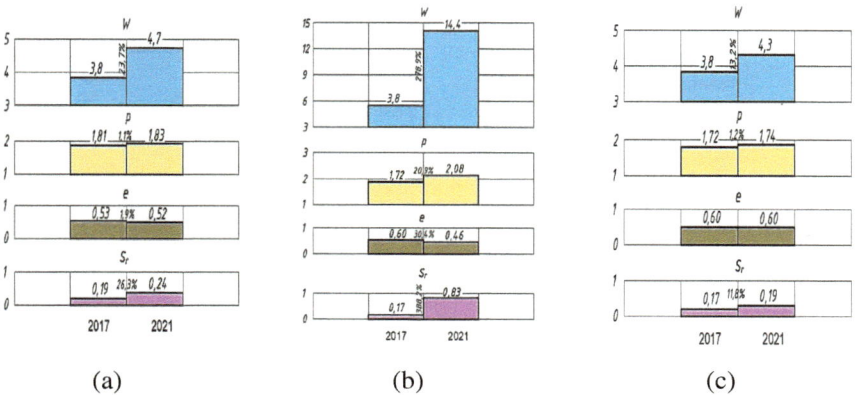

(a) (b) (c)

Fig. 4 Diagrams of changes in the physical properties of the foundation soils for the period 2017–2021, where **a** is GTE-2 sand; **b**—sand GTE-3; **c**—sand GTE-4

(GTE-2, GTE-4), the change in physical properties over this period is insignificant; for sand (GTE-3), the change in physical properties is significant: humidity increased from 3.0 to 14.4%; the water saturation coefficient increased from 0.17 to 0.83. The change in the properties of sand (GTE-3) indicates the formation of technogenic waters of the «top water» type (pit No. 3), associated with water leaks from the heating system, damaged after the resettlement of the house (Patrikeev 2014; Antipov and Ofrichter 2016; Vallero and Brasier 2008; Shashkin et al. 2022).

The results of the study of the structures of the foundations and walls of the building. The study of foundation structures was carried out in three pits inside the building in the basement. The foundations of the walls of the building are of the same type— tape, made of solid masonry of ceramic bricks and rubble stone "sandstone" in a complex solution. The width of the base of the foundations of the outer longitudinal and transverse bearing walls is 0.65–1.30 m. The depth of the base of the foundations of the walls from the floor level of the basement is 0.20–0.35 m «mirrors of water» on the section of the pit No. 3. The condition of the masonry of the wall foundation in the places where the pits were opened is satisfactory.

The results of the verification calculations showed the following: the condition for limiting the pressure on the foundation soils is fulfilled and the bearing capacity of the foundation soils—sands of medium size (GTE-2, GTE-3, GTE-4) is sufficient for the actual dimensions of the foundation and indicators of physical properties. Settlements of foundations according to verification calculations do not exceed permissible standard values.

The masonry of the outer walls of the building has cracks with an opening width of 1–15 mm. The largest crack opening width of 10–15 mm was noted in the load-bearing walls from the side of the adjacent territory of a new high-rise building.

General view of cracks in the outer load-bearing walls of the building «House of Cantonists» is shown in Fig. 5.

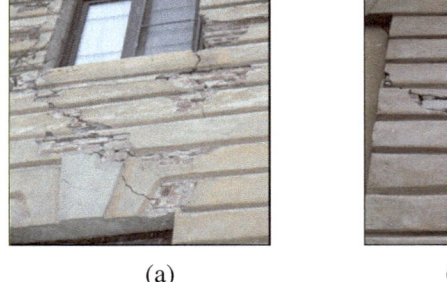

(a)   (b)

Fig. 5 View of cracks in the outer load-bearing walls of the building «House of Cantonists» from the side of a high-rise 22-storey residential building, where **a**—in the interwindow zone of the outer load-bearing wall; **b**—in the outer load-bearing wall

According to the results of the study of soil conditions and structures of the building «House of Cantonists», a diagram of the location of zones of influence of technogenic «top water» and additional load on soils from a high-rise building, is shown in Fig. 6.

In the Midas GTS software package, a numerical calculation was made of the limiting additional deformations of the foundation soils of the system «existing building «House of Cantonists»-soil foundation—new high-rise building with underground parking». The initial schemes for calculations are shown in Fig. 7. The calculation is performed in three stages: modeling of the initial stress state of the soil mass; modeling stages of construction of a high-rise building with underground parking; live load application.

The results of calculations in the PC Midas GTS showed that the following regulatory conditions are met:

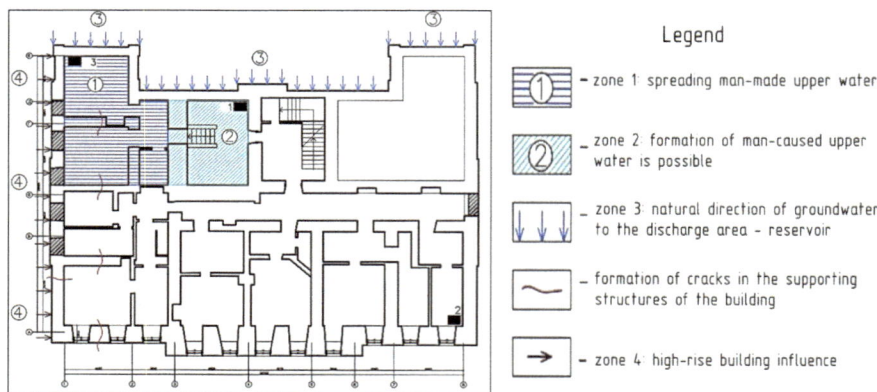

Fig. 6 Layout of zones of technogenic influence on soils of the foundation and building structure «House of Cantonists» on ground floor plan

Fig. 7 The initial scheme for calculating the limiting additional deformations of the soils of the foundation of the «building «House of Cantonists»—ground base—new high-rise building»

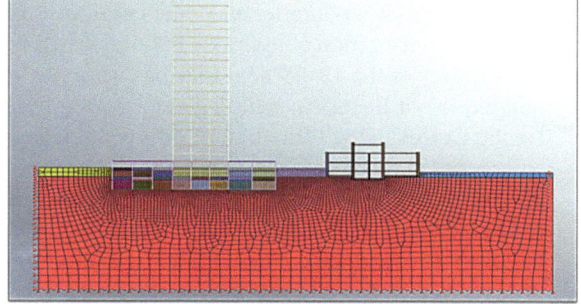

- limiting additional deformations of the soil base of foundations do not exceed the allowable value of 1.0 cm;
- the difference in sediment does not exceed the allowable value of 0.007.

Designs in the Midas GTS NX PC revealed the stabilization of the processes of the technogenic impact of a high-rise building on the foundation soils of the nineteenth century building.

4 Discussion

Modern man-caused processes that occur in the historical urban area in difficult engineering and geological conditions with high-rise infill development lead to a change in the state of soils and structures of the building—an object of cultural heritage. The formation of modern technogenic processes is very dynamic and in a short period of time causes changes in the properties of the foundation soils and structures of a historical building, leads to the loss of the functions of a residential building and complicates the process of modern adaptation. The monitoring of foundation soils of the system «historical building—ground base—new high-rise building» using numerical calculation methods in the Midas GTS NX software allows you to detail the process of the impact of changes in technogenic conditions on changes in the properties of foundation soils and choose the best options for restoring the conditions for the safe operation of historical buildings.

5 Conclusions

The analysis of the results of monitoring the soil conditions of the foundation base of the building—an object of cultural heritage allows us to draw the following conclusions:

- the influence of modern technogenic conditions of historical urban areas during the point construction of high-rise buildings on the soil conditions of the buildings foundations of the 18–19 centuries has been established;
- non-exploited historical buildings and currently not having modern adaptations can be factors in the formation of new technogenic processes associated with the emergency state of engineering systems;
- the results of the study can be used in the development of soil monitoring projects for foundations and structures of the building of the technical system «historical building—ground base—new high-rise building», established in the conditions of the urban historical territory.

References

Antipov VV, Ofrichter VG (2016) Modern non-destructive methods for studying the engineering-geological section. Bull Perm Natl Res Polytech Univ. Constr Arch 7(2):37–49. https://doi.org/10.15593/2224-9826/2016.2.04

Belostotsky AM, Akimov PA, Sidorov VN, Kaitukov TB (2018) On the development of adaptive mathematical models, numerical and numerical-analytical methods as the basis and component of monitoring systems for supporting structures of unique buildings and structures. In: Fundamental, exploratory and applied research of the RAACE on scientific support for the development of architecture, urban planning and the construction industry of the Russian Federation in 2017: sat. scientific tr. RAACE, T. 2. Izd-vo ASV, Moscow, pp 71–78. https://doi.org/10.22337/978 5432302663-71-78

Ilyichev VA, Kolchunov VI, Bakaeva NV (2019) Principles of strategic planning for the development of territories (on the example of the federal state of Bavaria). Vestn MSCU 14(2):158–168. https://doi.org/10.22227/1997-0935.2019.2.158-168

Karpova NV (2019) Sustainable development of urban settlements: theoretical postulates and their practical implementation. Econ Ecol Territ Entities 3(3):64–70. https://doi.org/10.23947/2413-1474-2019-3-3-64-70

Patrikeev AV (2014) The system of dynamic monitoring of an engineering structure as a key element of its technical safety. Vestn MSCU 3:133–140. https://doi.org/10.22227/1997-0935.2014.3.133-140

Shashkin AG, Shashkin VA, Bogov SG, Voronov AS (2022) Waterproofing of architectural monuments. Geotechnics XIV(2):28–41. https://doi.org/10.25296/2221-5514-2022-14-2-28-41

Telichenko VI, Roitman VM (2020) Analysis of the causes and consequences of major emergency situations in order to ensure the integrated safety of buildings and structures. Vestn MSCU 15(1):72–84. https://doi.org/10.22227/1997-0935.2020.1.72-84

Vallero DA, Brasier C (2008) Sustainable design: the science of sustainability and green engineering. Wiley, p 333. https://doi.org/10.1002/9780470259603

Zhuravlev PA, Marukyan AM (2020) Engineering protection of buildings, structures and territories as a factor in the innovative development of territorial planning. Vestn MGSU 15(10):1440–1449. https://doi.org/10.22227/1997-0935.2020.10.1440-1449

Carbon Sequestration in Managed Salt Wetlands (Carbon Polygon)

Victor V. Afanas'ev and A. B. Faustova

Abstract The results of studies conducted in 2021–2022 on the territory of the Sakhalin region showed that marine wetlands are giant carbon sink systems. Peat accumulation rates in estuary-lagoonal biomorpholithosystems reach 3.5 cm/year, and the total average annual growth of marshes in the lagoons of the Sakhalin Region alone is 50–100 hectares per year. In various subtypes of peat-gleyzems with an organic thickness of less than 50 cm, the average carbon reserves vary from 20.9 to 42.7 kg/m^2. According to the already dated sections of marsh peatlands, the value of the annual net carbon sink from 1 to 2 kg/m^2 or 3.7–7.4 kg/m^2 CO_2 was obtained. To develop the scientific foundations for the management of coastal wetlands through their accelerated programmable formation, it is proposed to create a pilot industrial carbon landfill. The development of experimental biopositive technologies of a nature-like type to increase the rate of peat accumulation is planned to be localized in the estuary-lagoon zone of the Tsunai and Susuya rivers.

Keywords Estuary · Salt marsh · Intertidal mudflat · Peat accumulation · Carbon sequestration · Increase in sedimentation rate

1 Introduction

Marching formations on the shores of Sakhalin are represented almost everywhere, but their geospatial characteristics, structure, features of formation and evolution have been of little interest until recently. The situation changed dramatically with the start of the implementation of the action plan to ensure the reduction of greenhouse gas emissions [Decree of the Ministry of Natural Resources of Russia dated June 30, 2017 N 20-r, ed. from 01/20/2021]. And the climate experiment that began in

V. V. Afanas'ev (✉)
Institute of Marine Geology and Geophysics, FEB RAS, Yuzhno-Sakhalinsk, Russia
e-mail: vvasand@mail.ru

A. B. Faustova
Sakhalin State University, Yuzhno-Sakhalinsk, Russia

T. Chaplina (ed.), *Processes in GeoMedia—Volume VII*, Springer Geology,
https://doi.org/10.1007/978-981-99-6575-5_28

the Sakhalin Region in accordance with the Federal Law of 03/06/2022 has dramatically accelerated the search for new opportunities to solve the problem of reducing GHG emissions (N 34-FZ "On Conducting an Experiment to Limit Greenhouse Gas Emissions in Certain Subjects of the Russian Federation").

It became clear that the region, where the length of the coastline exceeds 4 thousand kilometers, and the area of lagoon reservoirs is more than 2200 km^2 estuary-lagoonal geosystems (Najjar et al. 2018; Ouyang and Lee 2014; Spivak et al. 2019; Villa and Bernal 2018).

As a result of studies conducted in 2021–2022 on the territory of the Sakhalin Region, the geospatial parameters of marshes and silt lands formed on the coast of Sakhalin Island were determined. This work was based on the analysis of arrays of aerial photographs of 1952 and satellite images of 2014–2020, which was performed in the Quantum GIS geoinformation system. The calculations were made on the WGS84 EPSG:7030 ellipse. At the next stage of the study in the same Quantum GIS-environment, using standard procedures for analyzing remote sensing materials, we reconstructed the change in march areas for the period 1952–2019. Piltun, Chaivo, Nyivo, Nabil, Lunskoye, Nevskoye, Ainskoye lagoons. The age of the deposits was established based on the results of the reconstruction of the changes in the areas of marches for the period 1952–2019 and the method of radiocarbon dating of the deposits.

The results obtained make it possible to characterize the average annual rate of growth of marshes and, accordingly, the average annual increase in the volume of carbon in marsh soils for more than 75% of the total area of the lagoons of the island. Sakhalin (Afanas'ev and Faustova 2022a). It has been established that the average long-term growth of marches in the Sakhalin region is 70–100 hectares per year. And, for example, in the upper 30 cm of the section of the low march of the Nabil lagoon, the content of organic carbon is 35–40% (sampling at 5 cm intervals).

The published data and the authors' own results were used in calculating the carbon stocks in the sediments of marshes, silty desiccations, and bottom sediments. Drilling and sampling from the march deposits was carried out using a geoslicer and a set of Edelman soil hand drills (Eijkelkamp). Sediment analyzes were carried out in the laboratories of St. Petersburg State University and the Institute of Forestry. V. N. Sukacheva FRC KSC SB RAS.

One of the main results of the 1st stage of our work was the establishment of the fact of the highest sequestration capacity of salt marsh and intertidal mudflat in the northern part of the Aniva Bay. Over 60 years, marsh peatlands up to 80 cm thick formed on an area of about 2 km^2 (Afanas'ev et al. 2022). According to the already dated sections of marsh peatlands, the value of the annual net carbon sink from 1 kg/m^2 to 2 kg/m^2 or 3.7–7.4 kg/m^2 CO_2 was obtained. An analytical review of the problem showed that the maximum rates of vertical growth of lowland (marching) peatlands are observed in the first 20–30 years (Temmerman et al. 2004).

Given the fact that methane emissions, unlike freshwater marshes, estuarine-lagoon wetlands are minimal, such a potential allows us to form a technological offer corresponding to market demand in the form of technologies and climate projects for large companies in the Sakhalin Region (Steinmuller et al. 2020).

2 Managed Coastal Wetlands of the Sakhalin Region

As part of the Sakhalin experiment (34-federal law), it is proposed to comprehensively work out a system for regulating the absorption, accumulation and long-term disposal of organic carbon by managed PMVBU using nature-like technologies, which will create new opportunities for fulfilling the obligations of the Russian Federation under the Paris Agreement. Provisions of the Supplements on Wetlands 2013 The Guidelines of the IPCC 2006 National Greenhouse Gas Inventories on Accounting for the contribution of PVBU in National Inventories make it possible to take into account emissions and removals of greenhouse gases (GHGs) when using wetlands (Solomon et al. 2007; Hiraishi et al. 2014).

The movement towards the inclusion of marine swamps in the national inventories of many countries is already in full swing (Sapkota and White 2020; Lovelock et al. 2022). At the same time, there are even no estimated data for some regions of the world (Mcowen et al. 2017).

A lot of work has recently been done in the Sakhalin region to determine the geospatial parameters of estuarine-lagoon geosystems and to determine the carbon pool in their sediments (Afanas'ev and Faustova 2022a). Many processes are involved in the emission and absorption of greenhouse gases (GHGs), as a result of which the variability of fluxes in various blue carbon systems is very high. In our research approaches, we adhere to Geomorphological ideas that have already proven their effectiveness (Rosentreter et al. 2021; Hu et al. 2015; Curado et al. 2014).

We also acknowledge that the lack of a detailed facies analysis of sedimentation in the intertidal zone of the lagoons raises quite a few questions. However, the main tasks of this stage in the form of the analysis of geospatial information by modern methods and the assessment of the main biomorpholithic complexes of the lagoons have been completed. The geomorphological position, sequestration potential, and features of estuarine-lagoonal sedimentogenesis characterized by a high organogenic component were determined. Mainly thin silty and pelitic material brought to the silty dry land during the tide phase or by river runoff falls on its surface. Due to these sediments and weakly decomposed organic matter of plant origin, mainly the seagrass Zostera, this surface enters an altitudinal range already favorable for terrestrial vegetation. In our case, these are mainly sedges and pondweeds. The accumulation of peat-gley earth layers at the first stage of formation occurs at a very high rate, according to our data, up to 3 cm/year.

Further progress along the path of capitalization of environmental resources, the creation of systems for dynamic monitoring of blue carbon resources, systems of economic accounting and trading in carbon units is impossible without fundamental research, technological innovations in the field of blue carbon science.

It is supposed to form a technological proposal corresponding to market demand in the conditions of a full-scale pilot test site, the next stage of development of which will be the creation of carbon farms for carbon sequestration by the managed coastal wetlands of the lagoons of the Sakhalin Region. Figure 1 shows the location of the area with the already established rates of formation of peat-gley soils, where work is planned to develop scientific foundations and test nature-like technological solutions. According to the morpholithodynamic position, this area is in fact a crypto lagoon, because during the maximum low tides, a barrier form is exposed, separating this part of the Salmon Bay from the main water area (Figs. 2 and 3).

The creation and technological turnover of priority zones of accelerated organogenic coastal marine sedimentation is proposed. The biopositive nature-like technology for increasing the sequestration capacity of marine wetlands is based on a controlled process of growth of silt lands to altitudinal intervals, at which an active artificially accelerated growth of marsh halophyte vegetation begins. After 20–30 years, depending on the biomorpholithodynamic position, when the growth rate of the peat-gley strata slows down significantly, the deposits are removed and processed into humic and other products, or are buried in environments that prevent the release of GHGs.

Works on a detailed study of carbon turnover at the polygon require high-precision observations of the processes of sedimentation on silty dry lands and the accumulation of peat-gley soils of marshes. The technology for constructing high-precision photogrammetric digital multi-temporal surface models has been used by us for more than 7 years in studying the destruction of benches (Afanas'ev and Uba 2022b). Evaluation of quantitative parameters of organogenic sedimentation of marshes and silty drylands also includes a systematic geochemical analysis of deposits formed over a

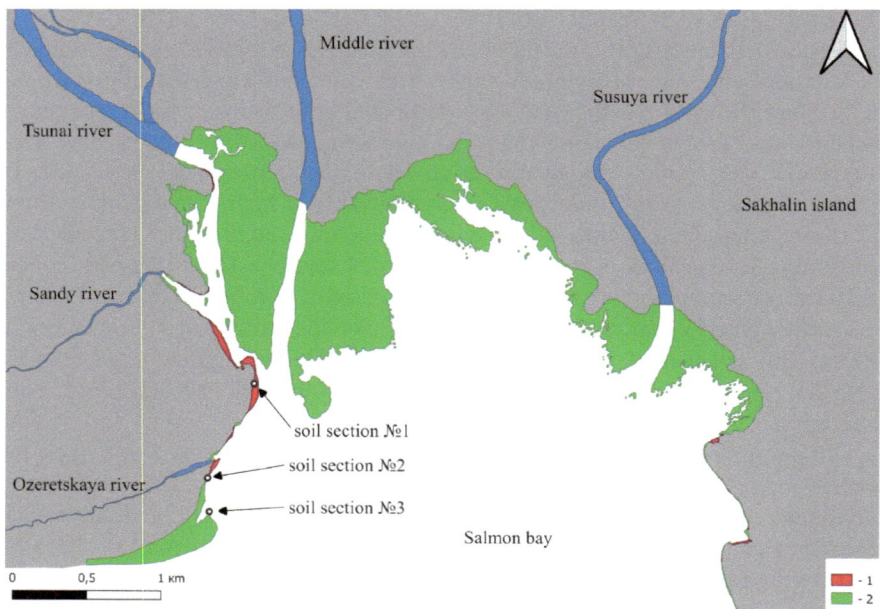

Fig. 1 Growth and erosion of marching terraces in Lososey Bay and location of soil sections (Aniva Bay, Sakhalin). 1—blur, 2—increase

Fig. 2 The surface of the march formed over the period 1952–2019 at the site of the polygon (image from a UAV)

Fig. 3 The surface of the march and silty drying in the area of the landfill

certain period of time. A geobotanical research program has been formed, supported by spectrozonal remote sensing from the Geoscan 401 UAV.

Thus, measurements of carbon stocks in coastal wetland ecosystems will mainly cover vegetation carbon pools (including above and below ground), soil and sediment carbon pools. It is proposed to use lead isotopes or cesium isotopes to date the stratospheric differences of soils and sediments.

However, without modern high-quality, reproducible data on gas exchange, the "cooling effect" of peat-gley soils of marshes, which are formed at a high rate, is very uncertain. The possibility of unambiguous conclusions can only be given by studies of the processes of soil gas exchange.

3 Conclusion

Our preliminary calculations showed that the sequestration potential of the managed marine wetlands on Sakhalin's potential is about 1.6 million tons of organic carbon per year.

When solving problems of coastal wetlands management by means of their accelerated programmable formation, heterogeneous hydro-biogeomorphological systems merge; absorption and transformation of existing and the emergence in their place of natural-anthropogenic systems with inherited or nature-like properties and structure.

Biopositive nature-like technology for the creation and technological turnover of zones of accelerated organogenic coastal-marine sedimentation is a controlled process of the growth of silty dries to altitude intervals, at which the active artificially accelerated growth of marching halophytic vegetation begins. After 20–30 years, depending on the biomorpholithodynamic position, when the rate of increase of peat-gley sediments significantly slows down, the deposits are removed and processed into humic and other products, or go to burial in environments that prevent the release of greenhouse gases.

References

Afanas'ev VV, Faustova AB (2022a) The first results of the study of sequestration properties of coastal marine biomorpholithosystems (sakhalin region). In: Physica land mathematical modeling of earth and environment processes. Springer Proceedings in Earth and Environmental Sciences. https://doi.org/10.1007/978-3-031-25962-3_46

Afanas'ev VV, Uba AV (2022b) Destruction of rocky coastes of cold seas. In: Processes in geomedia—vol 5. Springer Geology. Springer, Cham. https://doi.org/10.1007/978-3-030-85851-3_27

Afanas'ev VV, Latkovskaya EM, Uba AV, Levitsky AI (2022) Carbon Sequestration in the Coastal Marine Biomorpholithosystems of the Salmon Bay of Aniva Bay. In: Proceedings of physical and mathematical modeling of earth and environment processes. Springer Proceedings in Earth and Environmental Sciences. https://doi.org/10.1007/978-3-031-25962-3_49

Curado G, Rubio-Casal AE, Figueroa E, Castillo JM (2014) Plant zonation in restored, nonrestored, and preserved Spartina maritima salt marshes. J Coast Res 30(3):629–634. https://doi.org/10.2112/JCOASTRES-D-12-00089.1

Federal Law No. 34-FZ of 06.03.2022 (2022) On conducting an experiment to limit greenhouse gas emissions in certain subjects of the Russian federation

Hiraishi T, Krug T, Tanabe K, Srivastava N, Baasansuren J, Fukuda M, Troxler TG (2014) 2013 supplement to the 2006 IPCC guidelines for national greenhouse gas inventories: wetlands. IPCC, Switzerland

Hu Z, Van Belzen J, Van Der Wal D, Balke T, Wang ZB, Stive M, Bouma TJ (2015) Windows of opportunity for salt marsh vegetation establishment on bare tidal flats: the importance of temporal and spatial variability in hydrodynamic forcing. J Geophys Res 120(7):1450–1469. https://doi.org/10.1002/2014JG002870

Lovelock CE, Adame MF, Bradley J et al (2022) An Australian blue carbon method to estimate climate change mitigation benefits of coastal wetland restoration. Restor Eco:e13739. https://doi.org/10.1111/rec.13739

Mcowen CJ, Weatherdon LV, Van Bochove JW, Sullivan E, Blyth S, Zockler C, Fletcher S (2017) A global map of saltmarshes. Biodivers Data J (5). https://doi.org/10.3897/BDJ.5.e11764

Najjar RG, Herrmann M, Alexander R, Boyer EW, Burdige DJ, Butman D, Zimmerman RC (2018) Carbon budget of tidal wetlands, estuaries, and shelf waters of Eastern North America. Global Biogeochem Cycles 32(3):389–416

Order of the Ministry of Natural Resources of Russia dated June 30, 2017 N 20, ed. Accessed 20 Jan 2021

Ouyang X, Lee SY (2014) Updated estimates of carbon accumulation rates in coastal marsh sediments. Biogeosciences 11(18):5057–5071

Rosentreter JA, Al-Haj AN, Fulweiler RW, Williamson P (2021) Methane and nitrous oxide emissions complicate coastal blue carbon assessments. Glob Biogeochem Cycles 35:e2020GB006858. https://doi.org/10.1029/2020GB006858

Sapkota Y, White JR (2020) Carbon offset market methodologies applicable for coastal wetland restoration and conservation in the United States: a review. Sci Total Environ 701:134497. https://doi.org/10.1016/j.scitotenv.2019.134497

Solomon S, Qin D, Manning M, Averyt K, Marquis M (eds) (2007) Climate change 2007-the physical science basis: working group I contribution to the fourth assessment report of the IPCC, vol 4. Cambridge University Press

Spivak AC, Sanderman J, Bowen JL, Canuel EA, Hopkinson CS (2019) Global-change controls on soil-carbon accumulation and loss in coastal vegetated ecosystems. Nat Geosci 12(9):685–692

Steinmuller HE, Hayes MP, Hurst NR, Sapkota Y, Cook RL, White JR, Xue Chambers LG (2020) Does edge erosion alter coastal wetland soil properties? a multi-method biogeochemical study. Catena 187:104373. https://doi.org/10.1016/j.catena.2019.104373

Temmerman S, Govers G, Wartel S, Metre P (2004) Modelling estuarine variations in tidal marsh sedimentation: response to changing sea level and suspended sediment concentrations. Geology 212(1–4):1–19

Villa JA, Bernal B (2018) Carbon sequestration in wetlands, from science to practice: an overview of the biogeochemical process, measurement methods, and policy framework. Ecol Eng 114:115–128

Methodology for Determining the Content of BTEX and Oils in Water with Various External Parameters

V. P. Pakhnenko and T. O. Chaplina

Abstract The article considers the main methods used in environmental monitoring systems for determining oil products in waters and other environmental objects. The possibilities and limitations of gravimetry, IR spectroscopy, fluorimetry, and gas chromatography are described. Systematized information about instruments for determining the content of petroleum products in water and standard samples of the composition for their calibration. The results of using the BlueTrace fluorescent oil in water sensor to measure BTEX and oils in water are also shown.

Keywords Oil products · Emergency oil spills · Gravimetry · IR spectroscopy · Fluorimetry · Gas chromatography · Experimental work

1 Introduction

At present the great danger is represented by spills of oil and products of its processing, the volume of emergency discharges of which continuously grows and reaches critical values in man-made disasters (accidents at oil fields, ruptures of pipelines). With an annual volume of pollution of the World Ocean of 6–12 million tons the sources of pollution are distributed (in %) as follows: sea transport (washing water, docking, leakages, loading–unloading operations etc.)—35; industrial discharge—13; sea oil extraction—1.5; river discharge—32; inflows from the atmosphere—10; natural sources of oil inflows—about 8.5 (www.itopf.org). Large-scale pollution zones include not only the shelf, but also some areas of the open part of the sea. Average pollution of such seas as the Barents Sea, the Baltic Sea, the Black Sea, the Caspian Sea, the Mediterranean exceeds the maximum permissible concentration (MPC) of oil products in water several times.

Oil is a multicomponent energy carrier that includes substances of both organic and mineral origin. The polycyclic aromatic hydrocarbons (PAH) that make up petroleum

V. P. Pakhnenko (✉) · T. O. Chaplina
A.Yu. Ishlinsky Institute for Problems in Mechanics, Prospect Vernadskogo 101-1, Moscow 119526, Russia
e-mail: terkin95@mail.ru

© The Author(s), under exclusive license to Springer Nature Singapore Pte Ltd. 2023
T. Chaplina (ed.), *Processes in GeoMedia—Volume VII*, Springer Geology,
https://doi.org/10.1007/978-981-99-6575-5_29

products are highly toxic substances. Some of them, in particular anthracene, ovalene and benzpyrene have carcinogenic properties and also contribute to gene mutation. Other compounds in petroleum products also have adverse effects on the environment.

Petroleum products (PP) are divided into the following main groups: fuels, petroleum oils, petroleum solvents and lighting kerosene, solid hydrocarbons, oil bitumen, other PP (Othme 1992).

The water-soluble fractions of oil products which consist basically of aromatic hydrocarbons represent a special danger. They get into surface and ground waters during transportation, together with wastes of industrial enterprises, with rain water (Shagidullin et al. 2002).

In this regard, there is a need to constantly monitor the level of petroleum product concentration in water bodies. The maximum allowable content of petroleum products in water is regulated by standards and amounts to:

- for drinking water—0.1 mg/dm^3;
- for fish farming and domestic facilities—0.05 mg/dm^3.

The need to control the turnover of petroleum products and prevent their entry into the soil and water bodies is due to the high toxicity of these substances. In this regard, of great importance are measures aimed at identifying petroleum products in wastewater discharged by industrial enterprises.

This explains the need to control the actual concentration of petroleum products in water, as well as regulation of this value, carried out at the state level. Legislative acts of the Russian Federation established standards of maximum permissible concentrations (MPC) of oil and its derivatives in water for various purposes. To determine petroleum products in the environment a wide range of methods of analysis, devices and standard samples of composition for their calibration are used.

2   General Approaches to Determining the Presence of Petroleum Products in Water

At present, the technology of controlling the presence of BTEX, various oils, oil and its products in water mainly consists of periodic sampling of water for subsequent laboratory analysis.

Laboratory analysis, in turn, is usually carried out by one of the following methods: infrared spectrophotometry method; gravimetric method; gas chromatography; fluorimetric method.

To analyze any of these methods in the laboratory, you must first perform extraction of petroleum products from the sample. For this purpose, special chemicals—extractants are used. After extraction of components of oil from the studied sample with the help of various substances, determination of quantity of hydrocarbon compounds follows.

In the photometric method carbon tetrachloride is used, as well as physicochemical method using a column filled with aluminum oxide. When using the gravimetric method, an organic solvent and an aluminum oxide column are used. When analyzing by the fluorimetric method, the extractant is hexane.

After separation of petroleum products, the study in the photometric method, the sample is subjected to spectral (spectrophotometric) analysis, based on the absorption by petroleum hydrocarbons in some parts of the infrared spectrum which is irradiated with the sample. Gravimetric method is reduced to a simple weighing of the petroleum product separated from the sample. Gas chromatography is accompanied by the use of an auxiliary gas, a carrier gas, by which the test sample is fed into a special gas chromatographic column.

Such methods of control, which are reduced to periodic, even if quite frequent, sampling for analysis, have a number of clear disadvantages. In essence, it is just a point control, which does not provide an objective picture. Implementation of a system that ensures continuous monitoring of oil product discharges allows a company to monitor the content of discharges, as well as to plan and carry out various activities aimed at meeting the necessary requirements for controlling the discharge of oil products into the environment.

Determination of oil and petroleum products in water can be carried out by differential or integral methods. Differential methods include such methods as: gas, gas–liquid and high-performance liquid chromatography, chromatography-mass spectrometry. Integral methods include gravimetry, UV and IR—spectrophotometry, and luminescence. Moreover, integral methods are simpler and more convenient for carrying out observations of the state of oil pollution of water bodies and are mainly used in routine daily analysis. However, none of the listed methods does not allow to receive a full picture of qualitative composition of PP present in natural waters. For an exhaustive assessment of oil pollution, it is necessary to apply a combination of methods. In this case, for practical purposes application of any one integral method, for example IR—spectrophotometric or gravimetric is often quite enough (Oradovsky 1977). At the present time the area of analytical control of environmental objects pollution by oil products can be attributed to sufficiently well provided in the methodological plan section of analytical chemistry. In different books (Lurie 1984; Fomin 1995; Drugov 2000), ASTM standards (2000), guiding normative documents of Russia (1993) the procedures of sampling, sample preparation are considered in detail and methods of petroleum products determination in the air (ERD F 13.1:3.1-96), waters (ERD F 14.1:2:4.128-98 2002), soils (ERD F 16.1:2.2.22-98 2005) and bottom sediments (Oradovsky 1979) are stated.

3   Methods for Determining the Presence of Petroleum Products in Water

The technologies used for laboratory studies of the condition of water resources can give different results from each other, so the choice of method depends on the purposes of analysis. The methodologies of the examinations are regulated by federal environmental regulatory documents (ERD F).

3.1   Fluorimetric Method

Of all the methods currently used to determine the mass concentration of petroleum products in water, fluorimetric analysis is best suited to carry out continuous monitoring of this value in real time "online". The technique used in it deserves wider coverage in view of the appearance of devices functioning on its basis and raising the solution of the control problem to a qualitatively new level. The peculiarity of this technique is the use of radiation of the ultraviolet spectrum, in contrast to the photometric analysis, in which infrared radiation is used.

The fluorescence or fluorimetric method for determining the mass concentration of PP in water is based on the special properties of polycyclic aromatic hydrocarbons (PAH). In nature, these compounds are formed by pyrolysis of cellulose, so they are found in hydrocarbon deposits. For example: coal, gas and oil, which makes it very convenient to use them as a marker of the presence of petroleum products in water. PAH belong to the class of organic compounds whose molecular structure is characterized by the presence of condensed benzene rings.

The fluorescent properties of PAH are as follows. When these substances are exposed to radiation of certain wavelengths of the ultraviolet spectrum, PAH atoms, which are subjected to photon bombardment of UV radiation and received excess energy, begin to generate light emission of lower frequency. That is, having a longer wavelength than the original radiation. The luminescence of the substance irradiated by this method is called fluorescence. This process is caused by the fact that the electrons of the irradiated substance, receiving excess energy, make a transition to a higher energy level with a subsequent return to the old orbit. The transition from one state to another is accompanied by the release of released energy, released in the form of light radiation. This process does not stop as long as the substance continues to be irradiated. The intensity of the fluorescent glow is proportional to the mass of the substance irradiated with ultraviolet, which allows this method to be used for quantitative analysis of fluorescent compounds.

The advantages of the method include high sensitivity (the lower limit of the measuring range is 0.005 mg/litre), quick results, small volumes of the analyzed sample ($100 \, cm^3$) and the absence of significant interfering effects of lipids. The method of PP determination by fluorimetric method is described in the normative documents (Gladilovich 2001). Only aromatic hydrocarbons participate in formation of analytical signal. Since they have different conditions of excitation and emission registration, a change of the extract fluorescence spectrum depending on the wavelength of the excitation light is observed. When excited in the near UV, and even more so in the visible region of the spectrum, only the polynuclear hydrocarbons fluoresce. Since their proportion is small and depends on the nature of the PP, there is a very strong dependence of the analytical signal on the type of NP. Thus, the fluorimetric method of PP determination, which is based on emission registration in the visible spectral region, cannot be used for mass analytical measurements. Using the fluorimetric method, not only petroleum products as such, but also many other organic compounds of other origin are determined. Among polycyclic aromatic hydrocarbons (PAH) naphthalene and methylnaphthalene are the most common in light fractions of petroleum products.

This method is used to determine the content of hydrocarbons in water in a wide quantitative range from 0.005 to 50 mg/dm^3.

Practical realization of the fluorimetric method of water analysis was embodied in creation of a special submersible fluorescent sensor of oil products concentration in water. This device is designed for stationary placement in a monitored flow. The sensor is designed to work as a part of information-measuring system, controlling the state of the object according to different parameters, for which the sensors measuring different quantities are used. Such systems can have the widest application in various fields.

As an example, consider a sensor for determining the mass fraction of petroleum products in water from the GO Systemelektronik catalog (Fig. 1). The sensor is a thin cylinder made of stainless steel. In order to increase the corrosion resistance so that the device can work in aggressive environments, molybdenum is present in the material. The working side of the sensor designed to measure the mass concentration of petroleum products in waste water is its end surface, on which there is a transparent measuring window.

A special xenon lamp installed inside the sensor serves as a source of ultraviolet radiation with a wavelength of 285 nm. The receiving photodiode captures the fluorescent radiation generated by the PAH atoms, which has a wavelength of 325–375 nm. The device has a high sensitivity, the lower limit of determining the mass fraction of the petroleum product by this method is 3 ppm, which is 3 millionths of the substance in the total mass. At the same time, the device is very accurate, the measurement error in the process of analysis is 2%.

Fluorescence analysis is quite accurate, but there are nuances. The method is affected by external conditions: medium pH, temperature, viscosity, and other

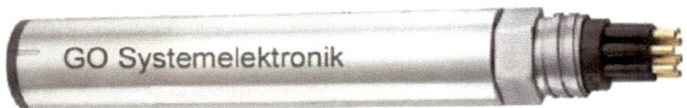

Fig. 1 GO Systemelektronik sensor

medium parameters. Since the quantitative version of this method is not easy to use, its accuracy is not good for samples with complex compositions, subject to degradation, overly viscous samples and others. In addition, let us pay attention to the following phenomena that interfere with quantitative fluorimetry: screening of molecules of the test substance by other compounds, stray light, light reabsorption by the studied molecules, non-uniform distribution of molecules over the volume of the sample. Nevertheless, this method of analysis shows itself very well when working with biological compounds, such as proteins and amino acids. At the same time, the modern equipment used for fluorimetry allows to level out most of the disadvantages of this approach.

The luminescence ability of the substances under study can also change with a change in dispersity, which in turn allows for a fairly accurate analysis if the original spectra are known. Qualitative fluorimetry uses information about the emission spectra of the substances under study. However, since not all chemical compounds are capable of emitting photons, this method is used as an auxiliary method in the spectrophotometry of multicomponent systems. A number of disadvantages of qualitative fluorimetry cannot outweigh its advantages, so this method of analysis is preferred in medical, biochemical and biological studies.

The main advantage of fluorimetry is its extremely high accuracy. It is believed that analysis methods based on the fluorescence properties of substances are a thousand times more accurate than other spectrophotometric methods. A certain error in these measurements is introduced by various effects mentioned above: screening, stray light, reabsorption and other phenomena of intermolecular and intramolecular interaction. The use of fluorimetry requires the construction of a calibration chart using solutions of the substance under study at the desired concentration, as well as the ability to select correction factors to account for various side effects. Modern instruments most often have software with built-in options for the automatic use of various correction factors.

The general principle for determining sample indices is to measure the detector response at a given wavelength and excitation intensity. As a result, the fluorimetry method measures the fluorescence intensity related to the concentration of the detected substance. This dependence has an exponential character, which requires the construction of a calibration graph. This graph is usually constructed using solutions of the desired substance with known concentrations and is a straight line whose slope tangent is described as follows:

$$tga = 2.3 \times I0 \times K \times jkB \times e \times l$$

where $I0$ is the intensity of the excitation radiation, K is the sensitivity coefficient of the device, jkv is the quantum yield of luminescence, and e is the molar extinction coefficient of the substance under study at a given excitation wavelength.

Fluorimetry is a sensitive and highly selective method of analysis. At the same time, it has a number of drawbacks. These disadvantages are levelled by modern equipment, but may hinder the study with the necessary accuracy.

The disadvantages of the method:

- noticeable dependence on environmental conditions;
- stringent requirements for the sample under study;
- inability to determine some substances that are not capable of fluorescence.

Advantages of the method:

- high accuracy;
- comparative ease of sample preparation (subject to the availability of suitable equipment);
- possibility of control of biochemical and biological processes.

Fluorimetry is mostly used as: auxiliary method (for other spectrophotometric types of analysis); method of control of processes of biochemical nature. Examples of laboratories in which fluorimeters are required would be biochemical and pharmacological institutions, that is, those laboratories where the accuracy of analysis and purity of substances would be of paramount importance.

3.2 Gravimetric Method

Gravimetric method (ERD F 14.1:2.116-97) is based on the extraction of petroleum products from analyzed waters with organic solvent, separation from polar compounds of other classes by column chromatography on aluminum oxide and quantitative determination by gravimetric method.

The gravimetric method (Rodin et al. 2007) is based on the extraction of PP from the analyzed samples with low-polarity solvents (chloroform, hexane, carbon tetrachloride, pentane, petroleum ether, freon (chladone)—(1,1,2-trichloro-1,2,2-trifluoroethane). This is followed by purification of the extract from polar substances by passing it through a column with a sorbent (aluminum oxide of activity level II (which contains 3% H2O), silica gel, floresil (basic magnesium silicate). This is followed by removal of the extractant by evaporation and weighing the residue to determine the amount of "petroleum products". Usually for the analysis take 0.1–3 L of water, acidify with HCl to pH < 5, then extract in two portions of 20 cm^3 of the solvent and combine the extracts. When the concentration of PP in water < 0.3 mg/litre of a 3-L sample will not be sufficient for their reliable determination by the gravimetric method. In this case it is recommended to extract the sum of nonpolar organic compounds from large volumes of water (10–20 L) using continuous liquid–liquid extraction or sorbents like active carbon, etc. The sorbed substances are then desorbed by carbon tetrachloride in a Soxhlet apparatus (Lurie 1984). Different degree of extraction corresponds to different components of oil: aliphatic hydrocarbons—up to 85%, for aromatic hydrocarbons—20%. As a rule, the method is applicable for the analysis of natural waters and industrial effluents at the concentration of petroleum hydrocarbons from 5 to 1000 mg/liter. In the process of sample preparation and analysis the hydrocarbons with the boiling point less than 100°C may be lost. In this case the direct extraction with hexane leads to underestimated results.

The error may be up to 30% if the analyzed water contains suspended particles. Production of carbon tetrachloride as an ozone-depleting substance has been discontinued in Russia since late 1999. The main advantage of the gravimetric method (one of the few "absolute" methods of analytical chemistry) is that it eliminates the need for standard samples of the same qualitative and quantitative composition as the test sample. There is also no need to calibrate the measuring instruments beforehand. Limit of PP determination in waters when using refrigerant is 2–4 mg/litre (Fomin 1995). The arbitrary gravimetric method for determining low concentrations of PP requires large volumes of analyzed water and solvents. In this connection, the fluorescent chromatographic method is recommended for routine work to control the PP content in drinking water and water bodies (Lurie 1984). The method is based on chromatographic separation of PP from polar hydrocarbons and impurities of water of not oil origin in a column with active aluminum oxide with use of extractants—chloroform and hexane and further definition of the isolated PP by a luminescent method. Only a part of hydrocarbons (aromatic high-molecular weight, especially polycyclic ones) is possessed the ability to luminesce under influence of UV light and, moreover, in different ways depending on excitation conditions. To obtain reliable analysis results, it is therefore necessary to have a standard sample solution containing the same luminescent substances and in the same relative amounts as in the test sample, which is very difficult to implement in practice. It is easier to set the "division value" of the applied instrument by comparison with the result obtained by one of the arbitration methods, and to make such comparison necessarily each time anew if there are indications that the qualitative composition of water has changed. However even in this case the received results should be considered less reliable, than those received by gravimetric arbitrage method described above. Light PP (gasoline, kerosene) cannot be determined by the luminescent method. The sensitivity of the method is 0.02 mg per sample. When the content of suspended substances less than 10 mg/litre one solvent—hexane can be used, which significantly simplifies and refines the determination, as it excludes the evaporation of chloroform and the loss of light oil products (Rodin et al. 2007).

3.3   Method of Infrared Spectrophotometry

The infrared spectrophotometry method is one of the powerful analytical tools widely used for research purposes and routine monitoring of production processes. The method makes it possible to determine the presence of petroleum products in the amount of 0.05–50 mg/dm^3.

This method uses carbon tetrachloride to extract dissolved and emulsified petroleum products. Then the desired substance is separated from other organic components on a column with a sorbent (aluminum oxide). The fraction of the compound is calculated by the degree of absorption intensity in the infrared region of the spectrum.

The most widespread method for oil HC monitoring is the IR-spectrometry, which allows to determine the amount of aliphatic HC and PAH. In this case, the content of both petroleum HC of anthropogenic origin and those produced by marine organisms is measured. The corresponding analysis techniques are based on the extraction of PP from the sample with an organic solvent (CCl4 or chladone 113). This is followed by purification of the extract from polar compounds by column chromatography on aluminum oxide and subsequent registration of the IR spectrum in the region of $2700-3200$ cm^{-1}. The method can be implemented both in the variant of registration of an absorption spectrum in the specified area by means of a traditional or Fourier spectrometer, and in a simpler version in which the analyzer measuring integral absorption in the area of $2900-3000$ cm^{-1}, in which the most intensive bands corresponding to asymmetric valence vibrations of CH3 and CH2 groups are observed, is used. IR-spectroscopy is applicable to the analysis of natural waters and industrial effluents at the concentration of petroleum hydrocarbons from 0.1 to 50 mg/liter (using a cuvette with the optical path length of 10 mm), does not require solvent distillation and heating of the extract, which eliminates the loss of HC with low boiling point. The advantage of the infrared spectroscopy method is lower losses of light fractions than in the determination of hydrocarbons by other methods. The lower limit of the measurement range is 0.05 mg/liter. It is worth noting that the main advantage of the method is a weak dependence of the analytical signal on the type of petroleum product, which forms the basis of the sample contamination. Among the disadvantages arising when using this method, it is worth noting the interfering influence of lipids and other polar compounds at their high content, at which the capacity of the chromatographic column used to purify the extract is exhausted. The main disadvantage of the method is its unecological nature, which is due to the highly toxic solvents used.

3.4 Gas Chromatography

The technology is suitable for the analysis of aqueous solutions with a petroleum product content of more than 0.02 mg/dm^3. Hydrocarbon compounds are extracted from the sample with an extractant and purified with a sorbent. The extract obtained is examined with a gas chromatograph. This method allows not only to detect the presence of petroleum products in water and their quantity, but also to identify the specific composition of impurities.

The gas chromatography method (GOST 31,953-2012) is used to determine the mass concentration of petroleum products in drinking water, including bottled water, natural (surface and underground) water, including water from drinking water supply sources, as well as waste water with a mass concentration of oil products not less than 0,02 mg/dm^3.

After extracting petroleum products from water samples with an extractant, the extract is purified from polar compounds with a sorbent. Then the eluate is analyzed on a gas chromatograph. The areas of chromatographic peaks of hydrocarbons in the retention time range equal and (or) more than n-octane are summarized and the content of oil products in water is calculated according to the established graduation dependence. This method allows to determine not only the total content of petroleum products, but also to identify the composition of petroleum products.

The GC method is suitable for analysis of samples that contain petroleum products at the ERD level. Duration of registration of the chromatogram is 20–30 min. There is a method of PP determination including extraction with hexane, evaporation of the extractant and gas chromatographic detection of analyzed components in the concentrate (Korenman and Fokin 1993). The given way of PP determination has rather low detection limit (0.01–0.03 mg/litre at luminescent detection) and allows to control reliably quality of natural and potable water. The method is recommended for sanitary and toxicological analysis of natural and treated wastewater.

Gas chromatography with a mass-spectral detector allows not only to determine the total content of PP (like other methods), but also to identify and quantify individual hydrocarbons that are part of oil products. The latter circumstance makes it possible to detect the source of oil pollution (to determine the type and brand of PP) and to take measures to eliminate the consequences of pollution. The gas-chromatographic method was developed and successfully applied in the system of control laboratories of Mosvodokanal in 1985–1995. (Smolyaninov et al. 1981), was certified by Gosstandart of the Russian Federation (MVI-05-94 1994) and is a reliable and informative way of PP determination in any natural and waste waters, as well as in drinking (tap) water.

4 Using the BlueTrace Sensor to Detect Petroleum Product Concentrations in Water

The BlueTrace Water Sensor is a compact fluorescent sensor for measuring BTEX and oils in water. The BlueTrace sensor is designed to be used in harsh environments, such as aggressive environments or high pressures environments.

The BlueTrace sensor has the following specifications: measuring principle is fluorescence; light source: <300 Nm; emission wavelength: 280 Nm; fluorescence detection: 300–400 Nm; measuring range: 0–30/100/300 ppm; measurement accuracy: 3% FS; detection limit: 0.1 ppm; measurement interval: ≥1 c.

This sensor was used by the authors to control samples when purifying surface and wastewater from oil and its products with a natural sorbent, namely, sheep's wool. A number of laboratory experiments on adsorption by wool of PP spilled on the surface of water at varying its thermodynamic characteristics—temperature and salinity—were carried out (Table 1).

Table 1 Values of oil concentration in water samples after three consecutive sheep wool cleaning operations (using BlueTrace sensor)

Sample	Oil concentration, mg/l
#1	395
#2	84
#3	4.2

5 Fluorescence Diagnosis of the Quality of Sheep's Wool Water Treatment: Direct Fluorimetry of Water Samples

To analyze water samples taken at different stages of the process of water purification from oil pollution, we used the procedure developed in the laboratory "Laser spectroscopy of aqueous media and laser biophotonics" of MSU named after M. V. Lomonosov for normalization of the PP fluorescence intensity by the intensity of Raman light scattering band by medium molecules—in our case, water or hexane (Chaplina et al. 2016).

In the standard prescription, the extraction technique includes the extraction of PP with carbon tetrachloride or hexane and determination of PP concentration by absorption IR spectroscopy (for carbon tetrachloride extraction) or fluorescence spectroscopy (for hexane extraction). A preconstructed calibration curve linking the measured optical parameter with the concentration of PP in the extract is used. We used the hexane technique with a fluorescence end. The calibration was performed using a solution of the same oil (brand "Siberian Light", density 845 kg/m^3 (36.50 API), sulfur content 0.57%). The solution of oil in hexane was prepared by adding 1 ml of oil in 10 ml of hexane and then calibrated diluted the obtained extract with hexane to the required concentration. Fluorimetry can be calibrated in a similar way in the case of a direct aqueous sample. Fluorescence measurements were performed on a FluoroMax 4 spectrofluorimeter (Horiba Jobin Yvon, France). To correctly apply the internal repeater method, we recorded sections of the optical response spectrum containing both the PP fluorescence band and the Raman band of the medium (hexane or water). To increase the accuracy of determining the Φ_0 parameter, we ensured the ratio of the intensities of these bands within plus or minus one order of magnitude relative to unity by diluting the solutions.

The determined value of oil concentration in the extract C_{ex} was recalculated to the concentration in the water sample C_w according to the formula:

$$C_w = C_{ex}(v/V)$$

Figure 2 shows the fluorescence spectra of water samples measured at different values of the excitation wavelength (λ_{ex}); the lower right segment of the figure shows the absorption spectra, or rather the optical density determined by both absorption and light scattering, including that on the oil emulsion. Absorption spectra were measured on a Lambda 25 spectrophotometer (PerkinElmer, USA).

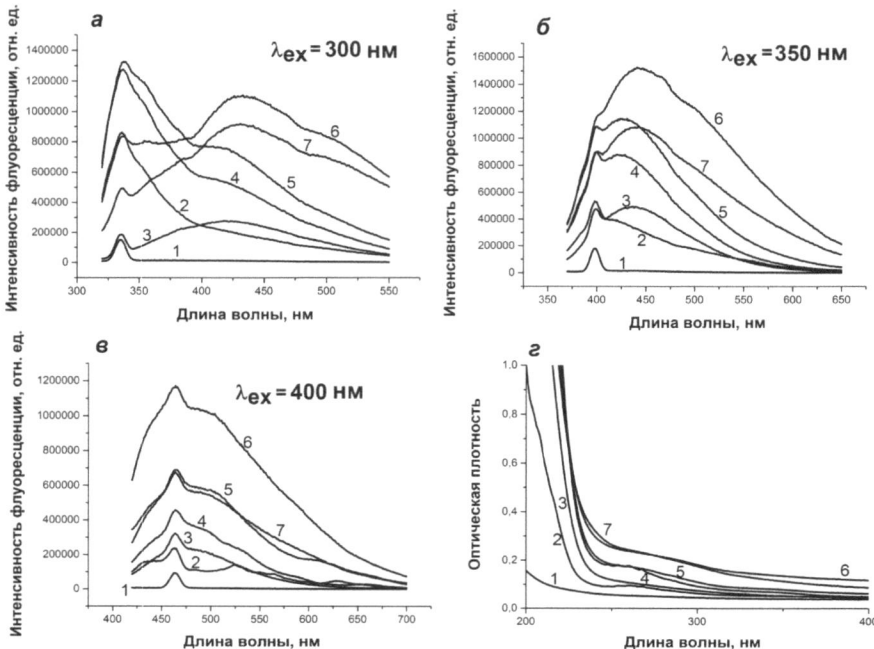

Fig. 2 Fluorescence and absorption spectra of water samples 1–7. Absorption spectra of these samples are shown in the right lower sector of the figure. The numbers in the figure indicate: 1—distilled water; 2—distilled water contaminated with oil; 3—tap water used for the experiment, hereafter "water"; 4—water contaminated with oil before cleaning; 5—water contaminated with oil after first wool cleaning; 6—water contaminated with oil after second wool cleaning; 7—water contaminated with oil after third wool cleaning

Sheep wool contributes to the purification of water from oil pollution, as evidenced by the decrease (on the scale of the water Raman band, which can serve as an internal repeater) of the intensity of spectral bands of fluorescence of different oil fractions (in Fig. 2 at excitation wavelengths below 320 nm two such bands are visible—with maximums at 370 and 410 nm; at higher values of the excitation wavelength only the long-wave band of oil fluorescence is shown).

The results given in Table 2 show that at ultra-high oil concentration in water (368 mg/l) a portion of sorbent (wool) with the weight of 1 g immersed in a water volume of 2 l, sorbs 76% of oil, but at oil concentration of 89 mg/l the same portion of sorbent leads to a higher degree of purification (95% of oil is collected). The suggested explanation for this difference is an insufficient amount of sorbent in the first case and a more favorable ratio of amounts of sorbent and oil in the second case.

This work was performed under Government Order 123,021,700,046-4 (FFGN-2023-0006).

Table 2 Values of oil concentration in water samples after three consecutive cleaning operations with sheep wool	Sample	Oil concentration, mg/l	Degree of purification, (%)
	#1	368	76
	#2	89	89
	#3	4.5	95

References

ASTM D5412-93 (2000) Standard test method for quantification of complex polycyclic aromatic hydrocarbon mixtures or petroleum oils in water

Chaplina TO, Voloshina OV, Stepanova EV, Fadeev VV (2016) Fluorescence quality control of water purification from oil pollution by sorbent on the basis of sheep's wool. Process Geomedia 2:81–92

Documents & Guides: the international tanker owners pollution federation limited, ITOPF. www.itopf.org

Drugov YS (2000) Ecological analytical chemistry. Anatolia, SPB, p 432

ERD F 14.1:2:4.128-98 (2002) Methods for measuring the mass concentration of petroleum products in natural, drinking and waste water samples by fluorimetric method on "Fluorat-02" liquid analyzer (instead of ERD F 14.1:2:4.35–95). NPF "Lumex"

ERD F 16.1:2.2.22-98 (2005) Methods for measuring the mass fraction of petroleum products in soils and bottom sediments by infrared spectrometry. Tyumen State University

ERD F 13.1:3.1-96 Methods for measuring the volume fraction of hydrocarbon components of oil in the mixture with air by gas chromatography and determination of the saturated vapor pressure of petroleum products. AO "KU-BAN-ECO", PNU "Orgneftekhimzavod"

Fomin GS (1995) Water, control of chemical, bacterial and radiation safety according to international standards. Encyclopedic Handbook, 2th edn, revised and supplemented. VNII standard, M, p 618

Gladilovich DB (2001) Fluorimetric method to control the content of petroleum products in waters. Partners Competitors 12:11–15

Guidelines for chemical analysis of seawater (1993) RD 52.10.243-92. Gidrometeoizdat, SPB, p 264

Korenman YI, Fokin VN (1993) Gas-chromatographic determination of petroleum products and volatile phenols in natural and treated wastewater. Chem Technol Water 15(7):530–533

Lurie Y (1984) Analytical chemistry of industrial wastewater. Chemia, M, p 448

MVI-05-94 (1994) Methods of PP in natural and waste waters by gas chromatographic method with flame ionization detector. GOST R Certification System. Center for Water Certification and Metrological Support of Environmental Monitoring-AO TSSV

Oradovsky SG (1977) Manual on methods of chemical analysis of sea waters. Gidrometeoizdat, L., p 208

Oradovsky SG (1979) Methodological guidelines for the determination of pollutants in marine bottom sediments, vol 43. Hydrometeoizdat, M

Othme K (1992) The Chemical Encyclopedia: in 5 vols: vol 3. The Big Russian Encyclopedia, Moscow, p 641

Rodin AA, Drugov YS, Zenkevich IG (2007) Ecological analyses in case of oil and oil products spills. BINOM, M Laboratory of knowledge, p 270

Shagidullin RR, Avvakumova LV, Do-Roshkina GM, Selyanina SG (2002) Determination of oil products in waters based on FT-IR spectral complex and measurement of integral intensities of absorption bands νCH. J Anal Chem 57(3):250–256

Smolyaninov GA, Filippov YS, Zelvensky VY, Senin NN, Sakodynsky KI, Artemova IM (1981) Gas-chromatographic determination of oil products in natural and wastewater. J Anal Chem 26(2):342–349

Abrasion-Denudation Cycle Levels as the Basis of Paleoreconstructions on Rising Coasts

Victor V. Afanas'ev and A. V. Uba

Abstract The paper proposes to consider the problem of marine cyclical levels on rising rocky shores from the standpoint of coastal morpholithodynamics and sedimentation features in the post-sea period of bench development. The data of facies analysis of deposits with high-precision geospatial reference of the geological section of coastal ledges and the results of dating of deposits overlying the bench make it possible, even in the absence of micropaleontological confirmation, to substantiate the marine origin of the basement terraces. The paper presents for the first time the results of high-precision measurements of the geospatial position of the elements of the geological section and the basement, as well as the determination of the C_{14} age from the coastal marine deposits of the so-called Boshnyakovskaya terrace, whose C_{14} age was previously determined as early-Middle Pleistocene.

Keywords Coastal cut platform · Marine terrace · C_{14} age · Neotectonic reconstructions

1 Introduction

Discussion about the C_{14} age and geospatial parameters of the Pleistocene terraces on the coast of the island. Sakhalin has recently calmed down somewhat (Svitoch 2004). Nevertheless, differences in views on the problem of marine cycle geomorphological levels have not been overcome here and are due, on the one hand, to the lack of results of dating the age of deposits, on the other hand, as Aleksandrova (1978) by the fact that no micropaleontological evidence in favor of a marine origin was found for terraces with a height of 50–80 m and higher (Aleksandrova 1978).

However, at the same time, the actual geomorphological "trace" in the formation of these surfaces remains practically unattended by researchers. A great influence on the

V. V. Afanas'ev (✉) · A. V. Uba
Institute of Marine Geology and Geophysics, FEB RAS, Yuzhno-Sakhalinsk, Russia
e-mail: vvasand@mail.ru

Sakhalin State University, Yuzhno-Sakhalinsk, Russia

T. Chaplina (ed.), *Processes in GeoMedia—Volume VII*, Springer Geology,
https://doi.org/10.1007/978-981-99-6575-5_30

formulation and presentation of such an approach to the analysis of the results of the high sea level in the Late Pleistocene was exerted by our studies of the formation and development of rocky coasts and, in particular, benches (Afanas'ev 2020; Afanas'ev and Uba 2022).

In the presented work, we draw attention to the fact that the socle terraces on the sea coast with neotectonic movements of a positive sign are nothing more than a paleobench overlain by marine, estuarine-lagoonal, alluvial, slope of various types, and eolian deposits. The thickness of these deposits at one cycle abrasion-denudation level, depending on the position in relation to the source and type of sedimentation, the features of the destruction of the coastal ledge, can vary in a very wide range. So, for example, even over a 3 km segment of the coastal ledge near the village of Novoselovo, the thickness of the proluvial-colluvial cover of the basement terrace varies from 3 to 20 m as the exposed section of the coastal ledge moves away from the source of slope deposits.

We analyzed the composition of the sediments overlying the paleo coastal cut platform carried out high-precision measurements of the geospatial position of the elements of the geological section and the bedrock base, and for the first time obtained C_{14} age determinations from the coastal marine deposits of the so-called Boshnyakovskaya terrace, whose age was previously determined as Middle Pleistocene (Kulakov 1973; Aleksandrova 1982; Korotkiy et al. 1997). The location of the study area is shown in Fig. 1.

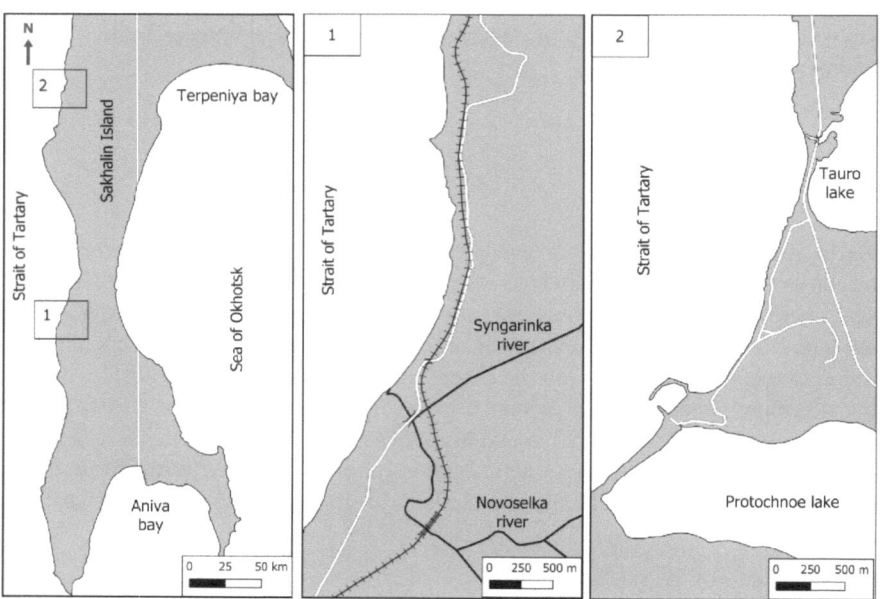

Fig. 1 Location of the study area. 1-marine basement terrace in the area of Novoselovo village, 2-sea basement terrace in the area of Shakhtersk village

2 Methods and Results

All measurements were made with SOUTH Galaxy G1 GNSS receivers (GPS, GLONASS, BEIDOU, GALILEO). Pickets at work sites were taken in RTK mode (real time kinematics) from a local base station (BS) and using a Topcon GTS236-N total station. The coordinates of local BSs were received in static mode from 2 permanent BSs of the EFT-CORS network located in Korsakov and Yuzhno-Sakhalinsk. At the same time, the RTK accuracy is within 1 cm in the plan and 2 cm in height, the statics are no more than 2 cm in the plan and 2 cm in height. The ITRF2014 (epoch 2010.0) coordinates of the BS Korsakov and Yuzhno-Sakhalinsk are taken as reference, the height above the WGS84 ellipsoid. The obtained picket heights were recalculated into heights relative to the EGM2008 geoid (Earth Gravitational Model 2008).

Determination of the C_{14} age of the samples by the radiocarbon method was carried out in the laboratory "Geomorphological and paleogeographic studies of the polar regions and the oceans" of St. Petersburg State University. The calendar age values are based on the calibration program "OxCal 4.4.4" (calibration curve "IntCal 20"). Christopher Bronk Ramsey (https://c14.arch.ox.ac.uk).

In a detailed study of the section of the coastal ledge and geological wells on the surface of the basement terrace in the area of the village of Novoselovo, under the soil layer or thin bulk soils, hard gravelly sandy loams with lenses of gravel loams with rare thin clay interlayers occur, which are replaced to the north by sandy loams and loams of soft-plastic and plastic consistency (Fig. 2).

The thickness of the upper layer varies within 4.0–4.5 m. On the southern flank of the site, the most technogenically modified, its thickness decreases to 1 m. In the outcrop of the coastal ledge, there are no water intrusions in this layer. The

Fig. 2 Orthophotoscheme of the coastal ledge in the Novoselovo village area: 1–3 geological sections with sampling for radiocarbon and facies analyses, 4–5 points for determining the geospatial parameters of the paleobench

groundwater level in the wells is below the base of this horizon. According to their characteristics, the deposits belong to the slope deluvium, overlying the alluvial marine deposits of the Upper Pleistocene age.

Below this layer, a 2.0–2.5 m layer of water-bearing flooded gravels can be traced in the outcrop of the coastal ledge. In wells located closer to the ancient coastal ledge, this layer has a lower thickness and a higher hypsometric position than in the coastal cliff. Probably, here we are dealing with a pack of marine beach deposits proper, and given the proximity of the river, these deposits are most likely alluvium, reformed by wave-cutting processes near the mouth zone.

Even lower, a subhorizontal horizon of coastal-marine deposits, estuary-lagoonal lithofacies complexes—clays and loams, less often bluish and gray sandy loams, in which mixed complexes of brackish-water and marine diatoms were identified and the interlayer lying between them intensely ferruginous gravel-pebble deposits with a thickness of 2.04 m. C_{14} age determinations were obtained from these deposits (Table 1).

The roof of this member in the section of description (r.d.) No. 3 of the coastal ledge lies at elevations of 36.09 m; No. 1 at 35.19 m, and in wells located 50–70 m from the outcrop of this contact in the coastal ledge towards the ancient coastal ledge, the heights of this layer are 36.89 m. 38.2 m. The thickness of the upper layer is 0.8–0.91 m. The bottom of the lower layer of silt deposits with a thickness of 1.38 m (r.o. No. 2) lies at a mark of 31.76 m.

Pebbles and gravels with sandy loamy and sandy aggregates, as well as very dense gravelly sandy loams 5.50 m thick, are noted below the clayey layer (N1 hl), Nevelsk (N1 nv), Chekhov (N1 ch), Upper Duya (N1 vd), Kurasi (N1 kr) formations. In point 5, the absolute elevations of the paleobench surface are 26.26 m, and in point 4, they are 25.64 m (B.S. 77)

The abrasion-denudation cyclic geomorphological level in the Shakhtersk region is worked out in siltstones, mudstones and fine-grained sandstones of the Kurasiyskaya suite (N1kr) of various strengths. The absolute marks of the basement surface lie in the range of 8.0–8.5 m (B.S. 77). Sandy-pebble deposits occur directly on the paleobench, the nature of the layering of which allows them to be attributed to

Table 1 Results of absolute age determination by radiocarbon method

Lab. number	Location, sample type	Altitude position, m	Radiocarbon age, years	Calibrated age (calendar), years
LU-10099	Section №. 1, wood	35.00–35.19	25,880 ± 340	30,200 ± 390
LU-10100	Section №. 1, algal peat	34.86–34.90	25,120 ± 290	29,440 ± 310
LU-10715	Section №. 3, wood	36.09–36.00	24,410 ± 430	28,630 ± 480
LU-10718	Section №. 3, wood	35.60–35.66	25,420 ± 280	29,650 ± 290
LU-10717	Section з №. 3, wood	35.18–35.24	25,590 ± 350	29,830 ± 390
LU-10716	Section №. 2, wood	32.56–32.70	25,790 ± 370	30,080 ± 430

Fig. 3 Orthophotoscheme of the coastal ledge in the area of the village of Shakhtersk

marine beach sediments. These deposits are overlain by aeolian fine-medium-grained sands 2.0–3.0 m thick. On the surface of the terrace, a fragment of a paleodune about 5 m high is observed (Fig. 3).

It is interesting that the top of the estuarine-lagoonal deposits that fill the paleoincision south of the considered part of the basement terrace lies at approximately the same elevations as the surface of the paleobench. The thickness of the eolian deposits covering the surface of the lagoon silts is 7–8 m.

3   Conclusion

Exact geospatial parameters of the basement and overlying sedimentary strata have been determined for the first time as a result of the study of the socle sea terraces of the western coast of Sakhalin. The age of marine regressive deposits of the marine terrace, determined by the predecessors as the early Pleistocene, is significantly rejuvenated as a result of radiocarbon dating of these deposits. Thus, the amplitude of neotectonic movements, for which the analysis of marine terrace levels is used, is significantly lower than expected (Bulgakov and Senachin 2019). The facies analysis of the deposits and the geospatial position and parameters of the terrace basement surface make it possible to attribute these formations to the category of paleobenches. A very interesting fact is the inclination of the surface of the benches to the lines of the general runoff, which were bays and straits during the period of sea level decrease. This phenomenon is also found on the pumice abrasion shores of the Holocene age (Afanas'ev et al. 2019).

References

Afanas' ev VV, Uba AV (2022) Destruction of rocky coastes of cold seas. In: Processes in geomedia, vol 5. Springer International Publishing, Cham, pp 243–252. https://doi.org/10.1007/978-3-030-85851-3_27

Afanas'ev VV et al (2019) Vetrovoy isthmus of iturup island–holocene strait. In: IOP conference series: earth and environmental science. IOP Publishing, vol 324(1), p 012029. https://doi.org/10.1088/1755-1315/324/1/012029

Afanas'ev VV (2020) Morpholithodynamic processes and development of the coast of the contact zone of the subarctic and temperate seas of the North Pacific. IMGiG FEBRAS, Yuzhno-Sakhalinsk, p 234. ISBN 978-5-6040621-8-0, https://doi.org/10.30730/978-5-6040621-8-0.2020-1

Aleksandrova AN (1978) Sea terraces of Sakhalin and Hokkaido. In: Geomorphology and paleogeography of the shelf. Nauka, Moscow, pp 123–131

Aleksandrova AN (1982) Pleistocene of Sakhalin. Nauka, M., p 292

Bulgakov RF, Senachin VN (2019) Marine terraces and hydroisostasy influence on the vertical movements of the Sakhalin. Geosyst Transit Zones 3(3):277–286 (In Russ.). https://doi.org/10.30730/2541-8912.2019.3.3.277-286

Korotkiy AM, Pushkar VS, Grebennikova TA (1997) Marine terraces and the Quaternary history of the Sakhalin shelf. Dal'nauka, Vladivostok, p 229

Kulakov AP (1973) Quaternary coastlines of the Sea of Okhotsk and the Sea of Japan. Nauka, Novosibirsk, p 189

Svitoch AA (2004) Marine pleistocene of the far eastern coasts of Russia. Pac Geol 23(3):76–93

Exact Analytical Expression for Maximum Heat Flux from Impulsive Sources of Charged Particles in Magnetic Field

Maxim Glebovich Ponomarev

Abstract Exact analytical expression is obtained for upper limit to heat flux (maximum heat flux) from impulsive (explosive) charged particle sources in magnetic field. This exact expression is based on the obtained distribution function as analytical solution for Boltzmann equation for charged particle flows in magnetic field. This can be considered as a saturation phenomenon, when we see significant slow-down of the heat flow increase versus the magnetic field still increasing at the same rate. It is instrumental when we need to estimate the saturation maximum of the heat flow, when we have the only 1 parameter to change: magnetic field.

Keywords Geomagnetic fields · Magnetic fields · Charged particle flows · Magnetic field effect on ionized flows · Plasma · Geomagnetism · Impulsive charged particle sources · Explosive charged particle sources · Maximum heat flux

1 Introduction

This paper deals with yet another example from extensive number of innovative physical models with applications to real world problems outlined in Carati and Ponomarjov (2003), Gun'ko and Ponomarev (1993, 1995a, b), May et al. (2006), Ponomarev 1995, 1996, 1997a, b, 1998a, b, 1999, 2000a, b, c, d, e, f, 2001a, b, c, d, 2002, 2022a; b; c, 2023a, b), Ponomarev et al. (2010, 2012a, b, 2013a, b, 2014, 2022), Iovea et al. (2012, 2013), Garrison et al. (2011), Ponomarjov and Gunko (1995), Ponomarjov and Carati (2006, 2003, 2004a, b, c, d, e), Webb and Ponomarev (2007).

Specifically, this paper deals with kinetic simulations of heat flow control from impulsive (explosive) charged particle sources in magnetic field. These kinetic simulations extend directly previous results by Ponomarev MG, published in Carati and

M. G. Ponomarev (✉)
Cambridge Research and Consultancy, Cambridge, UK
e-mail: maxim.pv@gmail.com

Adyghe State University, Maykop, Russia

© The Author(s), under exclusive license to Springer Nature Singapore Pte Ltd. 2023
T. Chaplina (ed.), *Processes in GeoMedia—Volume VII*, Springer Geology,
https://doi.org/10.1007/978-981-99-6575-5_31

Ponomarjov (2002), Gun'ko and Ponomarev (1993, 1995a, b), Ponomarjov (1995, 1996, 1997a, b, 1998a, b, 1999, 2000a, b, c, d, e, f, 2001a, b, c, d, 2002), Ponomarjov and Carati (2003, 2004a, b, c, d, e), Ponomarev et al. (2011), Webb and Ponomarev (2007).

It is well known that magnetic field can control charged particles flows. There are a lot of cases when there is a need to know the upper limit (saturation maximum) to heat flow for variable magnetic field. Namely, if we plot this heat flow versus this magnetic field, this paper shows analytically, that there is an upper limit to the heat flux (saturation effect).

No matter how big the magnetic field is, this limit to the heat flux can not be exceeded.

From the view point of saving energy, for magnetic field generation, this obtained heat flux limit allows to save the energy by applying corresponding limit on the requirements for the generation energy to generate the required magnetic field. So, if we consider this as magnetic field driving this heat flux (flow), the upper limit to this heat flux (flow) corresponds to the saturation effect.

2   Heat Flux Vector Measurements from Experiments

The heat flux vector corresponds to the net flux of kinetic energy of chaotic motion. As usual, it is splitted into $\vec{B_\tau}$, the component tangential to $\vec{B}$, and $\vec{B_n}$, the component normal to $\vec{B}$:

$$\vec{B} = \vec{B_\tau} + \vec{B_n} \tag{1}$$

$\left|\vec{B_\tau}\right|$ is expected to be significantly larger than $\left|\vec{B_n}\right|$ component, because the magnetic field resists heat flux in the normal direction. Heat flux can show that fast particles escape along the magnetic field lines from some injection region.

3   Moments of the Distribution Function, Partial Moments

Compression of data is only one case of the applications of moments. Even more significant are their contributions as the functions taking part in the fluid equations, especially those related to the conservation of energy, momentum, and mass. It should be noted here that in these applications, the exact type of the distribution function describing the moments is not significant.

As usual, the moments to be considered in the conservation equations need integration over the entire velocitiy space. But in some special cases there are problems that require moments calculated over just a part of the velocity space like Resonant Moment Method (Ponomarjov and Carati 2006).

The moment of order L can be defined by the expression

$$
M_{ijk..}^{L}(\vec{r}, t) = \int_{v} v_i v_j ... v_k f(\vec{r}, \vec{v}, t) dv_x dv_y dv_z \tag{2}
$$

where the velocity components $v_i v_j ... v_k$ appear L times in the integral operator.

The first three moments including density, the average velocity, the pressure dyad, momentum flow dyad were considered in previous papers (Carati and Ponomarjov 2002; Gun'ko and Ponomarev 1993; 1995a, 1995b; Ponomarjov 1995, 1996, 1997a, b, 1998a, b, 1999, 2000a, b, c, d, e, f, 2001a, b, c, d, 2002; Ponomarev and Carati 2003, 2004a, b, c, d, e; Ponomarev et al. 2011; Webb and Ponomarev 2007).

The 4th moment of the distribution function $f(\vec{r}, \vec{v}, t)$ will be considered here.

4 Triad of Thermal Energy Flux

We will use the notations of the average value of the function $G(\vec{r}, \vec{v}, t)$ over the distribution function $f(\vec{r}, \vec{v}, t)$ as following:

$$
\langle G(\vec{r}, \vec{v}, t) \rangle = \frac{1}{n(\vec{r}, t)} \iiint_{V} G(\vec{r}, \vec{v}, t) f(\vec{r}, \vec{v}, t) dv_x dv_y dv_z \tag{3}
$$

where $n(\vec{r}, t)$ is the number of particles in the volume $dxdydz$ around the point $\vec{r}$ at the time t.

Then tensor of thermal energy flux is described by the following triad

$$
\psi = \rho \langle \vec{v} \vec{v} \vec{v} \rangle \tag{4}
$$

where $\rho = mn$ is the mass density of the particles.

The components of this tensor are

$$
\psi_{ijk} = \rho \langle v_i v_j v_k \rangle \tag{5}
$$

In Cartesian coordinates, the thermal energy flux triad has the following form:

$$
\psi = \psi_x \vec{i} + \psi_y \vec{j} + \psi_z \vec{k} \tag{6}
$$

where dyads ψ_x, ψ_y, ψ_z have the following matrix representation:

$$\psi_x = \begin{pmatrix} \psi_{xxx} & \psi_{xyx} & \psi_{xzx} \\ \psi_{yxx} & \psi_{yyx} & \psi_{yzx} \\ \psi_{zxx} & \psi_{zyx} & \psi_{zzx} \end{pmatrix} \tag{7}$$

$$\psi_y = \begin{pmatrix} \psi_{xxy} & \psi_{xyy} & \psi_{xzy} \\ \psi_{yxy} & \psi_{yyy} & \psi_{yzy} \\ \psi_{zxy} & \psi_{zyy} & \psi_{zzy} \end{pmatrix} \tag{8}$$

$$\psi_z = \begin{pmatrix} \psi_{xxz} & \psi_{xyz} & \psi_{xzz} \\ \psi_{yxz} & \psi_{yyz} & \psi_{yzz} \\ \psi_{zxz} & \psi_{zyz} & \psi_{zzz} \end{pmatrix} \tag{9}$$

5 Heat Flow Vector

The stochastic or thermal energy flux through a surface corresponds to the normal component of the heat flow vector q_n. If the normal to the surface directs as the unit vector $\vec{n}$, the component of the heat flow vector in the direction of $\vec{n}$ is

$$q_n = \vec{q} \cdot \vec{n} \tag{10}$$

Takinkg into account the kinetic energy of chaotic motion of the particles

$$E_{kinetic} = \frac{mv^2}{2} \tag{11}$$

we obtain for the projection of the heat flow vector in this case:

$$q_n = \frac{\rho}{2} \langle v^2 \vec{v} \cdot \vec{n} \rangle \tag{12}$$

For the corresponding heat flow vector we have

$$\vec{q} = \frac{\rho}{2} \langle v^2 \vec{v} \rangle \tag{13}$$

In addition, from Eq. (12) we have a connection between q_n and the thermal energy flux triad:

$$q_n = \frac{\rho}{2} \left(\langle v_x^2 v_n \rangle + \langle v_y^2 v_n \rangle + \langle v_z^2 v_n \rangle \right) \tag{14}$$

Taking into account Eq. (6), we obtain

$$q_n = \frac{1}{2}\left(\psi_{xxn} + \psi_{yyn} + \psi_{zzn}\right) \tag{15}$$

5.1 Heat Flow from Impulsive (Explosive) Injector

So, we will investigate the heat flow (flux) $\vec{q}$ in the following form:

$$\vec{q} = \frac{m}{2} \iiint \vec{v}v^2 f\left(v_x, v_y, v_z\right) dv_x dv_y dv_z \tag{16}$$

For the impulsive point-like source (Ponomarjov 2000a) placed at position $\vec{r_s}$, we have

$$Q(\vec{r}, \vec{v}, t) = N\delta\left(\vec{r} - \vec{r_s}\right)\delta(t - t_0)W(\vec{v}) \tag{17}$$

and in case of the shifted Maxwell-particle velocity distribution near this source

$$\begin{cases} W(\vec{v}) = \left(\frac{\beta}{\pi}\right)^{3/2}exp\left(-\beta\left(\vec{v} - \vec{V}\right)^2\right) \\ \beta = \left(\frac{m}{2kT}\right) \\ \vec{V} = (V_1, V_2, V_3) \end{cases} \tag{18}$$

Equations (17) and (18) describe N particles placed very close to the $\vec{r_s}$ position with the mean velocity $\vec{V}$ at the initial time $t = t_0$ and $W(\vec{v})$ velocity distribution.

A Cartesian coordinate system $OXYZ$ is introduced in Ponomarjov (2000a) so that OX axis is directed in parallel to the magnetic field $\vec{B}$ and the point-like source located at the center of this coordinate system $OXYZ$ ($\vec{r_s} = \vec{0}$).

It is the problem which is described by the following Boltzmann equation in the centimetre-gram-seconds (CGS) unit system:

$$\frac{\partial f(\vec{r}, \vec{v}, t)}{\partial t} + \vec{v} \cdot \frac{\partial f(\vec{r}, \vec{v}, t)}{\partial \vec{r}} + \frac{e}{m}\left(\frac{1}{c}\left[\vec{v}, \vec{B}\right] + \vec{E}\right) \cdot \frac{\partial f(\vec{r}, \vec{v}, t)}{\partial \vec{v}} = Q(\vec{r}, \vec{v}, t) \tag{19}$$

In this paper we will use equivalent equation in the metre-kilogram-seconds (SI, System International):

$$\frac{\partial f(\vec{r}, \vec{v}, t)}{\partial t} + \vec{v} \cdot \frac{\partial f(\vec{r}, \vec{v}, t)}{\partial \vec{r}} + \frac{e}{m}\left(\left[\vec{v}, \vec{B}\right] + \vec{E}\right) \cdot \frac{\partial f(\vec{r}, \vec{v}, t)}{\partial \vec{v}} = Q(\vec{r}, \vec{v}, t) \tag{20}$$

This Eq. (20) is solved here analytically by a method of characteristics in a similar way to the previous cases for the first three moments including density, the average velocity, the pressure dyad, momentum flow dyad (published in previous papers Gun'ko and Ponomarev (1993, 1995a, b), Garrison et al. (2011), Ponomarjov (1995, 1996, 1997a, 1998a, b, 1999, 2000a, b, c, d, 2002), Ponomarjov and Gunko (1995), Ponomarjov and Carati (2006, 2001a, b, e). On the final stage, it required triple analytical integrations in velocity space after special 3-dimensional variable transformations (described in detail in Ponomarjov (2002)).

So after analytical solution of Eq. (20) and taking into account Eqs. (1) and (16), we have for $q_x = q_{\vec{B_\tau}}$, the heat flow (flux) along the OX axis, which is parallel to the magnetic field $\vec{B}$:

$$q_{\vec{B_\tau}} = \frac{mN}{2}\left(\frac{\beta}{\pi}\right)^{3/2}\frac{x}{\tau}\left(\left(\frac{x}{\tau}\right)^2 + \omega^2\frac{y^2+z^2}{4sin^2(\omega\tau/2)}\right)exp\left(-\beta\left(\frac{x}{\tau}-V_1\right)^2\right)$$

$$\frac{exp\left(-\beta\left(\omega^2\frac{y^2+z^2}{4sin^2(\omega\tau/2)} - \omega y(V_2cot(\omega\tau/2) + V_3) - \omega z\left(V_2 + V_3cot(\omega\tau/2) + V_2^2 + V_3^2\right)\right)\right)}{4\tau sin^2(\omega\tau/2)/\omega^2} \quad (21)$$

where $\tau = t - t_0$ is the time since the impulse (explosion) happen.

Consider the shift velocity $\vec{V}$ pointed along the magnetic field $\vec{B}$, that means $\vec{V} = (V, 0, 0)$ in the coordinate system $OXYZ$. In this case, according to Eq. 21, the heat flow vector has the following expression:

$$q_{\vec{B_\tau}} = \frac{mN}{2}\left(\frac{\beta}{\pi}\right)^{3/2}\frac{x}{\tau}\left(\left(\frac{x}{\tau}\right)^2 + \omega^2\frac{y^2+z^2}{4sin^2(\omega\tau/2)}\right)$$

$$\frac{exp\left(-\beta\left(\frac{x}{\tau}-V\right)^2\right)exp\left(-\beta\omega^2\left(\frac{y^2+z^2}{4sin^2(\omega\tau/2)}\right)\right)}{4\tau sin^2(\omega\tau/2)/\omega^2} \quad (22)$$

where $\tau = t - t_0$ is the time since the impulse (explosion) happen.

Let's introduce polar coordinates r, ϕ so that

$$\begin{cases} y = r\cos\phi \\ z = r\sin\phi \\ 0 \le r \le R \\ 0 \le \phi \le 2\pi \end{cases} \quad (23)$$

then the Eq. (22) takes the following form:

$$q_{\vec{B_\tau}} = \frac{mN}{2}\left(\frac{\beta}{\pi}\right)^{3/2}\int_0^{2\pi}\int_0^R\frac{x}{\tau}\left(\left(\frac{x}{\tau}\right)^2 + \omega^2\frac{r^2}{4sin^2(\omega\tau/2)}\right)$$

$$\frac{exp\left(-\beta\left(\frac{x}{\tau}-V\right)^2\right)exp\left(-\beta\omega^2\left(\frac{r^2}{4sin^2(\omega\tau/2)}\right)\right)}{4\tau sin^2(\omega\tau/2)/\omega^2}rdrd\phi \quad (24)$$

For integration of Eq. (24) consider division of the heat flow $q_{\vec{B_\tau}}$ into two parts $q_{\vec{B_{\tau1}}}$ and $q_{\vec{B_{\tau2}}}$ so that:

$$q_{\overrightarrow{B_\tau}} = q_{\overrightarrow{B_{\tau 1}}} + q_{\overrightarrow{B_{\tau 2}}} \tag{25}$$

Then, for the first part $q_{\overrightarrow{B_{\tau 1}}}$ we have

$$q_{\overrightarrow{B_{\tau 1}}} = \frac{mN}{2}\left(\frac{\beta}{\pi}\right)^{3/2} \int_0^{2\pi}\int_0^R \left(\frac{x}{\tau}\right)^3 \frac{exp\left(-\beta\left(\frac{x}{\tau}-V\right)^2\right) exp\left(-\beta\omega^2\left(\frac{r^2}{4sin^2(\omega\tau/2)}\right)\right)}{4\tau sin^2(\omega\tau/2)/\omega^2} r\,dr\,d\phi \tag{26}$$

After double integration of Eq. (26), the first part of the heat flow takes the following precise analytical form:

$$q_{\overrightarrow{B_{\tau 1}}} = \frac{mN}{2}\left(\frac{\beta}{\pi}\right)^{1/2}\left(\frac{x}{\tau}\right)^3 \frac{exp\left(-\beta\left(\frac{x}{\tau}-V\right)^2\right)}{4\tau}\left(1 - exp\left(-\beta\omega^2\left(\frac{R^2}{4sin^2(\omega\tau/2)}\right)\right)\right) \tag{27}$$

Initially, the second part of the heat flow $q_{\overrightarrow{B_{\tau 2}}}$ has the following form:

$$q_{\overrightarrow{B_{\tau 2}}} = \frac{mN}{2}\left(\frac{\beta}{\pi}\right)^{3/2} \int_0^{2\pi}\int_0^R \frac{x}{\tau}\frac{\omega^2 r^2}{4sin^2(\omega\tau/2)}$$
$$\frac{exp\left(-\beta\left(\frac{x}{\tau}-V\right)^2\right) exp\left(-\beta\omega^2\left(\frac{r^2}{4sin^2(\omega\tau/2)}\right)\right)}{4\tau sin^2(\omega\tau/2)/\omega^2} r\,dr\,d\phi \tag{28}$$

After double integration, the Eq. (28) takes the following precise analytical form:

$$q_{\overrightarrow{B_{\tau 2}}} = \frac{mN}{2}(\beta\pi)^{-1/2}\left(\frac{x}{\tau}\right)\frac{exp\left(-\beta\left(\frac{x}{\tau}-V\right)^2\right)}{4\tau}$$
$$\left(1 - \left(1 + \beta\omega^2\left(\frac{R^2}{4sin^2(\omega\tau/2)}\right)\right)\right) exp\left(-\beta\omega^2\left(\frac{R^2}{4sin^2(\omega\tau/2)}\right)\right) \tag{29}$$

It is important that from these precise analytical results (Eqs. 27, 29) we obtain exact upper limit for the heat flow, which can not be exceeded by applying very large magnetic fields ($\beta\omega^2 R^2 \to +\infty$):

$$q_{\overrightarrow{B_{\tau 1}}} \to \frac{mN}{2}\left(\frac{\beta}{\pi}\right)^{1/2}\left(\frac{x}{\tau}\right)^3\frac{exp\left(-\beta\left(\frac{x}{\tau}-V\right)^2\right)}{4\tau} \tag{30}$$

$$q_{\overrightarrow{B_{\tau 2}}} \to \frac{mN}{2}(\beta\pi)^{-1/2}\left(\frac{x}{\tau}\right)\frac{exp\left(-\beta\left(\frac{x}{\tau}-V\right)^2\right)}{4\tau} \tag{31}$$

That means the exact upper limit to the heat flow, maximum heat flux, $q_{max \overrightarrow{B_\tau}}$, which can not be exceeded by applying very large magnetic fields, is outlined below:

$$q_{\overrightarrow{B_\tau}} = q_{\overrightarrow{B_{\tau 1}}} + q_{\overrightarrow{B_{\tau 2}}} \rightarrow q_{max \overrightarrow{B_\tau}} = \frac{mN}{2} \left(\frac{\beta}{\pi}\right)^{1/2} \left(\frac{x}{\tau}\right) \left(\left(\frac{x}{\tau}\right)^2 + \frac{1}{\beta}\right) \frac{exp\left(-\beta\left(\frac{x}{\tau} - V\right)^2\right)}{4\tau}$$

(32)

In addition, for large time $\tau \rightarrow +\infty$ from Eq. (32) we obtain

$$q_{max \overrightarrow{B_\tau}} \rightarrow \frac{mN}{2} (\beta\pi)^{-1/2} \frac{x}{\tau} \frac{exp\left(-\beta V^2\right)}{4\tau}$$

(33)

That means that for the large time this upper limit of heat flow decreasing as $C\tau^{-2}$ where

$$C = \frac{mN}{8} (\beta\pi)^{-1/2} x \cdot exp\left(-\beta V^2\right)$$

(34)

and $\tau = t - t_0$ is the time since the impulse (explosion) happen.

6 Conclusion

Exact analytical expression is obtained for upper limit to heat flow $q_{max \overrightarrow{B_\tau}}$ (maximum heat flux) from impulsive (explosive) charged particle sources (injectors) in magnetic field. This exact expression is based on the obtained (via characteristics method) distribution function as analytical solution for Boltzmann equation for charged particle flows in magnetic field. This can be considered as a saturation phenomenon, when we see significant slow-down of the heat flow increase versus the magnetic field still increasing at the same rate. It is instrumental when we need to estimate the saturation maximum of the heat flow, when we have the only 1 variable parameter to change: magnetic field.

References

Carati D, Ponomarjov MG (2003) Statistical description of currents induced by two electron cyclotron counter-propagating waves. In: Proceedings of the 12th joint workshop electron cyclotron emission and electron cyclotron heating. Held 13–16 May 2002 in Aix-en-Provence, France. Edited by Gerardo Giruzzi (Association Euratom-CEA sur la Fusion, France). Published by World Scientific Publishing Co. Pte. Ltd., 2003. ISBN #9789812705082, pp 77–82

Gun'ko YF, Ponomarev MG (1993) The charged particles emission in a magnetic field. Vestnik Sankt-Peterburgskogo Universiteta. Ser 1. Matematika Mekhanika Astronomiya (2):89–94

Gun'ko YF, Ponomarev MG (1995a) The charged particle layer expansion in a magnetic field. Vestnik Sankt-Peterburgskogo Universiteta. Ser 1. Matematika Mekhanika Astronomiya (4):105–108

Gun'ko YF, Ponomarev MG (1995b) The charged particle distribution due to emission in a magnetic field. Astronomische Nachrichten 316(1):17–21

Iovea M, Neagu M, Stefanescu B, Mateiasi G, Halai H, Amos M, Ponomarev M, Kappatos V, Selcuk C, Gan TH, Gierl C (2012) Preliminary NDT investigation of sintered powder metallurgy parts by high-resolution TDI based X-ray digital radiography. In: European international powder metallurgy congress and exhibition (Euro PM 2012), vol 1, Basel, Switzerland, 16–19 Sept 2012 (Code 105676)

Iovea M, Neagu M, Stefanescu B, Mateiasi G, Clarke A, Nicholson I, Ponomarev M, Kappatos V, Selcuk C, Gan TH (2013) PM parts fast in-line X-ray digital radiography. In: International powder metallurgy congress and exhibition (Euro PM 2013). European PM Conference Proceedings, vol 1. European Powder Metallurgy Association (EPMA)

May J, Ponomarev M, Kuball S, Gallardo J (2006) A case for new statistical software testing models. In: Annual reliability and maintainability symposium, RAMS'06; Newport Beach, CA; United States; 23–26 Jan 2006; Category number CH3744; Code 69734 Proceedings—Annual Reliability and Maintainability Symposium, Article Number 1677399 349 353 https://doi.org/10.1109/RAMS.2006.1677399

Ponomarjov MG (1995) Disturbances of the ambient magnetoplasma due to its interactions with object surfaces: imaginary emission method, far-wake of objects moving through a rarefied plasma at different angles to the ambient magnetic field. Planet Space Sci 43(10–11):1419–1427

Ponomarjov MG (1996) Imaginary-emission method for modeling disturbances of all magnetoplasma species: reflecting and absorbing objects in motion through a rarefied plasma at different angles to the ambient magnetic field. Phys Rev E 54(5):5591–5598

Ponomarjov MG (1997a) Pressure of charged particle flows in ambient magnetic fields. Astron Nachr 318(3):187–192

Ponomarjov MG (1997b) New ways of protecting astronomical equipment and solar batteries of spacecrafts. Bull Am Astron Soc 29:1023. American Astronomical Society, DPS meeting No 29, id.27.02

Ponomarjov MG (1998a) Outer atmosphere and wake of space objects, kinetic simulation: disturbances of ambient magnetoplasma due to diffuse reflecting bodies in motion. In: APS division of plasma physics meeting. abstract id. U9Q.09

Ponomarjov MG (1998b) Outer atmosphere and wake of space objects, kinetic simulation. Disturbances of ambient magnetoplasma due to diffuse reflecting bodies in motion. American Physical Society, Division of Plasma Physics Meeting. New Orleans, LA. Abstract Id. U9Q 09, 16–20 Nov 1998

Ponomarjov MG (1999) Kinetic simulation of stratifications and flute structures of charged particle jets and wakes in the ambient magnetic field. Bull Astron Soc 31(4):1157 (id.53.12)

Ponomarjov MG (2000a) Space flows and disturbances due to bodies in motion through the magnetoplasma. Astrophys Space Sci 274(1):423–429

Ponomarjov MG (2000b) Simulation of oscillations in charged particle systems under the ambient magnetic field control. In: 2nd international conference control of oscillations and chaos (COC 2000). St. Petersburg, Russia, 5 July 2000 through 7 July 2000. Code 57512. Proceedings (Cat. No.00TH8521), vol 1, pp 167–170. https://doi.org/10.1109/COC.2000.873548

Ponomarjov MG (2000c) 3D collisional kinetic simulation of stratifications and flute structures of plasma flows and wakes in external magnetic fields. In: AAS, DPS Meeting. No 32. id.15.10. 2000. Bull Am Astron Soc 32:1022

Ponomarjov MG (2000d) 3D kinetic dynamical models of ionized HII clouds in external magnetic field. In: Ionized gaseous nebulae. Meeting abstract id. 44, Mexico City, 21–24 Nov 2000

Ponomarjov MG (2000e) 3D collisional kinetic simulation of stratifications and flute structures of plasma flows and wakes in external magnetic fields. In: 32nd annual meeting of the division for planetary sciences. Pasadena, California, USA. Session 15. Outer Planets IV-Aurorae and Magnetospheres. 2000/10/24. Abstract 15.10. 1022, 23–27 Oct 2000

Ponomarjov MG (2000f) 3D time-dependent kinetic simulation of turbulent plasma flows under the effect of external magnetic field. In: 42nd Annual meeting of the APS division of plasma physics combined with the 10th international congress on plasma physics. Québec City. Canada Meeting ID: DPP00. abstract id. No 1.007. American Physical Society, 23–27 Oct 2000. https://ui.adsabs.harvard.edu/abs/2000APS..DPPNO1007P/abstract

Ponomarjov MG (2001a) Acceleration and transport of particles in collisionless plasmas: wakes due to the interaction with moving bodies. Astrophys Space Sci 277(1):39–44

Ponomarjov MG (2001b) Kinetic simulation of magnetic field effects on stratifications and flute structures of space plasma flows and wakes of bodies. In: 2001 Joint assembly american geophysical union spring meeting. Boston, Massachusetts, USA. Abstracts, SM52B-04

Ponomarjov MG (2001c) Kinetic simulation of magnetic field effects on wakes of meteoroids imaginary emission method. Eur Space Agency (Special Publication) ESA SP 495:295–300

Ponomarjov MG (2001d) Kinetic modeling magnetic field effect on ion flows, disturbances and wakes in space plasma. In: Büchner J, Dum CT, Scholer M (eds) Space plasma simulation: proceedings of the sixth international school/symposium. Schaltungsdienst Lange o.H.G, Berlin, p 328. ISSS-6, Garching, Germany, 3–7 Sept 2001

Ponomarjov MG (2002) 3D time-dependent kinetic simulation of space plasma disturbances due to moving bodies with the ambient magnetic field effect. Adv Space Res 29(9):1397–1402

Ponomarev M (2022a) A novel physical model to enhance precision and performance of 3-dimensional force sensors. Process Geomedia 2(32):1589–1600

Ponomarev M (2022b) Experimental validation of novel physical model for improvement of sensing 3-dimensional fluid flow loads and responses in real sea conditions with South Western Mooring Test Facility (SWMTF). Process Geomedia 2(32):1579–1589

Ponomarev M (2022c) A novel physical model to enhance precision and performance of multidimensional force sensors. Proceedings of 8th international scientific conference-school physical and mathematical modeling of earth and environment processes. Springer, pp 347–364. ISBN 978-3-031-25961-6

Ponomarev M (2023a) Physical foundations of mechanics part 1. Newtonian mechanics. World of Science LLC, pp 1–93

Ponomarev M (2023b) Physical foundations of mechanics, part 2. Relativistic Mechanics. World of Science LLC, pp 1–83

Ponomarjov MG, Carati D (2003) Kinetic simulations of relativistic electron flows in time-dependent electromagnetic fields. In: Proceedings of the conference electron cyclotron emission and electron cyclotron heating. World Scientific Publishing Co Pte Ltd, pp 137–142

Ponomarjov MG, Carati D (2004a) Search for optimal 3D wave launching configurations for the acceleration of charged particles in a magnetized plasma: Resonant Moments Method. In: 46th Annual meeting of the division of plasma physics, Savannah. GA. Meeting Id: DPP04. abstract id. RP1.042. American Physical Society, 15–19 Nov 2004

Ponomarjov MG, Carati D (2004b) Acceleration of charged particles by crossed cyclotron waves, Resonant Moments Method. In: 35th COSPAR scientific assembly. Paris, France, p 2851, 18–25 July 2004

Ponomarjov MG, Carati D (2004c) Search for optimal 2D and 3D wave launching configurations for the largest acceleration of charged particles in a magnetized plasma. Resonant Moments Method. https://arxiv.org/format/physics/0411075 arXiv:physics/0411075

Ponomarjov MG, Carati D (2004d) Acceleration of electron populations by crossing EC waves in an external magnetic field. In: 13th Joint workshop on electron cyclotron emission and electron cyclotron resonance heating. Nizhny Novgorod, Russia. https://ec13.iapras.ru/papers/Ponomarjov.pdf

Ponomarjov MG, Carati D (2004e) Enhanced acceleration of charged particles by crossing electromagnetic waves in a magnetized plasma, resonant moments method. https://hal.archives-ouvertes.fr/hal-00001926v1

Ponomarjov MG, Carati D (2006) Enhanced acceleration of electrons populations by crossing electron cyclotron waves in an ambient magnetic field, resonant moments method. Adv Space Res 38(8):1576–1581

Ponomarjov MG, Gunko YF (1995) Kinetic modeling of charged particle cloud expansion and emission in magnetic and electric fields. Planet Space Sci 43(10–11):1409–1418

Ponomarev M, Johanning L, Parish D (2010) Enhancing precision and reliability of tri-axial load cells for mooring load measurements. In: Proceedings of the 3rd international conference on ocean energy, 6 Oct 2010. (ICOE2010). ISBN 978-84-693-5467-4

Ponomarev MG, Garrison BJ, Vickerman JC, Webb RP (2011) A molecular dynamics study of a 5 keV C60 fullerene impact on a two-component organic molecular sample. Surf Interface Anal, Wiley Online Library 43(1–2):107–111

Ponomarev MG, Kappatos V, Selcuk C, Gan TH, Amos M, Halai H, Gierl C, Iovea M (2012a) PM dimensional control: development of a digital radiographic inspection technique for production friendly quality assessment of powder metallurgy parts. In: Proceedings of the international euro powder metallurgy congress and exhibition (Euro PM 2012), vol 1. Switzerland, Basel, 16–19 Sept 2012 (Code 105676)

Ponomarev MG, Selcuk C, Gan TH (2012b) Construction of pattern recognition system optimized for X-ray inspection of plastic electronics and OLED displays. In: 51st Annual conference of the British institute of non-destructive testing (NDT 2012) BINDT 2012. Northamptonshire, United Kingdom, 11–13 Sept 2012 (Code 107048 354 360)

Ponomarev MG, Kappatos V, Selcuk C, Gan TH, Amos M, Halai H, Gierl C, Iovea M (2013a) Digital radiographic inspection technique for production friendly quality assessment of PM parts. Powder Metall 56(2):92–95

Ponomarev MG, Selcuk C, Gan TH, Amos M, Nicholson I, Iovea M, Neagu M, Stefanescu B, Mateiasi G (2013b) Non destructive testing: a defect detection and classification system for automatic analysis of digital radiography images of powder metallurgy parts. In: International powder metallurgy congress and exhibition (Euro PM 2013). Gothenburg. Sweden, 15–18 Sept 2013 (Code 110976)

Ponomarev MG, Selcuk C, Gan TH, Amos M, Nicholson I, Iovea M, Neagu M, Stefanescu B, Mateiasi G (2014) Defect detection and classification system for automatic analysis of digital radiography images of PM parts. Powder Metall 57(1):17–20

Ponomarev M, Johanning L, Parish D (2022) Experimental validation of novel physical model for improvement of sensing multidimensional fluid flow loads and responses in real sea conditions with South Western Mooring Test Facility. In: Proceedings of 8th international scientific conference-school of the physical and mathematical modeling of earth and environment processes, Springer, pp 365–376. ISBN 978-3-031-25961-6

Webb RP, Ponomarev M (2007) Molecular dynamics simulation of low energy cluster impacts on carbon nanotubes. Nuclear Instrum Methods Phys Res Sect B Beam Interact Mater Atoms 255(1):229–232